THE BOTANIST
IN IRELAND

THE BOTANIST
IN IRELAND

BY

ROBERT LLOYD PRAEGER

D.Sc.

EP PUBLISHING LIMITED
1974

Republished 1974
EP Publishing Limited
East Ardsley, Wakefield
Yorkshire, England

by kind permission of the copyright holders

Reprinted 1977

Originally published by Hodges, Figgis & Co., Dublin, 1934

ISBN 0 85409 955 7

Please address all enquiries to EP Publishing Limited
(address as above)

Printed in Great Britain by
The Scolar Press Ltd. Ilkley
West Yorkshire

PREFACE.

IRELAND is a pleasant country for the botanist. It is true that its flora offers less variety than that of England, and still less than that of equal Continental areas. But this comparative paucity of species is richly compensated by the occurrence, often over a wide range, of plants of high interest and often of great beauty, whose analogues in England or Scotland are few and of very restricted distribution, while in the parts of Europe next adjoining they are almost all unknown. These peculiar features are due mainly to the position of Ireland on the extreme western edge of the Old World, a fact which in itself, for the student of European or Eurasian geography or botany, gives to this country a special interest.

Again, the extraordinary diversity of rocks in Ireland produces a corresponding variety of land-forms, with repercussions on the character of the flora and fauna. Geological diversity increases as one passes across Europe from east to west, from the interior of the great land-mass to its unstable oceanic edge. In Russia, areas far larger than Ireland are built up of a single type of rock, stretching featureless from horizon to horizon. Within the fretted coast-line of Ireland, almost every sedimentary rock which has gone to building up the crust of the Earth is represented, from Archæan to Neolithic, as well as many of igneous origin. The Irish mountains are exceeded in elevation by the loftiest summits in Wales and in Scotland, but they offer within a smaller area an equal variety of rock and consequently of scenery. The Scottish Highlands have their analogue in Donegal and Connemara, the Welsh Silurian mountains and the English Lake District find an echo in the Galtees and the Wicklow foot-hills, Devon is matched and surpassed by Kerry, as the limestone clints of Ingleborough are by the "pavements" of Clare and Galway. If Ireland lacks the Chalk downs, it has the remarkable Tertiary basalt plateau of the north-east with its characteristic flora, and if it has not got the Norfolk Broads it has a great and varied array of large lakes with many interesting plants on their shores and islands.

Another charm of Ireland is the freedom with which one may wander over its hills and plains. Seldom is one arrested by a notice forbidding access, and a friendly word will work wonders on ground preserved for game or kept for private purposes.

To the botanist studying the vegetation of a country as a whole, or the distribution within a country of any plant or group of plants, a Flora arranged on the usual plan, under which the species are enumerated according to their botanical affinities and the information relative to each is given under the plant's name, is the most convenient presentation of the subject. For Ireland as a whole two such works have been compiled since its flora became well known. In "Cybele Hibernica" (2nd ed., 1898, by N. COLGAN and R. W. SCULLY; 12s. 6d., now 4s. 6d.), under each plant its range is shown with reference to twelve Districts into which the country is divided, as in the previous edition of the same work; specially detailed treatment, geographical and historical, being awarded to the rarer species. In "Irish Topographical Botany," compiled by the present writer (1901, 12s. 6d.), distribution is shown with reference to forty Divisions or Vice-Counties of about the area of those used by H. C. Watson for Great Britain, and a more uniform treatment is awarded to the plants, whether rare or common, the latest known record alone being usually given, even in the case of rare species, with a brief indication of the scope of other records. No Supplement to "Cybele Hibernica" has as yet appeared. "Irish Topographical Botany" has been brought up to date or nearly so by the publication of three Supplements— the first for the period 1901–5 (Proc. Roy. Irish Acad., **26** Sect. B, 1906), the second for 1906–28 (*ibid.*, **39**, Sect. B, 1929), and the third for 1929–34 (*ibid.*, **42**, Sect. B, 1934). For shorter general accounts of the flora the five geographical volumes ("Ireland," "Ulster," "Leinster," "Connaught," "Munster," reduced to 2s. 6d. per volume), issued in 1921-2 by the Cambridge University Press under the editorship of George FLETCHER, may be consulted. With information as to the distribution of the Higher Plants in Ireland, therefore, the student is well equipped.

But for the botanist resident in this country, or visiting it and finding himself here or there within it, a question very likely to arise is—What interesting plants grow in this or that neighbourhood, or in the vicinity of this or that town or hotel, or on this or that island or mountain or lake? It is to supply information on this basis that the present book is compiled, plants being arranged under localities instead of localities under plants. This scheme of treatment leads to a restriction as to the species dealt with, all those of general distribution being necessarily omitted (just as, conversely, the stations for them are omitted from the works previously mentioned). The commoner species, indeed, appear only when some plant association or other natural group requires for its description a list of the species occurring. In the Introduction, however (**79-207**), opportunity is taken briefly to indicate the whole range, not only within Ireland but outside it, of a selection of the more interesting plants which are included in the Irish flora. At the end of the book a Census List of the Irish Flora (species, subspecies, and hybrids) is added, the distribution of each plant, native or naturalized, being shown according to the forty "Divisions" used in "Irish Topographical Botany." This is desirable especially because the Irish range of many of the rarer species has been extended, and nomenclature has suffered many changes, since the publication of "Cybele Hibernica" and "Irish Topographical Botany."

The book, therefore, aims at giving, with a different orientation, the bulk of the information supplied by the works mentioned, with the inclusion of the large amount of additional matter, published or unpublished, which has accumulated within the last thirty years or more. In "A Tourist's Flora of the West of Ireland" which appeared in 1909, the plan of the present work was used in the "Topographical Section." But that book dealt with only eleven out of the forty Irish County-divisions, and within that area, as well as outside it, a good deal of botanical exploration has been done in recent years. So far as this western district is concerned, the matter in the "Tourist's Flora" has been included, revised and considerably enlarged.

The works which have appeared dealing with the flora of restricted areas in Ireland are limited in number and of varied date, and all are arranged in the localities-under-species plan of the typical Flora. A few of them are of recent date and of comprehensive scope, and form invaluable companions to the botanist visiting the districts with which they are concerned. This applies especially to COLGAN'S "Flora of the County Dublin" (1904, 12s. 6d.) and SCULLY'S "Flora of County Kerry" (1912, 12s. 6d.). HART'S "Flora of the County Donegal" (1898, 7s. 6d., now out of print) is somewhat earlier; STEWART and CORRY'S "Flora of the North-east of Ireland" (1888, reduced price, 2s. 6d.), dealing with the counties of Down, Antrim, and Londonderry, has been brought up to date by the issue by the Belfast Naturalists' Field Club of two Supplements, the first in 1895, the second (a "Second Supplement and Summary") in 1923. HART'S "Flora of Howth" (1887, o.p.) deals with a quite small but rich and interesting area near Dublin. The remaining Irish local or general Floras are of earlier date; the information which they supply is naturally incomplete, and all of importance has been incorporated in later books.

An inconvenience of the plan of the present work is that names of finders and other authorities for the records are of necessity omitted. This hiatus is met to some extent by a list of the principal workers at the Irish-flora, with a note of their chief publications, and the names and dates of their most important finds (**78**). Further information may be obtained from the works quoted above, as well as from the numerous subsequent papers and notes referred to in the pages which follow. Cross-references (by paragraph numbers) are inserted freely, to facilitate the bringing together of information relative to both plants and places.

In the matter of botanical nomenclature, the eleventh edition (1925) of the "London Catalogue of British Plants" is for convenience followed, save in the case of a few genera where revision has been lately carried out.

It has not been found possible to include in this book maps adequate for use over so wide an area as Ireland, but six of the most important districts—northern Donegal, eastern Antrim, the Sligo neighbourhood, western Galway-Mayo, the Wicklow mountain region, and southern Kerry—are illustrated by excerpts from Messrs. Bartholomew's quarter-inch map. For field work, the one-inch map of the Ordnance Survey (contoured edition) is indispensable. For more general purposes the beautiful contour-coloured half-inch Ordnance map, and the already quoted similar quarter-inch map issued by Messrs. Bartholomew, supply all that is required.

All that I have to say at the conclusion of fifty years' field-work in Ireland, during which I have explored the flora of every county, of every important mountain-range, lake, river, and island, is embodied in condensed form in the present work. The last four seasons have been devoted to study of those areas and those plants which a general survey of the field of work showed to be most in need of further investigation. So while much remains to be done by future generations of field botanists, it is hoped that the present account will be found to offer a tolerably balanced view of the flora of Ireland as it is at present known. The claims of ecology as contrasted with purely floristic work have been recognized, and an endeavour has been made all through to correlate vegetation with geography and geology. The human element in local botany has also been kept in view, and in the topographical notes certain non-botanical features of the land, such as its more important antiquities, have not been altogether ignored.

As to the illustrations, their chief value lies in the series of excellent photographs of rare native plants in the field taken by Mr. R. J. Welch during visits with me to many places, and reproduced here. For these and some others illustrating habitat and scenery my best thanks are due to him.

For the loan of blocks illustrating geographical features I am indebted to the Syndics of the Cambridge University

Press; for help in various directions thanks are tendered to Dr. R. W. Scully and Messrs. J. P. Brunker, A. Farrington, T. Hallissy, R. A. Phillips, A. W. Stelfox, and a number of other friends.

As the book is intended for use in the field as much as in the study, it has been made as portable and compact as possible.

R. Ll. P.

19 Fitzwilliam Square, Dublin.

November, 1934.

Note on 1974 reprint

The vice comital list (Census List) which was included in the original edition (pages 492–539 inclusive) has been omitted in this reprint because it was out of date and is now replaced by a new publication:

M. J. P. Scannell and D. M. Synnott
Census Catalogue of the Flora of Ireland
Irish Stationery Office.

Obtainable from Government Publications, Sales Office, G.P.O. Arcade, Dublin 1.

CONTENTS.

The four coloured maps included in the original edition have been omitted from this reprint. Large scale maps of the whole of Ireland are now readily available from bookshops.

ILLUSTRATIONS.

The figure shows the order in which the various parts of Ireland are dealt with in the Topographical Part. The sequence is an attempt to group together areas of floristic affinity, and may be summarized as follows, the numbers indicating the sections in the letterpress devoted to each area:—

Dublin (227–237), Central Plain, SE portion (238–259), SE Coastal Region (260–289), South-west (290–332), Western Limestones (333–375), Central Plain, western portion (353–382), Galway-Mayo Metamorphic Area (384–415), North-west (416–439), Donegal (440–452), Basaltic Plateau (455–470), NE Silurian Region (472–485), Central Plain, NE portion (485–491).

INTRODUCTION.

1. Ireland presents several features which give it a special interest for the botanist. It is the most westerly portion of Eurasian land, extending to 10° 29′ W longitude. Portugal and N Spain come next, falling short by some 65 miles of the Kerry headlands. But this approximation in longitude does not find a parallel in the general character of its vegetation. Ireland, being separated by two sea-channels from the Continent, suffers all the botanical penalties of insularity, in the form of a flora considerably reduced when compared even with England, and more when contrasted with the Peninsula. And again, the prevalence in Ireland of W and SW winds, blowing in off the warm Atlantic, results in a climate of extraordinarily small annual range in temperature, and of high humidity (**22–26**), unlike the comparatively Continental climate of the greater part of Spain and Portugal.

The flora more nearly akin to that which prevails over the greater part of Ireland will be found by moving eastward into N England, and thence across Holland into N Germany, though the more Continental climate encountered as one goes E introduces increasing divergence in the vegetation as compared with this country. A NE course across Scotland into Scandinavia is productive of a greater amount of diversity, though it tends to retain in the flora a number of species which are lost along the eastern line.

2. The island itself is unusual in several respects: notably in having the central portion occupied by a low plain of limestone (**10, 240–5**)—the largest continuous area of Carboniferous limestone in Europe—of only some 200 ft. average elevation, while mountains, of 1500 to 3000 ft. in height, occupy much of the margin, and only occasionally rise far from the sea. The surface, from sea-level to the tops of the hills, is occupied by peat-bog to an extent not found elsewhere in Europe (**28, 52**). The most marked characters of the Irish flora are its reduction due to insularity, and the presence, due to mildness of climate, of southern plants, which have in past times crept along the warm coast from the Mediterranean and the Peninsula, and which are not found so far north anywhere else (**35, 37**). A less conspicuous but more startling feature is the presence of some North American plants, which find their European headquarters in Ireland, and die out in west Scotland and the NW parts of the Continent (**35, 37**). All of these features, it is important to note, characterize the fauna also (**37**).

GEOLOGICAL HISTORY IN RELATION TO PRESENT SURFACE.

3. The nature of the vegetation of any part of the Earth's surface is governed primarily by conditions of temperature and precipitation, resulting in tropical rain-forest or daisy-starred grassland or cactus desert or other types as the case may be. But within any area of uniform climatic conditions, such as Ireland, vegetation is influenced chiefly by the nature of the substratum—the soil in which the plants are fixed. The soil results mainly from the wasting of the rocks of which the Earth's surface is locally composed; this detritus, mixed with the disintegrated remains of past generations of plants, produces the clayey or loamy or sandy material which is the first essential of a vegetation, and the basis of all agriculture and horti-culture. Hence, if we are to understand the distribution of the flora of a restricted area, we must first direct our attention to the rocks of which it is built up. This is the more essential in the case of Ireland, since its geological features are remarkably varied, its surface displaying a greater diversity of rocks, as regards both age and com-position, than is to be found in almost any similar area in Europe. It is true that, climatic conditions being the same, the operation of the soil-forming processes tends to produce similar soils from totally dissimilar materials. But, generally speaking, the soils of Ireland have not yet reached an advanced stage of development. Like the soils of all glaciated countries they are immature, and retain many of their original lithological features.

4. That the present surface is so varied in its lithology is due to several causes, the most important of which is the frequent fluctuations of level to which it has been subject. Again and again this area has sunk below the sea, and beds of mud, sand, &c. (the detritus of adjoining lands), now hardened into slate or shale, limestone or sandstone, have been laid down upon it. Again and again the surface has risen above the sea, and its burden of deposits has been exposed to the forces of denudation—rain, frost, sun, wind—and in part or whole destroyed and carried by rivers back into the ocean. An equally potent factor has been lateral earth-movements to which the crust has been subjected at intervals, causing extensive crumpling of the rocks. These foldings, by elevating certain regions, mostly in the form of ridges, have exposed them to especially severe denudation, resulting in the laying bare of underlying deposits; by depressing other areas, the beds occupying them have been protected; the great pressure produced has often melted the rocks partly or even completely, and entirely altered their character; and the weakening of the Earth's crust resulting from these movements has sometimes allowed molten material to rise from below into the arching folds, where it has solidified as granite, or to break through and pour as lava over the surface of the country. The result of the interplay of all these natural forces is that the present surface of much of Ireland resembles a patchwork quilt in which a score or more of patches, many of them frequently recurring in varying size and shape, have each a quite different tale to tell. Even in the great limestone plain, which is the largest continuous area of any one kind of rock occurring in the country, there is no point from which a walk of twenty miles in a chosen direction will not bring one to some area large or small occupied by a different geological formation.

The surface rocks of Ireland, then, are extremely varied, and represent an immensely long geological history—running back, it may be, a hundred million years or more. And though this history as told by Irish rocks is very fragmentary, there are few complete gaps of any magnitude, save when we come to those comparatively late formations which are well developed, for instance, in SE England, the product of submergences which Ireland escaped.

5. The oldest rocks which now lie exposed anywhere on the surface of Ireland occupy extensive areas in the N and W, and are of late pre-Cambrian age, belonging to the series known as Dalradian. In the N they form an almost continuous mass extending from Lough Neagh and Lough Foyle westward and south-westward to the sea at

Donegal Bay. To their presence is due most of the mountainous country which occupies a great deal of the counties of Londonderry, Tyrone, and Donegal. The rocks consist of gneisses, schists, greenstones, quartzites, pebble-beds, slates, and intrusive granites, of very unequal hardness. Their strongly folded character, and the heavy and long-continued denudation, both aerial and marine, which they have suffered have produced, alike inland and along the coast, highly picturesque features.

In the W, the great Galway-Mayo buttress (**384**), which protects the limestones of the Central Plain from the fury of the Atlantic, is formed mainly of rocks of the same age and of the same types. These ancient rocks, folded and metamorphosed, uplifted and denuded, are connected with much of the finest scenery in Ireland. They support a calcifuge vegetation, of grassy or more usually heathy type. Bands of primitive limestone and other basic rocks occur, but not in quantity sufficient to modify appreciably the flora; one such band, however, is responsible for the appearance of the Maidenhair (*Adiantum Capillus-Veneris*) in its only Connemara station (to a similar cause is probably also due the presence of this fern in Achill and on Slieve League in Donegal (see **106, 193**), though this has been a subject of controversy). A small area of these Dalradian rocks S of Fair Head in Antrim (**460**) produces a change of coastal scenery, and causes the flora to lose some of the calcicole elements which characterize the surrounding basalts.

6. The succeeding Cambrian period has left but small impress on the Ireland of to-day—merely a few patches of slates and quartzites along the SE coast, but these are of local importance, as they form those two bold sentinels of Dublin Bay, the high promontory of Howth (**231**) on the N, and on the S the precipitous Bray Head (**267**) with the Little and Great Sugarloaf (**1660** ft.) behind it. The heathery upland of Howth, formerly an island but now joined to the limestone mainland by a Neolithic raised beach, is classic ground for the botanist and furnishes the most varied flora of any area of its size in Ireland.

7. Next in order of age among Irish strata come the Ordovician and Gotlandian rocks, representing the Silurian Period (in the wide sense of the term). Like the older beds just mentioned, they were long buried under subsequent deposits of various ages, but these have been removed to a considerable extent by denudation, so that the slaty rocks which are so characteristic of the Silurian system now occupy two rather large tracts in E Ireland as well as a number of smaller areas. In the N, they cover a wedge-shaped area extending from Belfast Lough S-W to the Shannon, and back to Drogheda. They form the low rolling lands of Down, and much of Louth, Armagh, Monaghan, Cavan, and Longford. Disintegration yields a light friable soil, which even on the mountains supports a grassy rather than a heathery vegetation. This is an undulating fertile area, with more tillage than is found in most parts of Ireland, and many small lakes set among the fields; it is occupied by a characteristic calcifuge vegetation (**472**). South of Dublin, Ordovician strata lap up against both flanks of the granite core of the Leinster Chain, attaining a considerable development on the E side, in Wicklow and Wexford; much of the most beautiful scenery of Wicklow has been carved out of these rocks (**261**). In the southern midlands, on many of the ridges which rise out of the Limestone Plain, rain and frost have removed the Carboniferous strata from the summits, exposing the core of Gotlandian slates and Devonian conglomerates, elsewhere buried deeply. This is the characteristic structure of all the south-central hill-masses— Slieve Bloom, Devil's Bit, Keeper, Slieve Bernagh, Slieve Aughty, Slievenaman, the Galtees, Knockmealdowns, and Comeraghs. The change of rock thus produced, from lime-stone to sandstone and slate, is accompanied by a marked change of flora, from calcicole to calcifuge.

8. At the close of Silurian times, extensive earth-move-ments began, resulting in crumplings of the crust which threw the Silurian and older rocks into a series of folds running NE from Ireland towards the Arctic Circle. (Long previously, Irish rocks had been subjected to intensive folding (Huronian) with a similar trend, as may be traced in western Scotland; but in Ireland this pheno-menon has been obscured by the superimposed folds of the Caledonian series, with which we are now dealing.) The results of these stresses and of subsequent uplift are seen in the uplands which characterize not only Donegal and Mayo, Wicklow, and the area around Newry, but the Highlands of Scotland and the great raised mass of Scandinavia. In Mayo, Galway, and Donegal, denudation has laid bare over wide areas the ancient pre-Cambrian strata already referred to (**5**), which now form a heathery

mountainous surface. In the Leinster Chain, weathering
has revealed a massive core of granite, which rose in a
molten mass under the Silurian rocks as they became arched
by lateral pressure, so that we have in Wicklow a broad
heathery granite range lapped round by the denuded edges
of the former overlying Ordovician dome. The uniform
nature of the granite has caused it to wear down into
rather similar rounded hills, now deeply peat-covered;
while the slates which encircle them present a more
diversified surface, with picturesque glens and a less
uniform flora.

9. The Devonian Period, which was ushered in by these
mountain-building movements, has left a series of thick
beds of conglomerate, slate and sandstone which are a
conspicuous feature of S Ireland, occupying large areas
of Waterford, Cork, and Kerry. These rocks were formed
not in ancient seas, like most of those with which we have
been dealing, but in large lakes and deserts. Extensive
"inliers" of the same rocks, resulting from denudation of
upthrusted hills, have been already referred to (**7**) as
occurring over the southern part of the great limestone
area of central Ireland (to be dealt with immediately); in
Tyrone Devonian rocks cover an area of several hundred
square miles stretching NW from Enniskillen. The soils
derived from these rocks usually take the form of red
clays; their flora is distinctly calcifuge.

10. We now approach that period—the Carboniferous—
which has left an impress on Ireland and its flora greater
than any other of the many epochs of geological history.
A large area of Europe, including that part which we now
call Ireland, sank below the sea, and, remaining submerged
for a very long time, received on its surface first beds of
mud (now hardened into slate), then thick deposits of a
very limy nature composed largely of the shells of marine
animals (now limestone); next, as the sea-bottom rose,
layers of sand and mud (now appearing as sandstones and
shales) and finally other deposits (the well-known Coal-
measures) laid down under swamp conditions, where beds
of river-mud and sand alternated with layers of vegetable
matter, the debris of ancient forests, now forming strata
of shale, sandstone, and coal. Elevation continued, and,
to premise, the "Ireland" thus produced has in the main
remained a land area throughout almost the whole of the
immense period intervening between Carboniferous times

and our own day. As a consequence, this great series of strata has been exposed almost continually to the agents of denudation, and frost, wind, and water have played havoc with it. The first beds to go were the uppermost, the soft Coal-measures, and with them perished the possibility of Ireland's ever becoming a great manufacturing country. The largest area of these deposits remaining lies in N Kerry (**330**) and the adjoining part of Limerick (**332**), but here they do not contain coal of workable amount. Outlying patches extend northward into S Clare (**340**). Another relic of these once wide-spread deposits forms the Leinster coalfield (**255**) in Co. Kilkenny, now a plateau of about a hundred square miles, lying between the ancient granite chain of Wicklow and the newer fold of Slieve Bloom. In this area a small amount of coal is mined, but its rural character remains unaffected. The shales weather down into a heavy clayey soil, supporting a rather monotonous and uninteresting rushy calcifuge flora. Smaller areas of Coal-measures lying on the hills around Lough Allen (**428**) in Leitrim and at Coalisland in Tyrone need only be mentioned here.

Next, the underlying series of sandstones and shales, constituting the Millstone Grit, Yoredale and Pendleside beds, were attacked and to a great extent removed. In the areas mentioned they crop out from under the remaining Coal-measures, and extend to a considerable degree the area of the calcifuge flora.

Over the wide tracts in the Central Plain from which the Coal-measures and the beds just referred to have been removed, the Carboniferous limestone now extends. Ever since it was laid down, earth-movements have not disturbed, save to a small amount, its horizontal beds. The limestone now forms a slightly undulating plain stretching from Dublin to Galway and far to the N and S. In these latter directions its edges get more and more broken up by patches of older or newer rocks, but outlying fragments of the limestone reach Cork on the S and Donegal in the N. The most characteristic part of the plain lies between two east-and-west lines drawn respectively from Dublin to Galway Bay and from Skerries to Westport, an area of close on 4,000 square miles. Across this one may pass from the Irish Sea to the Atlantic without rising more than 300 ft. (Only occasionally, indeed, as in Sligo and N Clare, does the limestone form hills of any noticeable height.)   Heavily covered with drift in the E, it is occupied there by rich pasture with many trees; in the midlands it is often marshy, with great swelling bogs; while in the W, the rock often lies quite bare (**240–245**).

Since the presence or absence of lime · in the soil determines more than any other chemical factor the nature of the vegetation which the soil supports, this mass of limestone constitutes from a botanical standpoint the most important geological feature of the country.   Its floral characters are dealt with later (**41, 59**).

11. After the end of Carboniferous times, an important epoch of earth-movement, known as the Hercynian, resulted, as at the close of the Silurian Period, in a general uplifting and folding of the Irish area.   These foldings ran normally E and W, and in southern Ireland have given us the striking scenery of Kerry and Cork, and the bold ranges of the Galtees and Knockmeeldowns a little to the N.   Further N the folding was deflected by the older Caledonian axes such as the Leinster Chain, so that Slieve Bloom and other hill-ranges formed at this time have a NE and SW trend. In Cork and Kerry, denudation of the folded Carboniferous and Devonian rocks has left a series of high ribs of slate and sandstone, while the more soluble limestone has been preserved only in the intervening valleys.   In Slieve Bloom and Devil's Bit we have ridges formed of the older rocks (Devonian and Ordovician) from which the limestone covering has been denuded.   In the Leinster coalfield adjoining we see how a syncline or down-fold has preserved to us the beds which lay over the limestone.

With the close of Carboniferous times both the sedimentation and the folding which have produced the Ireland of to-day had to a great degree been accomplished, and during almost the whole of the time which has intervened Ireland has remained above the sea, and at the mercy of the forces of denudation, which have worn down mountain and plain alike, exposing older and older rocks.

The Permian period, which followed the Carboniferous, has left scarcely a trace in Ireland; only a few very limited patches of limestone, etc., in the N, supply evidence of a local submergence.

12. The whole of the Secondary (Mesozoic) and Tertiary (Cainozoic) epochs, during which the SE half of the England of to-day was built up of bed after bed of marine sediments, has left an imprint only in one corner of

Ireland, which, like Scotland and Wales, has remained essentially a Palæozoic area. But that corner, the NE (**455**), continues for us the geological history. There we still have, very locally developed, the red sandstones and gypsum-bearing marls of the Trias, the fossiliferous Rhætic, the grey shales and clays of the Lias, the glauconitic rocks and white Chalk of the Cretaceous. The Antrim Chalk differs from that of southern England in being a much harder rock, and it forms nothing approaching the English Chalk Downs, with their interesting and peculiar flora. There is reason to believe that the Chalk sea extended far over Ireland, but its deposits are now limited to Antrim and the immediate vicinity. Even there, it and the other Mesozoic strata would probably have perished completely owing to long-continued sub-aerial waste, had it not been for the intervention of volcanic outbursts, which poured lavas over the old Chalk land, and thus preserved it; in Eocene times this corner of Ireland was involved in eruptions which buried the Mesozoic rocks under hundreds of feet of dark basalt, forming a high plateau. This plateau, though broken-backed and much denuded, still persists, a peat-covered moorland with a rugged scarped edge which overhangs Belfast and extends round the coast to Lough Foyle; and from under it the buried Secondary rocks peep out. The basalt weathers into a deep heavy soil rich in calcium carbonate, and supports a varied flora which includes a number of plants elsewhere in Ireland found only on the limestone (**455**). The same convulsions caused depressions of the surface which produced the deep hollow, now mostly filled with clay, in which Lough Neagh (**463**), the largest lake in the British Isles, is situated. Near by, in the S part of the adjoining county of Down, a great mass of granite rose under the Silurian rocks, to form the Mourne Mountains (**479**).

13. The next event of importance (after a long interval), and one which profoundly affected the whole country, especially as regards its biological history, was the oncoming of the Pleistocene glacial period. Ice, extending from the Poles, accumulated over the area till, at the time of maximum glaciation, almost the whole country was deeply buried. Once or more than once the ice-front retreated and advanced again. When the land was at last freed, its surface was greatly altered. Hills had been lowered, valleys filled up or deepened, and a vast amount of detritus, including the pre-existing soft covering of the rocks, re-distributed over the country. In some places limy gravels were left spread over areas of acid rocks, or lime-free material over calcareous rocks. The sands and gravels which had previously accumulated on the bed of the Irish Sea, full of marine shells, were gouged out by a great glacier coming down channel from Scotland, and plastered up against the basalts of the NE, the Cambrian rocks of Howth, the granites of the Dublin Mountains, and pushed over much of the low grounds. The relation between the "solid geology" and the "drift geology" was no longer direct, especially along the boundary-lines of the former, a matter of importance to the botanist in Ireland, where the leading feature of the flora is the contrast between the limy and non-limy areas : between a calcicole and a calcifuge flora. The greater part of the lowlands was left covered by a thick deposit of tough clay, mostly full of ice-scratched stones ; and in the final stages of retreat the ice left behind, upon the surface of these clays, great dumps and ridges of gravel stretching especially across the middle of Ireland. The face of the country to-day bears everywhere the impress of this period of glacial erosion and deposition.

14. When the climate returned towards the conditions at present prevailing, a period rather colder and wetter than to-day favoured the growth and accumulation of peat, which attained so wide a development that it forms still one of the most conspicuous features of the surface of the country. In Neolithic times, a depression and subsequent re-elevation of the northern half of Ireland left beds of gravel (raised beaches) fringing the sea, and in the estuaries flat areas of low land formed of muds which accumulated in shallow water. The final act of the drama of our country's evolution has been something more than a mere geological episode. In the latest changes that have taken place in the Irish area civilised man has played the most important part. As a result of his agricultural and other activities native woodlands, formerly extensive, have diminished, lakes and marshes have been drained, bogs cut away for fuel, and the land tessellated by a multitude of fences and hedges, and traversed by a network of roads, canals and railways.

It will be seen that the present surface of Ireland is the result of a very long and varied history. Persistent emergence from the sea following periods of depression, and long-maintained land-surfaces, have allowed of very heavy denudation, so that a mere patchwork of rocks remains, fragments of once wide-spread formations. The ancient Caledonian and Hercynian periods of stress, which by some chance affected the margins more than the centre of our island, impressed themselves so profoundly that the sculpture of the whole country is still dominated by the foldings which they produced.

The ultimate product of the action and interaction of these forces is the Ireland of to-day.

Consult G. A. J. COLE and T. HALLISSY: "Handbook of the Geology of Ireland." London: Murby, 1924. 8s. 6d. Also section "Geology" in the series of provincial geographies (Cambridge University Press) referred to in **3**.

TOPOGRAPHY.

15. Ireland is separated from Great Britain by a channel of very variable width, the narrowest points being $13\frac{1}{2}$ miles between Antrim and Kintire, $23\frac{1}{2}$ miles between Down and Wigton, and 47 miles between Wexford and S Wales. The depth of this channel along its centre varies form 45 to 150 fathoms; the deepest and coldest water found along the coast lies off the Antrim shore. This sea-barrier may be contrasted with that which exists between England and France, which narrows to a width of 23 miles and a depth of 20–25 fathoms at the Straits of Dover, widening rapidly northwards into the broad but shallow North Sea, and more gradually southward along the deeper St. George's Channel. It is important to bear in mind that all these waters lie on a broad shelf, which surrounds Great Britain and Ireland (fig. 1). Along the coasts of Scandinavia, of Spain and Portugal, and of Africa the sea-bottom falls rapidly to oceanic depths—1000 to 2000 fathoms. But if we trace the isobaths of 100 or 200 fathoms northward from the Peninsula, we find they project in a broad curve which passes some 40 miles westward of Ireland, to approach the Continent again on the Norwegian coast. Ireland, like Britain, is essentially a portion of European land, and a slight uplift would re-join both to the Continent.

16. Ireland is an island roughly elliptical in outline, with a longer axis (NNE and SSW) of about 300 miles, and a shorter axis (WNW and ESE) of about 185 miles, and an area of 32,524 square miles. Its limits of latitude are 51° 26′ N and 55° 23′ N (about the same as Holland, N Germany and Poland), and of longitude 5° 26′ W and 10° 29′ W. It is on the whole a very flat island, saucer-shaped owing to its marginal mountain-ranges, and devoid of any distinguishable median watershed (fig. 2). The Limestone Plain stretches across the centre from horizon to horizon, encumbered with great swelling peat-bogs and

Fig. 1.—The Continental Shelf.

embosoming many extensive lakes—a feature unlike anything found in Great Britain. The mountain regions, on the other hand, often recall those of Wales or the Lake District or W Scotland, the most unusual Irish features being the cliff-walled limestone plateau of the Ben Bulben region in Sligo and the basaltic plateau of Antrim; also the scarped mountains in the W, which drop into the

Atlantic in grand precipices, like Brandon in Kerry,
Croaghaun on Achill Island, and Slieve League in Donegal.

17. The marginal position of most of the high grounds
affects the topography in many ways.   In most islands a
central backbone, more or less pronounced, causes the rivers
to flow radially towards the sea, from near the centre to
the margin.   In Ireland the streams that rise on the

FIG. 2.—LAND OVER 500 FEET ABOVE SEA-LEVEL.

seaward side of the mountain masses have short and mostly
steep courses, while those whose sources lie on the inland
side travel far before they reach an outlet.   Most of the
larger Irish rivers—Blackwater, Suir, Nore, Barrow, Liffey,
Bann, Erne, Shannon—rise from 10 to 60 miles inland, but
none of them flows towards the nearest sea; after devious

courses of 100 to 200 miles, they debouch at quite other points (fig. 3). The Shannon, longest river in the British-Irish area, with a course of 214 miles, rising on the hills near Lough Allen 25 miles from Donegal Bay, flows S from that lake for 130 miles with a fall of only 51 feet, forming

FIG. 3.—RIVER-BASINS.

at intervals great lake-like expanses. Then it cuts through hills of Old Red Sandstone and Silurian rocks, and drops 97 ft. in 18 miles to reach tidal waters at Limerick. Thence it continues its course for 55 miles more through a broadening submerged valley to meet the Atlantic at Loop

Head. The course of this river, and those of the Barrow, Nore, and Suir in the SE, give testimony of the great denudation to which the Carboniferous limestone of the Central Plain has been subjected, for after descending to the plain from their mountain sources they flow not towards what are now the lowest areas, but towards the southern hills of sandstone and slate, and cut through them in deep gorges which they have worn in these resistant rocks as the surface of the Limestone Plain was being lowered by denudation.

18. The southern rivers, Blackwater, Lee, and Bandon, with marked W–E courses changing abruptly to S near their mouths, furnish evidence of having undergone considerable natural development since they originated as "consequent" streams. In the early stages of their history they flowed S along the slope of the uplifted peneplain in channels unrelated to the geological sub-structure. But soon after, the drainage became complicated by the development of "subsequent" tributary streams. E-and-W tributaries, working along the strike of the softer Carboniferous rocks which, in consequence of folding and subsequent denudation, alternate with the more resistant beds of Old Red Sandstone, rapidly deepened their channels and eventually became predominant. Only in their lower reaches have the streams managed to retain their original N–S channels across the anticlinal ridges. In Ulster, Lough Neagh and the present course of the Bann owe their origin to a collapse caused by the outpourings of basalt in that region. Further W, the Erne has a singularly sluggish course—a mere succession of island-filled lakes—till at Belleek it plunges over a lip of limestone and rushes down to meet the sea below Ballyshannon. In the E, the Liffey, rising on the inland side of the Leinster Chain, performs a semicircular course of 72 miles before it enters the sea at Dublin, only 14 miles from its source. Among the larger streams, only the Boyne, Lagan, and Foyle have courses that can be described as normal.

19. The Erne is not alone among Irish rivers in being characterized by lake-like expansions. Owing to the soluble nature of the limestone, several of those flowing in the Central Plain show this peculiarity, the lakes being often of considerable size. Thus we have Lough Ree and Lough Derg on the Shannon, and Loughs Carra, Mask, and Corrib on the river Corrib. The quite different origin of Lough

Neagh in the north has been referred to already (**12**). That lake covers an area of no less than 153 square miles, and has a very uniform depth of 50 ft. or less. Around its margin, especially in the SW, a broad band of fresh-water clays of great depth testifies to a considerable former extension of its waters and to a long-continued subsidence of its floor.

20. The coast of Ireland in the N, W, and S is in general bold, deeply indented by drowned river-valleys and broader bays, and the influence of the Atlantic everywhere makes itself felt. In Donegal, Mayo, and Galway the ancient folded pre-Cambrian rocks produce an exceedingly broken coast-line, sometimes low, sometimes projecting in bold headlands and magnificent cliff-ranges, everywhere so exposed that the maritime flora is much reduced, though its vertical limit is greatly increased. Wherever, on the other hand, limestone prevails, the land is usually low, and the sea has penetrated deeply, as in Donegal Bay, Sligo Bay, Killala Bay, Clew Bay, and Galway Bay. In Clare the horizontal shales and sandstones of the Upper Carboniferous still withstand the onslaught of the Atlantic and form precipices which in the Cliffs of Moher rise vertically to a height of over 650 ft. In Kerry and W Cork the coast attains its boldest development. Here the great ribs of Old Red Sandstone project far into the ocean, the summits of the mountains often continuing seaward as high craggy islets. Between these Devonian buttresses the erosion of the weaker Carboniferous rocks of the troughs, coupled with submergence, has allowed the sea to flow far into the former river-valleys in deep wedge-shaped bays, producing very lovely scenery. Throughout E Cork and Waterford a less magnificent but high and broken coast-line prevails, with inlets at the river-mouths due to sunken valleys.

In the E, and there alone, the coast is usually low. The shore of Wexford and Wicklow is characterized by gravel beaches and stretches of sand, where the flora belonging to such ground attains its maximum for Ireland, and several species have their only station (**270**–9). In Dublin and Louth also sandy beaches prevail, broken by occasional rocky headlands. The coast of Down is low, with alternating stretches of sand and of jagged slaty rock. N of Belfast Lough (the drowned valley of the Lagan) the basaltic plateau of the NE presents to the sea a lofty and picturesque scarp.

21. As compared with the greater part of Europe, the most striking feature of Ireland (which it shares with Scotland, Wales, and Scandinavia) is the great age of the main features of the country and of the rocks which produce them. This is especially true of most of Ireland's mountain-ranges. Compared with the heathery granite domes of Wicklow and the ridges and peaks of Donegal, Connemara, and Kerry, the Mourne Mountains are happenings of yesterday : yet the rising of the Mournes heralded the great and long-continued earth-movements from which sprang the Pyrenees, the Alps, the wide Carpathian ring, the Balkans, the snowy Caucasus, and the lordly Himalayas themselves.

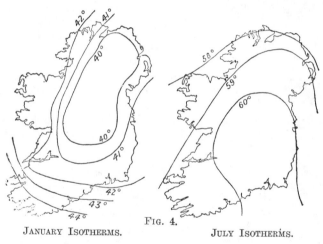

JANUARY ISOTHERMS.   FIG. 4.   JULY ISOTHERMS.

CLIMATE.

22. Ireland lies in about the same latitude as Berlin, Kamtschatka, Alberta, and Labrador. A comparison with the climatic conditions prevailing in any of these areas shows the extraordinary effect of the warm seas and sea-winds which envelop Ireland. The mean annual temperature over the whole island is above 50° F. (10° C.). The annual range is singularly small. The warmest area (the centre and SE) in July (the hottest month) has a mean of barely over 60° F. (15·5° C.), and the coldest area (the centre and NE) in January (the coldest month), a little under 40° F. (4·4° C.) (fig. 4). This may be compared with the 23° F. average range prevailing in England, itself possessing an insular climate in contrast to Continental countries. Most marked are the winter temperatures in the W. The isotherms there, running parallel with the coast, show a January mean of 40° to 42° F. in Donegal, 42° to 43° F. in the Mayo-Galway projection, and over 44° F. in SW Kerry and W Cork—warmer than Bordeaux or Rome.

The January isotherm of 42° F. (5·5° C.), which passes through the whole length of W Ireland from N to S, fringes the Welsh coast, cuts across Devon, runs along the W and S coasts of France, and thence through N Italy and Greece. The July isotherm of 59° F. (15° C.) runs likewise across W Ireland from S to N; then along the Cheviots, up the backbone of Scandinavia, and across the head of the White Sea into Siberia. Frost is rare in the W, and almost unknown on some of the islands. These high winter temperatures, combined with great humidity all the year round, have a marked effect on the vegetation, particularly in the W and S.

Fig. 5.—*Daucus Carota*, from Inishkea, Co. Mayo, showing effect of exposure. ½.

23. But there is also a high degree of exposure. Winter on the W coast might be described as a succession of westerly gales with westerly winds between. The influence of these conditions on the vegetation is clearly seen. In exposed ground, as in S Clare, not a native tree is to be found for miles inland; and everywhere in the western district (as indeed more or less over the whole of Ireland) the bending of trees towards the E is very noticeable (Plate 1 and fig. 17). On the western islands the vegetation is often extremely wind-shorn. On Achill, *Eryngium maritimum*, with spreading rhizomes and roots 3 to 4 ft. long, produces annual shoots from 3 to 4 inches high. On Inishkea, *Daucus Carota* grows quite stemless, with a sessile umbel nestling among a rosette of leaves (fig. 5). See also the description of *Plantago* sward (**63**) for an example of dwarfing by exposure.

PLATE I.

EFFECT OF WESTERLY WINDS ON TREES.

Crataegus Oxyacantha on flat country several miles from the sea near Doonbeg, Co. Clare.

G. Fogerty, Photo.

PLATE 2.

NORTHERN AND SOUTHERN PLANTS IN THE WEST OF IRELAND.

The alpine-arctic *Dryas octopetala* and the Mediterranean *Habenaria intacta* at Ballyvaughan in May.

R. Welch, *Photo.*

On the other hand, the bitter easterly winds which retard the spring flora along the Irish Sea are scarcely felt on the W coast, and exercise no influence on the vegetation. But even in the Dublin area, *Cochlearia danica* and *Ranunculus Lenormandi* commence flowering in January[1]—the latter at 1000 ft. elevation; and on a mid-winter walk, blossoms of 30 to 50 wild flowers may be seen, except after an unusual spell of frost.

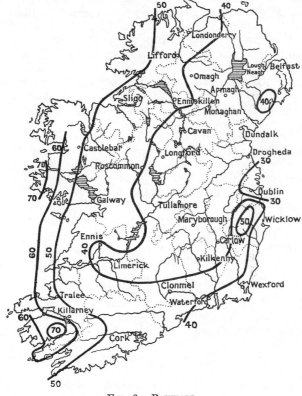

Fig. 6.—Rainfall.

24. Rainfall in Ireland increases from E to W. Along the E coast generally it corresponds to that prevailing on the British side of the Irish Sea. Along the line of the

[1] See also Journ. Bot., 1904, 89.

Shannon (roughly) it has risen from about 30 to about 40 inches (say 750–1000 mm.); and thence to the W coast it rises to 70 to 90 inches (1750–2250 mm.), while in the wettest spots in the western mountains (*e.g.*, Galway, Kerry) precipitation as high as 100 inches (2500 mm.) may occur (fig. 6). The average for sixteen years at Kylemore in Connemara is given as 81·79 inches. The S end of the Upper Lake at Killarney, 75 ft. above sea-level, showed an average of 87·36 inches for 17 years. The station at 1760 ft. on Mangerton in Kerry gave an average for 15 years[2] of 97·40 inches—minimum 63 inches in 1887, maximum 140·9 inches in 1903. But it is the high humidity prevailing in the W, and large amount of showery or drizzly weather, rather than heavy downpours, which produce the almost tropical luxuriance which characterizes the vegetation of sheltered spots, especially in Kerry.

The constancy of moist atmosphere on the W coast may be judged from the fact that, on the north side of Slievemore on Achill (**408**) *Hymenophyllum peltatum* forms an abundant ingredient of the sward covering the slope, growing fully exposed and open to the sky, among *Sphagnum* and stunted *Calluna*; while *Nymphœa alba* grows hard by in a marsh as a terrestrial plant, with short-stalked aerial leaves. The perennial nature of the precipitation in Ireland is seen in the rarity of parched vegetation during summer, causing that intense greenness which astonishes the American or S Continental visitor. In a normal season parching is confined to wall-tops, dry gravels, and places where the rock is a very few inches below the surface. On a well-drained hill-top in Westmeath in August of an unusually dry summer, it was noticed that rank pasture on loamy soil gave no indication of the presence of foundations and slabs of stone buried six inches below the surface—suggesting that in Ireland aerial survey will not prove an efficient means of discovering buried monuments, as it has in drier countries.

25. The general effect of climate and soils in Ireland is to render the country specially suitable for pasture. The forests of oak which once covered much of its surface have under the influence of man given way, not to tillage, but to grass-lands, often very rich, which, apart from mountain

---

[2] The years in question were not quite consecutive, owing to breaks in the sequence of returns.

grazing, occupy more than half of the entire area of the country (fig. 10). An important effect of the exposure is a reduction of the elevation up to which agricultural operations can be carried on. While in sheltered inland districts tillage may range as high as 1000 ft., in W Mayo and W Galway its limit does not exceed 400 ft., and is usually only about 200 ft.; on sheltered slopes in the E it may ascend to 1200 ft.

26. While the exposure has a dwarfing effect on vegetation, and on trees in particular, the mild, moist climate, and also the rich, warm soil that prevails on the limestone, produce a remarkable luxuriance of vegetation where shelter from wind is afforded. Thus, in Burren (**346**), *Asplenium marinum* may be seen with fronds 3 ft. in length; *Ceterach*, exuberantly crenated (fig. 16), attains a length of 1 ft.; *Adiantum*, nearly 2 ft. On Aran (**352**), *Allium Babingtonii* attains a height of 6 ft. On the islands in Lough Erne (**435**), *Solidago Virgaurea* grows 4 ft. in height, *Campanula rotundifolia* 3 ft., and *Polygala vulgaris* forms clumps with a hundred upright stems a foot high. On islands in L. Mask (**371**) *Campanula rotundifolia* grows 3 ft. 3 in. high, and *Orchis maculata* attains a height of 3 ft. By Lough Gill (**420**), *Digitalis* and *Heracleum* have been measured nearly 10 ft. high, *Agrimonia odorata* 7 ft., *Cœloglossum viride* 14 in.; and at Portlaw *Polypodium vulgare* on a wall with fronds over 2 ft. 3 in. long. H. C. Hart records from the banks of the Suir *Apium nodiflorum* 6 ft. high with leaves 3 ft. long, and *Œnanthe crocata* 7 ft. high. In the Killarney woods *Hymenophyllum tunbridgense* forms on rocks sheets of greenery up to 50 sq. ft. in area.

Another effect of the moist conditions is the abundance of epiphytes in sheltered places in the W, unexpected plants like *Saxifraga spathularis* (*umbrosa* auct.) perching on boughs 20 or 30 ft. above the ground (Blackwater near Kenmare). In more than one place in Kerry *Scrophularia aquatica* may be seen growing 6 to 8 ft. high on the top of 10-ft. walls.

GENERAL HISTORY OF THE FLORA.

27. In the history of Europe, Ireland appears to have possessed—save for the interlude of the Ice Age—a far longer period of continuous vegetation than most parts of the Continent, stretching back in all probability (unless interrupted by the Chalk sea) to the uplift that took place in late Carboniferous times (**11**). In this place we are concerned only with the plants which occupy the country at present and their immediate predecessors; and the history of these began in the Pliocene period. But of the Pliocene flora in Ireland we know as yet nothing. The subsequent changes in the vegetation owing to the on-coming of the cold of the Pleistocene period have not so far been traced, nor have deposits yet been found which throw light upon the flora of the mild Interglacial period or periods which alternated with the Glacial phases. Recent research indicates that the Aurignacian mitigation of climate that preceded the final advance of the ice-sheet was of long duration, and that it was accompanied by a considerable uplift of the land. This would have offered opportunity for extensive re-migration into Ireland of plants driven out by the main advance of the ice, and probably many elements arrived at that time. In spite of geological evidence of the presence of ice in every part of Ireland, and the frequent assumption that at the period of maximum glaciation this ice was confluent over the whole country, leaving no portion unexposed, it does not seem by any means certain that the Interglacial immigrant plants did not find in the country survivors of the Glacial and Preglacial floras—rather the reverse. Nor is it certain how far the subsequent and final less severe glaciation may have reduced or exterminated the reconstructed Interglacial vegetation. The geological and the biological interpretations of the evidence relating to these questions are not as yet in accord on many points, and widely different views are still expressed regarding both the time and the mode of arrival of our present plants and our present animals.[2] Controversy

[2] The following mostly recent papers on the Ice Age and its relations to the fauna and flora will illustrate some of the points of view:—E. FORBES: ''On the Connexion between the Distribution of the Existing Fauna and Flora of the British Isles, and the Geological Changes which have affected their area, especially during the epoch of the Northern Drift.'' Mem. Geol. Survey Gt. Britain **1** 336–432. 1846. C. REID: ''The Relations of the present Plant Population of the British Isles to the Glacial Period.'' Brit. Assoc. Report, 1911, 573–577. J. K. CHARLESWORTH: ''Some Geological Observations on the Origin of the Irish Fauna and Flora.'' Proc. Roy. Irish Acad., **39** B 358–390. 1930. M. L. FERNALD: ''Some Relationships of the

on these points has raged particularly with regard to certain ingredients of the Irish flora which have their home in the Pyrenean and Mediterranean region on the one hand, and in North America on the other. These are specially referred to below (**37**), when the plants composing them are considered.

It is not disputed that a full reconstructed flora and fauna, little different from those which occupy the country at present, were domiciled in Ireland in early Postglacial times. Probably belonging to the beginning of this period, one may note the occurrence of *Naias marina* at L. Gur in Limerick[3] in beds which yield remains of the extinct Great Irish Deer, which is assumed to have browsed a vegetation of tundra type. This plant is extinct in Ireland, and has now only one station in Britain (Norfolk). At Ballybetagh in Dublin *Salix herbacea* has been found in beds similarly associated with *Cervus giganteus*, at 750 ft. elevation, its nearest existing stations (in Wicklow) being at over 2000 ft.[4] A high land-level would seem to have persisted after the final passing away of the ice, allowing free re-colonization from the Continent. The break-down of the Irish-British land-connection prior to the disruption of a similar land-bridge between England and the Continent accounts for the absence from Ireland of many common British plants (see **32**) and animals—they were able to migrate overland into England, but found the way to Ireland blocked by the newly-formed Irish Sea.

28. Since the separation of Ireland from England there have been minor fluctuations both of land-level and of

Floras of the Northern Hemisphere.'' Proc. Internat. Congress Plant Studies 2 1487–1507. 1929. J. R. MATTHEWS: ''The Distribution of certain members of the British Flora. III. Irish and Anglo-Irish Plants.'' Annals of Bot. 1926, 773–797. R. LL. PRAEGER: ''Phanerogamia and Pteridophyta'' (Biol. Survey of Clare Island). Proc. Roy. Irish Acad. **31**, part 10. 1911. R. F. SCHARFF: ''The History of the European Fauna.'' London. 1899. G. C. SIMPSON: ''Past Climates.'' Manchester Memoirs 1929–30, 1–34. O. STAPF: ''The Southern Element in the British Flora.'' Englers Bot. Jahrb. **50** (Suppl. Band) 509–525. 1914. A. J. WILLMOTT: ''Concerning the History of the British Flora.'' *In* Contribution à l'Etude du Peuplement des Iles Britanniques 163–193 (Société de Biogéographie III). 1930. These and other papers are briefly summarized and discussed in PRAEGER: ''Recent Views bearing on the Problem of the Irish Flora and Fauna,'' in Proc. Roy. Irish Acad. **41** B 125–145. 1932.

[3] C. REID, Irish Nat. 1904, 162.
[4] A. W. STELFOX in Nature 1927, 781.

temperature. The forests of *Pinus* and *Betula* which characterized early Postglacial times gave way to *Quercus* when a slightly higher temperature prevailed. A few records of plants belonging to these more recent times have been given by Erdtman.[5] The main growth of the peat bogs represents a somewhat colder and wetter phase than that which prevails at present. The white marl which so often underlies the bogs shows already a flora of present-day type.[6] An amelioration of climate in Neolithic times allowed the forests to extend up the flanks of the mountains several hundred feet above the present tree-limit, and permitted the growth of woods on western coasts and islands where now their descendants survive—if at all—as mere bushes. In quite recent times *Quercus* has supplanted *Pinus* completely; but previous to this *Pinus* would appear to have become rare owing to the encroachment of the all-conquering peat, the deterioration of the climate, and finally to grazing in pre-historic and historic times.[7] The peat bogs in their turn have now passed their zenith. On the mountains they are being denuded by wind and rain, and in the plain their increase has usually slackened or ceased, and the operations of man are steadily reducing their area, just as his influence has much lessened the area of lake and marsh in the country, owing to drainage.

CHARACTER OF THE FLORA.

From a European point of view, Ireland, on account of its western position and insularity, is the home of an intensely "Atlantic" flora.

The Atlantic Flora.

29. "Atlantic Flora" is a relative term, dependent on the area which is being dealt with by the user of it, but signifying in all cases a flora developed under conditions of

---

[5] G. Erdtman: "Traces of the History of the Forests of Ireland." Irish Nat. Journ. **1** 242–5. 1927.

[6] C. Reid, Irish Nat. 1895, 131–2.

[7] A. C. Forbes: "Some Legendary and Historical References to Irish Woods, and their Significance." Proc. Roy. Irish Acad. **41** B 15–36. 1932.

greater humidity and more uniform temperature throughout the year than those which prevail in adjoining areas. To
the botanist of eastern Europe the flora of all the western
European countries is essentially Atlantic. The central
European worker might exclude E Spain, E France and
much of Sweden from his Atlantic area, the indefiniteness
of boundary being recognised by the addition of such
terms as "Subatlantic" and "Pseudoatlantic." [8]   For
Britain, Watson's "Atlantic" type signifies "species chiefly
seen in W England" [9]; half of the 62 species to which he
awards full "Atlantic" status do not occur in Ireland,
these being mostly of S English range, and the remainder
have a quite irregular distribution in this country.   The
Irish botanist using the term "Atlantic" would in turn
apply it to those plants which had in Ireland a range
corresponding to the maximum climatic influence, within
the island, of the neighbouring ocean—in other words, a
W and S distribution.   The focus of this flora is in Kerry,
to which area many of the most interesting are confined
(**35**); while some of its members of wider S and W range
are :—

Thalictrum minus	Juniperus communis
Arabis Brownii	sibirica
Draba incana	Taxus baccata
Saxifraga spathularis	Neotinea intacta
Drosera longifolia	Sisyrinchium angusti-
Galium sylvestre	folium
Asperula cynanchica	Naias flexilis
Hieracium iricum	Eriocaulon septangulare
Arctostaphylos Uva-ursi	Sesleria cœrulea
Gentiana verna	Adiantum Capillus-Veneris
Euphrasia salisburgensis	Asplenium viride
Bartsia viscosa	Polystichum Lonchitis
Euphorbia hiberna	

This is by no means a group of "Atlantic" facies in
a European sense. Half of these plants, while a few may
be rather western inside Europe, extend beyond European

[8] Consult *e.g.* M. TROLL: "Ozeanische Züge in Pflanzenkleid
Mitteleuropas," in Festgabe Erich von Drygalski, 1925; H.
CZECZOTT: "The Atlantic Element in the Flora of Poland." Bull.
Acad. Polon. Sc. & Lettr. (Cl. Sc. Nat.), 1926.

[9] Cyb. Brit. 1 43 (1847), 4 409 (1859).

confines into Asia and Africa; some are circumpolar. Three are members of the Lusitanian-Mediterranean migration along the W European coast (**37**); two belong to the N American immigration (**37**). On the British standard, the list contains only two "Atlantic" species, and while Watson's Atlantic plants are mostly southern in Britain, this list contains more of a northern than of a southern facies. From the point of view of distribution it is clear that the group in the list above is homogeneous only in one respect—its western range *within Ireland.* To many of the plants, the bare limestone "crags" of the Galway Bay area are the attraction; to others, the mountains; others again, truly Atlantic in a sense, are members of the Lusitanian and American immigrations.

30. The most genuine Atlantic group in the European (not American) sense consists of those whose world-range is confined to W Europe (a few extending to the Atlantic islands). Irish members of this group include:—

Ranunculus hederaceus	Carum verticillatum
Corydalis claviculata	Conopodium majus
Cerastium tetrandrum	Erica cinerea
Hypericum elodes	Wahlenbergia hederacea
Ulex europæus	Digitalis purpurea
Vicia Orobus	Scilla verna
Saxifraga hypnoides	

This list, selected as representative of the plants exclusively W European in range, shows in Ireland, as might be expected, no general "Atlantic" tendency. They are equally spread on the E and on the W sides of this island. In Britain, 4 of the 13 belong to Watson's "Atlantic" type. This means that Atlantic conditions (in the Continental sense) being prevalent over the whole of Britain and Ireland, these plants can grow equally in E and W, so far as climate is concerned. But that even in Ireland some of them are near the limit of their temperature-range is shown by the behaviour of *Ulex europæus* and *Digitalis purpurea.* The former extends N to Denmark, the latter to the Trondjem fiord in Norway. During hard winters in Ireland both suffer severely. In 1894–5 and again in 1916–7, in Wicklow and Down for instance, large quantities of both species were killed.

THE BRITISH AND IRISH FLORAS COMPARED.

31. The present vegetation of Ireland, as has been seen, was derived largely by immigration from W Europe through Britain subsequent to the passing of the ice. We may now consider the results of this and any previous immigrations, as represented by the flora of to-day. The flora of Britain is so well worked and so widely known that comparative notes as between the two islands are the simplest way of bringing out the salient features of the Irish flora.

The representation of total floras by numbers tends to be misleading on account of varying practice as to the validity of critical plants as species, subspecies, or varieties, and also as regards the inclusion or exclusion of alien plants of various degrees of permanency. On the conservative estimate used in '' Cybele Hibernica '' and '' Irish Topographical Botany,'' the flora of Ireland (species and subspecies of Phanerogams and Vascular Cryptogams, native or naturalized) may be taken as a little over 1000. The figure for Great Britain on the same basis is in the neighbourhood of 1500. On the basis of the '' London Catalogue of British Plants,'' 11th edition, the Irish total is slightly under 1300, the corresponding figure for Great Britain being 2300. The difference of proportion in these aggregate and segregate figures shows the greater extent to which critical plants have been worked out in the larger island.

British Plants Absent from Ireland.

32. Great Britain extends through a range of latitude nearly three times greater than does Ireland (50° to 61° instead of 51½° to 55½°) and N Scotland and especially the dry gravelly soils of S England support many plants unknown in Ireland, so it is not surprising to find in the latter area a smaller diversity of plants than in the larger island; but nevertheless the reduction in the Irish flora is due largely to its insularity : the Irish Sea has proved a serious barrier to immigration from the E, whence almost the whole of our flora is derived. This is clear from the fact that many common plants which extend from S England far up into Scotland have nevertheless failed to reach Ireland. Such for instance are:—

Genista anglica	...	...	89 (or 80 per cent.)			
Ononis spinosa	...	...	71	,,	63	,, ,,
Astragalus glycyphyllos		...	68	,,	61	,, ,,
Lathyrus sylvestris	...	...	67	,,	60	,, ,,
Chrysosplenium alterniflorum			79	,,	71	,, ,,
Scabiosa Columbaria		...	72	,,	64	,, ,,
Paris quadrifolia	...	...	77	,,	69	,, ,,
Convallaria majalis	...	...	67	,,	60	,, ,,
Avena pratensis	...	...	81	,,	72	,, ,,

The numerals following the names show in how many of the 112 vice-counties into which H. C. Watson divided Great Britain each species is found.[10] The majority of these plants have a rather wider range in England than in Scotland, thus approximating to the " British " range of the bulk of the Irish flora, from which they are nevertheless absent.

Common British Plants Rare in Ireland.

33. A more subtle problem is presented by the case of a number of species which have succeeded in arriving in Ireland, but for some reason have not made good, being much rarer on the Irish than on the English side of the Channel. Some no doubt are recent arrivals ("recent" on a plant-migration time-scale), and will in due course extend their range. Others are more or less manifestly relict species, which for one reason or another are decreasing, and in the absence of reinforcements (one of the disadvantages of insularity) will die out. A few of the more striking cases are cited below. The numbers appended to the species express the number of Irish divisions (40 in all) and British vice-counties (112 in all) in which each occurs.

[10] The statistics for Great Britain in this and subsequent lists are taken from Druce's Comital Flora—not a very accurate guide, as there are errors and misprints, but the best available.

Plants rare in Ireland, widespread in Britain.

A. Plants extending from N to S in Britain.

	No. of V.C.'s.		Same in % of V.C.'s.	
	Ireland.	*Britain.*	*Irish.*	*British.*
Trollius europæus	3	53	7	47
Corydalis claviculata ...	6	96	15	88
Helianthemum Chamæcistus ..	1	93	2	83
Geranium pratense	1	97	2	87
Hypericum hirsutum ...	4	92	10	82
Astragalus danicus	1	47	2	42
Ornithopus perpusillus ...	5	86	12	77
Spiræa Filipendula	2	65	5	58
Adoxa Moschatellina .	1	101	2	90
Galium Cruciata	2	95	5	85
Calamagrostis epigejos ..	3	72	7	64
Cryptogramme crispa ...	7	55	17	50
Lycopodium inundatum ...	5	60	12	54

B. Plants of Northern Range in Britain.

Geranium sylvaticum ...	1	54	2	48
Carex pauciflora	1	32	2	29

C. Plants of Southern Range in Britain.

Ranunculus fluitans... ...	1	59	2	53
Teesdalia nudicaulis ...	4	82	10	73
Trifolium subterraneum ...	1	46	2	41
glomeratum	2	21	5	19
Poterium officinale ...	4	69	10	62
Serratula tinctoria	1	68	2	61
Senecio erucifolius ...	5	69	12	62
‡Picris hieracioides	5	62	12	55
Hypochœris glabra	1	55	2	50
Campanula Trachelium ...	5	59	12	53
Limosella aquatica	2	53	5	47
Lamium Galeobdolon ...	4	63	10	56
Colchicum autumnale ...	3	53	7	47
Phegopteris Robertiana ...	1	29	2	26

The majority of the plants in these lists have in Ireland a continuous if limited range. In a few cases some reason—edaphic, or climatic, or both—can be suggested for their restriction in this country, as in the case of three E coast species — *Ornithopus, Trifolium subterraneum,* and *T. glomeratum*; but most are puzzling. A few show in Ireland a widely discontinuous range—*Poterium officinale* (Mayo and the NE counties), *Lycopodium inundatum* (Kerry, Cork, Wicklow, Galway, Mayo). Such discontinuities of range are not uncommon in the flora in general, and are difficult to explain. In some cases they undoubtedly represent relict distribution.

Plants Commoner in Ireland than in Britain.

34. Interesting also is the case of plants which, found both in Britain and in Ireland, have achieved greater success here than in the neighbouring island. The facts in regard to some of the more striking cases may be displayed as follows:—

		Distribution		*No. of V.C.'s.*		*Same in* $°/_\circ$	
		Irel.	*Brit.*	*Irel.*	*Brit.*	*Irel.*	*Brit.*
Spergularia rupicola ...	...	G^{10a}	G	20	31	50	28
Lathyrus palustris ...	...	n^{10a}	s	14	21	35	18
Rubia peregrina ...	...	S	SW	17	25	42	22
Andromeda Polifolia ...	...	G	s	27	35	67	31
Orobanche Hederæ ...	...	s	S	27	23	67	21
Euphrasia salisburgensis	...	W	N	10	2	25	2
Pinguicula lusitanica	...	G	W	34	31	85	28
Utricularia intermedia	...	W	G	23	8	57	7
Euphorbia hiberna ...	...	S & W	SW	11	3	27	3
Scirpus filiformis ..	...	W & S	W & S	26	31	65	28
Rhynchospora fusca ...	...	SW & S	SW & S	20	12	50	11
Trichomanes radicans	...	W	W	15	5	38	4
Lastrea æmula ...	...	G	G	37	36	92	32
Equisetum trachyodon	...	N	N	16	1	40	1
Chara aculeolata ...	...	G	G	27	18	67	16
desmacantha ...	...	G	G	24	13	60	12

[10a] G = General, N = northern, n = rather northern, and so on.

In the case of a few of these plants a reasonable suggestion can be advanced to explain their greater abundance in Ireland. Thus the higher humidity of air and soil may be held to account for the much wider range here of *Trichomanes* and *Lastrea œmula,* and the wide distribution of bog for *Andromeda, Pinguicula lusitanica, Utricularia intermedia, Rhynchospora fusca.* In other cases the advantages which Ireland offers as a habitat as compared with Britain are not obvious.

Irish Plants Absent from Britain.

35. One of the main interests that Ireland offers to the botanical geographer lies in the presence of a number of plants, of diverse origin, which are absent from the neighbouring island. To put the matter in proper perspective we must add to these a few others, clearly of similar class, which are not wholly absent from Britain, though very rare therein—such as *Euphorbia hiberna* and *Naias flexilis;* and also one or two belonging to the same migrations, which have colonized Britain but failed to reach Ireland—notably *Erica ciliaris* and *E. vagans.* The Irish plants absent from Britain and their companions just referred to group themselves as follows:—

1. Plants of Pyrenean-Mediterranean Facies.

	Ireland.	*Britain.*	*Continent.*
Saxifraga spathularis (umbrosa *auct.*)	S, E, W & NW	—	SW
Geum	SW & W	—	SW
Arbutus Unedo	SW & W	—	SW & S
Erica mediterranea	W	—	SW
Mackaii	W	—	SW
ciliaris	—	SW	SW
vagans	—	SW	SW
Dabeocia polifolia	W	—	SW
Sibthorpia europæa	SW	SW & S	SW
Pinguicula grandiflora	SW	—	SW & Alps
lusitanica	G	SW & W.Scot.	SW
Euphorbia hiberna ... ... S, W & NW	SW	SW
Peplis	S	SW & S	SW & S
Neotinea intacta...	W	—	S
Simethis planifolia	SW	SW	SW & S
Glyceria Foucaudii	SW & NE	S	S

The terms S, SW, etc., are here employed in a quite strict sense. SW Europe for instance means at most Spain, Portugal, and the W coast of France.

2. Plants of N American Facies.

		Ireland.	Britain.	Continent.	N. Amer.
Spiranthes gemmipara	...	SW	—	—	NW
stricta	... ...	NE	W. Scot.	—	N
Sisyrinchium angustifolium		SW–NW	—	—	G
Juncus macer	... ...	S, W, & E	Introd.	Introd.	G
Dudleyi[11]	... ...	—	Scot.	—	N
Naias flexilis	... ...	SW–NW	Scot., N. Eng.	N	G
Eriocaulon septangulare	...	SW–NW	W. Scot.	—	N

Hiberno-Lusitanian Plants.   Hiberno-American Plants.

Fig. 7.

DISTRIBUTION OF THE PLANTS IN TABLES 1 AND 2.

Pinguicula lusitanica is omitted in the left-hand map, as it has a wide range in Ireland.

[11] Recently detected in Perthshire (Fernald in Journ. Bot., 1931, 365); a plant of northern N America, unknown elsewhere in Europe.

3. Other Species.

		Ireland.	Britain.	Continent.
Arenaria ciliata ...	..	W	—	Arctic-alpine
Inula salicina ...	...	Shannon	—	Widespread, also Asia.
Chara tomentosa...	...	Shannon and Westmeath	—	Widespread, also Asia, N. Africa.

36. There are also a number of critical Irish plants which as at present known are absent from Britain, and some of them apparently endemic in Ireland; but the progress of field-work may yet reveal them in other places. Among these are :—

Arabis Brownii	Hieracium Scullyi
Alchemilla colorata	Stewartii
Saxifraga Drucei	subintegrum
Sternbergii	Orchis majalis.
hirta	Equisetum Moorei
affinis	Wilsoni
incurvifolia	Chara denudata

37. As regards the Pyrenean-Mediterranean group above, their curious range, mostly along the W coast, and their discontinuous distribution, have led many phytogeographers to assign a very early date for their arrival—either Preglacial, Interglacial, or early Postglacial. By others this is not admitted, and even quite recent arrival across existing seas has been advocated. It is impossible here to go into the pros and cons of a very debatable question; some of the leading contributions to the discussion of this subject have been given in a previous paragraph (**27**); but it should be borne in mind in all consideration of this matter that a fauna, small but varied, of similar distribution, is associated with the plants, and cannot be left out of account when the origin of the latter is being discussed. This fauna includes Mollusks (*Helix pisana, Geomalacus maculosus*), Beetles (*Otiorrhynchus auropunctatus*), False-Scorpions (*Obisium carpenteri*), Woodlice (*Metoponorthus melanurus, M. cingendus, Philoscia couchii, Eluma purpurascens, Trichoniscus vividus*), and Earthworms (*Lumbricus friendi*). The migration of these species across water-barriers offers much greater difficulties than in the case of the plants, and their Hiberno-Pyrenean or Hiberno-Mediterranean range suggests early overland migration.

The American list can be materially reinforced in Scandinavia, where a number of additional northern North American plants, absent from W Siberia but present in Greenland, etc., are found; such are *Ranunculus Cymbalaria, Rhododendron lapponicum, Campanula uniflora, Pedicularis flammea, Carex scirpoidea, C. nadina.* The occurrence of the American *Dulichium spathaceum,* now unknown in Europe, in deposits of the last Interglacial period in Denmark, and also in Poland, suggests that some at least of these plants were early arrivals. *Naias flexilis* also is a NW European Interglacial fossil.

The presence of the American group, which also includes at least one animal (the fresh-water sponge *Heteromeyenia ryderi*), offers great difficulties to any theory of recent or of trans-marine dispersal, and points to early migration, presumably *via* Iceland and Greenland, when the intervening seas were at least much narrower than they are at present. It should be remembered that in North America a corresponding group of European organisms is found along the NE coasts, pointing to reciprocal migration eastward and westward. For different views by geologists and biologists as to the time and mode of migration of these groups, the literature quoted in **27** should be studied.

38. Another element which makes its presence felt, especially on the W coast, which is the head-quarters of all the peculiar groups in the Irish flora, is the Arctic-alpine. The alpine flora of Ireland is in general poor, but in the W especially certain plants characteristically alpine on the Continent become abundant down to sea-level (**66–69**). *Dryas octopetala* (Plate 20), *Arctostaphylos Uva-ursi, Gentiana verna* (Plate 23), *Euphrasia salisburgensis* (Plate 27) may be mentioned as examples.

And so, when studying the flora of W Ireland, we are dealing with a startling mixture of types, not to be found elsewhere in the British Islands or in Europe. The pool from which we gather the American *Eriocaulon* is fringed with Pyrenean *Ericæ*. The cracks which are filled with the delicate green foam of *Adiantum* (Plate 24) are set in *Arctostaphylos* and *Gentiana verna*; *Neotinea intacta,* far from its Mediterranean home, sends up its flower-spikes through carpets of *Dryas* (Plate 2); and *Dabeocia* and *Juniperus sibirica* straggle together over the rocky knolls.

Relict and Incipient Species.

39. Assuming, as seems justified, that very limited range mostly signifies either incipient or relict distribution, we may analyse the single-station plants of Ireland, and by taking cognisance of their general range in Europe, attempt to determine where are the head-quarters from which they are advancing or the last refuges to which they are retreating; the single-station plants we may believe to be either the latest comers, or else the last remaining indication of former wider range. For this purpose we leave out of account plants in any way critical, of which the full range may not yet be determined: also any which are under suspicion of owing their presence to man. We find that of 21 plants so selected, twelve are northern in their general range, namely :—

Arenaria ciliata	Saxifraga nivalis
Geranium pratense	Epilobium alsinefolium
sylvaticum	Pyrola rotundifolia
Astragalus danicus	Carex pauciflora
Lathyrus maritimus	fusca
Rubus Chamæmorus	magellanica

Only four are southern :—

Trifolium subterraneum	Euphorbia Peplis
Erica Mackaii	Simethis planifolia

And five have a general range :—

Ranunculus fluitans	Adoxa Moschatellina
Helianthemum	Scirpus triqueter
Chamæcistus	Phegopteris Robertiana

If we take plants of somewhat wider Irish range, which either occupy a rather limited continuous area or have not more than two or three separated stations, we add 35 more. Of these nine have a distinctly northern range outside Ireland :—

Arabis petræa	Carex paradoxa
Potentilla fruticosa	elongata
Alchemilla alpina	Calamagrostis epigejos
Pyrola secunda	Poa alpina
Melampyrum sylvaticum	

while no less than 16 are of southern type :—

Matthiola sinuata	Erica mediterranea
Helianthemum guttatum	Dabeocia polifolia
canum	Microcala filiformis
Arenaria verna	Sibthorpia europæa
Trifolium glomeratum	Pinguicula grandiflora
Saxifraga Geum	Colchicum autumnale
Diotis maritima	Carex divisa
Arbutus Unedo	Glyceria Foucaudii

leaving six which are neutral in this respect :—

Elatine Hydropiper	Ajuga pyramidalis
Spiræa Filipendula	Asparagus maritimus
Inula salicina	Glyceria Borreri

The first northern list, and to a less extent the second, is rich in alpine plants which are clearly relict from Glacial or Subglacial times. '' Disintegration of area '' is in some cases—*e.g.*, *Potentilla fruticosa*—evident far beyond their Irish territory.[12] Presumably the Neolithic climatic optimum seriously affected them, and many are now in Ireland on the verge of extinction—for instance, *Lathyrus maritimus, Rubus Chamæmorus, Saxifraga nivalis, Carex fusca* (? already gone)—whose solitary Irish station can in each case now be reckoned in square yards. Other plants whose Irish station or stations are far removed from the nearest adjoining one, whether in Ireland or in Britain—*e.g., Astragalus danicus, Epilobium alsinefolium, Calamagrostis epigejos*—are no doubt also relict. But some others, whose Irish stations adjoin those in Britain, are probably incipient natives of recent arrival. The best examples are a NE group of Scottish immigrants —*Geranium pratense, G. sylvaticum, Melampyrum sylvaticum,* etc. (See also **43**.) The Dublin area supplies good similar examples in *Senecio erucifolius* and *Lamium Galeobdolon,* and some of the south-eastern plants, such as *Juncus acutus,* also suggest incipient colonization.

The same explanations suggest themselves as regards the southern groups. Those members of the SW or Lusitanian group which are included in the table in **35** are to my mind clearly relict. Vanishing species also are

---

[12] See for instance O. DRUDE in New Phytologist, 1913, 245.

Euphorbia Peplis (apparently gone), *Matthiola sinuata, Diotis maritima.* Probably relict, in view of the wide gap between their Irish and next stations, are also *Helianthemum Chamæcistus, H. guttatum, H. canum, Glyceria Foucaudii,* etc. Beyond these few suggestions it does not seem very safe to go.

In viewing these facts, it should be remembered that a number of other species have a wider range in Ireland than in Britain, whence they must have come to us. A list of some of these is given elsewhere (**34**). Whether the present disproportion is due to decrease in England or to more suitable conditions prevailing in Ireland is difficult to guess. Possibly they should rank in Ireland as incipient species in an advanced stage of success, possibly as decaying species in Britain, possibly as both.

40. SUMMARY.—We may summarize the foregoing paragraphs as follows :—the leading feature of the Irish flora as compared with that of Britain is its great reduction, about one-third of the flora of Britain failing to re-appear in Ireland. Among those plants which have colonized Ireland, there are notable examples of species which have not succeeded, and appear near extinction, or which, arriving recently (in a comparative sense), have not yet had time to extend their range. There are others which have achieved conspicuous success in Ireland, so that they are here actually or relatively more abundant than in Britain. Finally, there is a number of species in Ireland not found in Britain. These belong especially to southern (mostly Pyrenean) and to north-western (N. American) immigrations, both of which have affected also to a slight extent the flora of Britain and the adjoining parts of Europe.

DISTRIBUTION OF PLANTS WITHIN IRELAND.

41. In Ireland two general types of distribution govern the flora.[13] The first is primarily edaphic in character, caused by the general geological structure of the country— a central plain of limestone surrounded by groups of hills formed of non-calcareous rocks. This produces a Central and a Marginal type. The Central plants are largely calcicole species, inhabiting dry ground, marshes, and lakes;

[13] R. LL. PRAEGER: ''On Types of Distribution in the Irish Flora.'' Proc. Roy. Irish Acad. 24 B 1–60. 1902.

a few belong to the great peat bogs. Characteristic members of this group are :—

Ranunculus circinatus	Gentiana Amarella
Stellaria glauca	Teucrium Scordium
Rhamnus catharticus	Orchis morio
Myriophyllum	Ophrys apifera
verticillatum	muscifera
Sium latifolium	Potamogeton coloratus
Carlina vulgaris	Chara aculeolata
Crepis taraxacifolia	tomentosa
Andromeda Polifolia	

An interesting feature in connection with this group is the tendency of a number of its members to extend their boundaries in a NE direction, forming an arm as far as Lough Neagh. This appears to be due to an absence of hill-barriers in that direction, combined with NE extension of the Carboniferous limestone almost to the shores of that lake. Similarly a few penetrate SW into Kerry.

The plants of Marginal distribution (other than halophytes, which have no choice), while of homogeneous range, are of heterogeneous habitat; they comprise many alpine and montane species, and many of calcifuge tendency; also some xerophytes for which the light or sandy soils found near the sea are an attraction. Representative species are :—

Cerastium tetrandrum	Filago minima
Hypericum elodes	Hieracia
Radiola linoides	Jasione montana
Erodium moschatum	Scleranthus annuus
Trifolium striatum	Salix herbacea
arvense	Carex dioica
Vicia sylvatica	Milium effusum
Sedum roseum	Lycopodium alpinum
Callitriche intermedia	Nitella translucens

42. The other general type of distribution, which is partly due to topography, partly in all probability to the general direction of Postglacial immigration, and largely climatic, tends to produce isophytic boundaries running NE and SW, as suggested by fig. 8. In this connection fig. 4 should also be borne in mind, where both summer and winter isotherms are seen to follow similar directions, at least on the W coast. Precipitation also, in a broad sense, follows corresponding lines (fig. 6). See **47**.

North-western Plants.

The plants of general northern and alpine range (Watson's "Scottish" and "Highland" types, the former especially), while concentrated in northern Ulster, run down the W coast in much larger numbers than down the E, though the E possesses plenty of high ground and a lower winter temperature. Examples of such range are :—

	Watsonian type.	*Irish range.*
Silene acaulis ...	H	Derry, Don., Leitr., Sligo, Mayo.
Saxifraga oppositifolia	H	Derry, Don., Leitr., Sligo, Mayo, Galway.
aizoides ...	H	Antrim, Don., Ferm., Leitr., Sligo.
Callitriche autumnalis	S	Mainly in the N. and NW.
Circæa alpina ...	S–H	Ulster (exc. Cavan and Mon.), Louth, Leitr., Sligo.
Hieracium lasiophyllum	—	Down, Antrim, Derry, Don., Ferm., Galway.
Pyrola media ...	S	Down, Antr., Derry, Tyr., Ferm., Don., Sligo, Leitr., Mayo, Clare.
Lamium molucellifolium	S	Mainly in the N. and NW.
Salix phylicifolia ...	S–H	Antr., Derry, Don., Leitr., Sligo, Mayo.
Potamogeton filiformis	S	N. half of Ireland, mainly western.

Very few "Scottish" or "Highland" plants can be found which prefer the E coast.

43. The presence in Ulster only (with head-quarters in Antrim) of a number of species mostly "Scottish" in their distribution in Britain, strongly suggests direct and probably recent colonization from Scotland. Examples are :—

	Watsonian type.	*Irish range.*
Ranunculus fluitans ...	English	Antrim.
Trollius europæus ...	Scottish	Donegal, Fermanagh.
Geranium sylvaticum ...	Scottish	Antrim.
pratense	British-Eng.	Antrim.
Rubus Chamæmorus ...	Highland	Tyrone.
Ligusticum scoticum ..	Scottish	Don., Derry, Antrim, Down.
Pyrola secunda ...	Scottish	Ferm., Antrim, Derry.
Melampyrum sylvaticum	Scottish	Antrim, Derry.
Equisetum pratense ...	Scottish	Ferm., Don., Antrim.

On the E side of Ireland we find no corresponding group of "English" type plants of limited range save species confined to the coast. The bulk of the "English" and "British" plants have swept right across Ireland. The "British" plants are everywhere, each of the forty Irish Divisions possessing 90 to 100 per cent. of the species present in this country. The "English" plants, also very widespread, show some tendency to decrease towards the NW, especially in inland areas.

44. As regards the maritime flora, it is worthy of note that a number of plants which are by no means confined to the coast in Britain are essentially or exclusively maritime in Ireland. These are largely but not entirely "English" plants, widespread especially on the light soils which prevail in S England. In Ireland they are found mainly on the E side ("E" in the last column of the succeeding table), but some range round the whole coast ("EW"). Those marked * have also one inland station, and those marked ** two, mostly on lake shores. The second column indicates the Watsonian type of distribution in Britain.

*Cerastium semidecandrum ...	...	...	BE	EW
*Sagina subulata ...	...	...	...	SB	W
Trigonella ornithopodioides	...	...	E	E
Trifolium subterraneum	...	...	...	E	E
* arvense ...	...	...	...	BE	EW
striatum ...	...	...	...	E	E
scabrum ...	...	...	...	E	E
glomeratum ...	...	...	...	E	E
**fragiferum ...	...	...	...	E	EW
Ornithopus perpusillus[14] ...	...	...	BE	E
Vicia lathyroides ...	...	...	...	BE	E
Œnanthe Lachenalii	...	...	...	EB	EW
Plantago Coronopus	...	...	...	B	EW
**Carex distans	...	...	...	B	EW

[14] One station also on the W. Cork coast.

South-eastern Plants.

45. There is a group of East Coast species, largely maritime, which tends to range S rather than N. Exclusive of these, there is a fairly well-marked group with headquarters in the S, which tends to range up the E coast more than the W. Examples of the latter group are :—

	Watsonian type.	*Irish type.*
Linum bienne	A–E	Kerry to Boyne, Limk., Clare.
Wahlenbergia hederacea ...	A	Kerry, Cork, Wexf., Wickl., Dubl., Mayo.
Erythræa pulchella ..	E	Cork, Waterf., Wexf., Dubl., Leix.
Orobanche major ...	S–A	Cork, Wexford, Wicklow.
Salvia horminoides ...	G–L	Coast Cork to Louth, Kilk., Clare.
Juncus acutus	E–A	Cork, Waterf., Kilk., Wexf., Wickl.

It might be thought, in view of what one might call super-Atlantic conditions which find their maximum in Kerry and W Cork, whence they continue with slowly diminishing intensity up the W coast, that a marked group of plants of corresponding range would be found. This is true only in a general sense. The Lusitanian and American groups as a whole show this range, but individually they display considerable diversity of distribution. Conditions all along the W coast are sufficiently similar to allow a wide dispersal of the W coast species, and as a whole the plants southern in Ireland tend to range up the E rather than the W coast.

46. If we analyze the Irish flora according to Watson's well-known Types of Distribution we do not get far. His classification, based on the range of plants in Great Britain, fails in many respects if applied to Ireland. His "English" group is irregularly spread in this country, showing a general SE trend; "Germanic" plants are almost non-existent here, as might be expected from the definition; the "Scottish" plants are in Ireland northern and western, the "Highland" species the same, but more western and less northern, while "Atlantic" species range all round the coast of Ireland, being rather more plentiful in the S half. But if we take Watson's "English" type as representing broadly the southern element in the

Britannic flora, and his ''Scottish'' plants as representing the northern, we find that the limits of these groups in Britain and Ireland may be represented by rather suggestive curves as shown in the accompanying diagram (fig. 8).

Fig. 8.—Isophytic Lines in the Britannic Flora.

AA. Northern limit of the ''English'' flora.   BB. Southern limit of the ''Scottish'' flora.   From Praeger, Types of Distribution, &c., *loc. cit.*

47. This general NE-SW trend represents a feature which recurs in many connections. It is the direction of the ancient ''Caledonian'' folding which produced the leading topographical features of Scotland and of much of Ireland; and a glance at a geological map shows that it suggests also the main boundaries of the rocks of the southern half of England. These physical features are reflected in the distribution of the vegetation. Meteorological maps will show that it has besides a considerable climatic significance (see figs. 4, 7). It represents also the main front along which invasion advanced from the SE after the passing away of the ice, spreading W and N from a land-connection in the region of the Straits of Dover, and another across the Irish Sea. Its prevalence in the distribution of the flora is well shown for Britain in the maps illustrating J. R. Matthews' papers,[15] and for Ireland in a paper of my own.[16]

VEGETATIONAL SUBDIVISIONS.

Woodland.

48. The *Pinus* forests that long dominated the Irish landscape gave way to *Quercus,* and the *Quercus* woods under human influence have been replaced largely by grass-land, and now are seen only where the ground is unsuitable for pasture. But in certain areas—for instance in Kerry at Killarney and Loo Bridge, in Mayo about Pontoon, in Sligo on Lough Gill, in Wicklow in the Vale of Clara—the aboriginal Oak woods still occupy considerable areas, and may be studied to advantage. The native trees which mix as equals with *Quercus* (usually *sessiliflora*), which is often completely dominant, are *Betula* (frequently) and less often *Fraxinus.*[17] A number of smaller trees often form a lower stratum in the Quercetum—*Ilex, Corylus, Sorbus Aucuparia*; and more rarely *Cratægus, Populus tremula, Taxus,* and about Killarney *Arbutus* (Plate 13); where the *Quercus* is dwarfed, these may equal it in stature and form a mixed wood.

In damp places *Alnus* and *Salices* abound. On the western limestone "crags" dense *Corylus* scrub is often strongly developed, as along the eastern base of the Burren hills in Clare; it may be accompanied by *Euonymus, Cornus,* and the two species of *Rhamnus* (**348, 350**).

In addition to *Quercus, Betula pubescens* alone ever forms pure woods, as at Correl Glen in Fermanagh (**438**) and near Pontoon in Mayo (**374**), Mixed woods usually include a good deal of planted trees, particularly *Pinus sylvestris* and *Fagus sylvatica.* The native woods of the hills are being replaced largely by *Pinus, Picea,* and *Larix*

[15] J. R. MATTHEWS: ''The Distribution of certain portions of the British Flora.'' I–III. Annals of Bot. **37, 38, 40** (1923, '24, '26).

[16] R. LL. PRAEGER: ''On Types of Distribution in the Irish Flora.'' Proc. Roy. Irish Acad. **24** B, 1–60, 1902.

[17] *Ulmus montana,* the only other large indigenous tree, can seldom be claimed as native except on cliffs and rough ground, mostly limestone.

under the ægis of the Forestry Department. Some of the exotic trees, particularly *Pinus, Fagus,* and *Acer Pseudo-platanus,* seed freely and tend to extend their boundaries. The ground-flora varies, that of acid soils being the more pronounced, often with a great abundance of *Scilla non-scripta, Vaccinium Myrtillus, Luzula sylvatica, Melampyrum pratense,* and a profusion of ferns. The saprophytic and parasitic flora of old forests is poorly represented in Ireland. Many even of the English members of these groups are missing, and *Neottia, Monotropa,* and *Lathræa* are here in sole possession. Among the few other essentially old forest species of Ireland are *Cephalanthera ensifolia, Milium effusum, Festuca sylvatica, Pyrola media, P. minor.*

Descriptions of different types of native woodland will be found below, as in **266, 312, 313, 320–2, 348–9, 362, 372, 374, 420,** etc.

48 a. The history of Irish forests has not yet been fully worked out. For the earlier (Postglacial) periods, Erdtman has supplied some information by means of pollen analysis. While this method is invaluable for historical purposes, in tracing the changes of vegetation in successive strata, its results, when applied geographically, have to be interpreted in a broad sense, since pollen may be and is carried for long distances by wind. For instance, *Tilia* pollen has been obtained in Ireland, though it is very doubtful if any species of Lime was ever a native of Ireland. Augustine Henry has made useful contributions to the subject as regards medieval times (Louth Archæol. Journ., **3,** 237–245). A good discussion of Irish woods in relation to early human influence will be found in the paper by A. C. Forbes already quoted (**28**). For the study of the more recent developments in this matter, and especially the question of the introduction of exotic trees of various kinds into Ireland, the best material is another paper by A. C. Forbes.[18] He finds that the Church was probably the earliest planter of trees in Ireland, as far back as the 15th century, *Taxus* being the tree mostly used. The laying down of plantations did not take place on an appreciable scale, he demonstrates, until the 18th century, and then chiefly in conjunction with the creation and

---

[18] A. C. FORBES: "Tree Planting in Ireland during Four Centuries." Proc. Roy. Irish Acad. **41** C 168–199. 1933.

improvement of demesnes. The present tendency in tree planting is that of using Western American conifers in place of European species. This, he states, will undoubtedly alter the appearance of the country during the present century.

While State planting is not proceeding nearly so quickly as was envisaged by the Irish Forestry Committee of 1908, and while owing to the breaking up of estates under the Land Commission and unsatisfactory general financial conditions private planting has steadily declined, nevertheless in many parts of the country dense areas of conifers are replacing natural *Quercus* and *Betula* woods, and tracts of rough pasture and heath, all to the detriment of the indigenous flora.

49. The tree-limit is extremely variable, on account of the accentuated effect of aspect caused by the prevalence of strong westerly winds. Along the exposed W coast it is at sea-level, but shelter at once raises it to from 600 to 800 ft. even in Kerry, as about the Upper Lake of Killarney. On sheltered slopes in the E it may lie at about 1200 ft. or more, but the natural *Baumgrenze* is seldom seen on account of cutting, planting, enclosing, and grazing. As mentioned above (**28**), it has shifted both eastward (in the west of Ireland) and downward since the climatic optimum of Neolithic times, and may be descending still. In the extreme west, intensive grazing has combined with deterioration of climate to reduce arboreal vegetation to the condition of rare stunted bushes; and in some districts it is only on cliffs and on islets in lakes that trees are seen at all.

Grassland.

50. There is a good deal of evidence to show that considerable areas in Ireland were never under forest, but have been occupied by a grassy vegetation throughout the Postglacial period. This appears to be true especially of the centre and west. A progressive change in the composition of the grassland flora may be noted as one passes across Ireland from the rich pastures of Dublin and Meath, growing on deep soil derived from a thick mantle of calcareous Boulder-clay, to the rock-pastures which fringe Galway Bay. In the former areas—for instance, around the famous Hill of Tara, the ancient residence of the kings of Ireland—the grass is dense and luxuriant, and harbours a quite limited number of other plants; while in the west the grasses of the cragland are mixed with a non-gramineous flora rich in numbers and variety, and of high botanical interest (**344, 346, 352,** etc.).   Over large areas in the centre the subsoil is wet, as along the Shannon, and deep meadow-land prevails, with an admixture of paludal species—Orchidaceæ, Juncaceæ, and so on—often a quite rich flora.   The vegetation of a very flat area liable to flooding, near Monasterevan, is described in **239.**   The basalts of Antrim weather into a deep rich clay, with heavy grass; the Coal-measures likewise, but poor, and their rushy calcifuge vegetation is mostly monotonous in the extreme.   The Silurian slates of the NE yield a light loam which produces a short sward, with much *Agrostis tenuis, Aira caryophyllea,* etc., and a flora limited in variety.

51. An average low-level old grassland, well-drained, not particularly calcareous nor the reverse, gave the following list of plants :—

Ranunculus acris
  bulbosus
Lotus corniculatus
Trifolium pratense
  repens
Potentilla Anserina
  sterilis
Heracleum Sphondylium
Pimpinella Saxifraga
Galium verum
Chrysanthemum
      Leucanthemum
Achillea Millefolium
Senecio Jacobæa
Cnicus palustris

Cnicus pratensis
Hypochæris radicata
Centaurea nigra
Veronica Chamædrys
Prunella vulgaris
Plantago lanceolata
Rumex Acetosa
Orchis maculata
Briza media
Anthoxanthum odoratum
Cynosurus cristatus
Dactylis glomerata
Holcus lanatus
Festuca rubra
Agrostis tenuis

This was a pasture with a foot of soil overlying flattish esker material containing a good deal of limestone, resting on Silurian rock, a mile west of Dunlavin in Wicklow.

Peat-Bog.

52. More than one-seventeenth of the whole area of Ireland is returned as being covered with peat-bog.   While mountain bog accounts for a fair share of this, the greater part lies on the lowlands, forming the "red bogs" of the Central Plain. The question of the date and growth of the peat-bogs has already been mentioned (**28**). While on the mountains decay is often manifest, and the peat and its flora are in places being destroyed by wind and rain, on the plain, while growth has often practically ceased, the only enemy is man; turf-cutting is slowly reducing the characteristic flora, first by drainage and then by the obliteration of the bog by turf-cutting. In areas where bogs were few, as in Dublin and Down, the bog flora is almost extinct; but over the greater part of Ireland it is still abundant and highly characteristic of the country.

The mountain bogs differ from the "red bogs" chiefly in their better drainage and its effects, the most conspicuous of which is the stronger growth of *Calluna*; but where the mountain bogs are flattish, with pools, the difference in flora is less, and plants like *Andromeda, Oxycoccus, Rhynchospora alba, Schœnus* and *Carex lasiocarpa,* typical members of the vegetation of the lowland bogs or swamps, may come up from the plain and colonize them at 1000 to 1500 feet.

53. The great red bogs are very interesting, and form a remarkable feature in the Central Plain. Rising from the edges to the centre, like inverted saucers, they present brown, treeless, smooth tracts, 15 to 25 ft. higher than the surrounding farm-land. Their edges have almost invariably been nibbled into by man, and a floor of peat is often left, used for drying and stacking turf, on which the bog flora has been temporarily destroyed. *Rumex Acetosella, Holcus mollis, Agrostis tenuis* colonize this ground; and on rougher parts *Digitalis* and other characteristic calcifuge species, *Betula alba* (*verrucosa*)*; Lastrea spinulosa, Osmunda* are often abundant.[19] Turf-cutting drains to a greater or less extent the fringe of the bogs, and encourages the growth of Calluna, while injuriously affecting most of the bog species. Beyond this influence, we get the characteristic bog flora; *Calluna,* though stunted, is usually dominant; *Erica Tetralix, Scirpus cœspitosus, Narthecium, Eriophorum vaginatum, E. angustifolium* are abundant, with much *Sphagnum,* among which ramble the straggling stems of *Andromeda* and

---

[19] See J. M. WHITE: "Re-colonization after Peat-cutting." Proc. Roy. Irish Acad. **39** B 453–476, Pl. V–IX. 1930.

Oxycoccus; and often bosses of *Racomitrium uliginosum*. The shallow sinuous pools and depressions are full of *Sphagnum*, *Rhynchospora alba*, *R. fusca*, the three species of *Drosera*, *Menyanthes*, *Utricularia minor*, *Carex limosa*.[20]

The bogs display here and there, in the midst of their typical vegetation, colonies of relict plants revealing a former condition of the now uniform flora. Thus thin *Phragmites*, *Cladium*, *Juncus subnodulosus*, or *Carex lasiocarpa* growing on the bog betray the sites of former lakelets. Sometimes a marked change of vegetation occurs over a limited area which shows perhaps the site of a swallow-hole in the limestone under the bog or other cause of better drainage and less acid and saturated conditions. For instance, among the stunted flora of the large bog where *Sarracenia* flourishes near Termonbarry in Roscommon (**380**) is a sudden patch of tall *Calluna* and *Lastrea spinulosa* 3 ft. high, with *Salix aurita*, *Rubus idæus*, *Menyanthes*, *Cnicus palustris*, *Athyrium*, *Lastrea Filix-mas*, *L. aristata*, *Andromeda*, *Eriophorum*, *Anthoxanthum*, *Holcus lanatus*—a rather peculiar assemblage.

54. Lists showing the flora of typical small areas of some selected bogs in different parts of the country may be of interest.

Extensive Bog one mile north of Newbridge, Kildare.

This bog represents the most easterly extension of the typical Central Plain "red bogs." It had a smooth wet surface with many shallow *Rynchospora*-filled depressions.

Calluna dominant, with much *Eriophorum vaginatum*, giving a grassy surface. Also:—

Sphagna, c.	Aulocomnium palustre, f.
Rhynchospora alba, c. in pools	Drosera anglica, r.
	Eriophorum angusti-
Erica Tetralix, f.	folium, r.
Narthecium ossifragum, f.	Scirpus cæspitosus, r.
Drosera rotundifolia, f.	Cladonia, r.

The cut-away fringe was invaded by luxuriant *Juncus effusus*, *Lastrea aristata*, *L. spinulosa*, and *Pteris*.

[20] See also Irish Top. Bot., pp. xxx–xxxii.

It will be noted that two of the most abundant plants of the Central Plain bogs—*Andromeda* and *Oxycoccus*—are absent here in the east.

Bog near Killashee, Longford.

Very smooth surface, with great abundance of *Rynchospora alba*.

Rhynchospora alba, v.c.	Drosera rotundifolia
Calluna vulgaris, v.c.	anglica
Narthecium ossifragum, v.c.	Andromeda Polifolia
Rhynchospora fusca, c.	Scirpus cæspitosus
Erica Tetralix	Carex diversicolor
Eriophorum angustifolium	Lycopodium Selago.

Bog near Frankford, Offaly.

Flora poor and stunted, apparently dying out.

Calluna vulgaris	Eriophorum angustifolium
Scirpus cæspitosus	Narthecium ossifragum
Erica Tetralix	Drosera rotundifolia
Eriophorum vaginatum	Rhynchospora alba

Monmor Bog, Clare.

Extensive bog. Surface smooth, with low grassy vegetation. Fully exposed to the Atlantic.

Calluna vulgaris, c.	Erica Tetralix, f.
Narthecium ossifragum, c.	Drosera, 3 species.
Rhynchospora alba, c.	Menyanthes trifoliata
Eriophorum vaginatum, c.	Orchis maculata
Cladonia rangiferina, c.	Scirpus cæspitosus
Campylopus fragilis, c.	Sphagnum spp.

No pools, and hence absence of *Rhynchospora fusca,* etc. Great uniformity of vegetation. Occasional patches of *Myrica* and *Molinia.*

Large Bog by Cashen River, N. Kerry.

Myrica Gale ⎱ dom.	Erica Tetralix, f.
Rhynchospora alba ⎰	Orchis maculata
Narthecium ossifragum,	Schœnus nigricans
v.c.	Molinia cœrulea
Drosera, 3 species, c.	Cladonia
Eriophorum, 2 species, f.	Sphagna

The frequent great abundance of *Myrica* is a characteristic feature of the Kerry bogs.

Analysis of bog vegetation in Connemara, where *Molinia* moor is the prevailing type, will be found elsewhere (**387**): and an extended study of the same is in the paper quoted in **385**. Some notes on the bog flora of the Dublin mountains appear in PETHYBRIDGE and PRAEGER: '' The Vegetation of the District lying south of Dublin '' (Proc. Roy. Irish Acad., **25** B 124–180, pl. vii–xi, coloured map). Other useful notes will be found in A. G. TANSLEY and others: '' The British Vegetation Committee's Excursion to the West of Ireland '' (New Phytol., 1908, 253–260); and R. Ll. PRAEGER: ''The Flora of Achill Island '' (Irish Nat., 1904, 272–3).

55. Where bogs become the site of colonies of breeding gulls a remarkable change in the flora may occur, the trampling and guano of the birds killing out the natural vegetation, their place being taken by a rank vegetation brought by the birds in their crops or on their feet. A study of this as exemplified by a great colony of Black-headed Gulls on a bog near Tullamore in Offaly will be found in '' Irish Naturalist,'' 1894, 175.

56. The flora of bogs and of their vicinity is sometimes altered by the interesting phenomenon of bog-slides or bog-bursts, whereby a portion of a bog may suddenly discharge its lower semi-liquid portion, leaving the upper crust much fissured and drained, if not actually removed; the ejected material being deposited in a viscid layer at lower points. The chief effect of this occurrence is to encourage the growth of *Calluna* on the affected area of the bog owing to the improved drainage, and the sites of former flows can be sometimes plainly seen in flattish saucer-shaped depressions with a rough surface and increased amount of Ling. Where farm-land has been covered by the flowing peat, this has been cut for fuel as soon as it has consolidated, and the land restored to its former use.

Ireland is pre-eminently the country of bog-flows, a greater number being recorded from Ireland alone than from the rest of the world. An account of the most catastrophic slide which has happened, with a summary of previous occurrences of the kind, will be found in PRAEGER and SOLLAS : '' Report of the Committee . . . to investigate the recent Bog-flow in Kerry '' (Sci. Proc. Roy. Dublin Soc., n.s. **8**, 475–508, pl. xviii–xix, 1897). Reference may also be made to PRAEGER : ''A Bog-burst seven years after'' ('' Irish Nat.,'' 1897, 20) and '' Report on the recent Bog-flow at Glencullin, Co. Mayo,'' by A. D. DELAP, A. FARRINGTON, R. LLOYD PRAEGER, and L. B. SMYTH, in Sci. Proc. Roy. Dublin Soc. **20**, 181–192. 1932.

Marshes.

57. Before systematic draining dried the land, marshes and swamps occupied much of the surface of Ireland. The deposits of the Ice Age, laid down on the flatter areas irrespective of the previous drainage system, tended to the production of much standing water in the form of shallow lakes. The silting up of these, owing to the formation of marl, accumulation of detritus brought by streams, and vegetable growth, produced many areas of marsh, of which a considerable part still remains, especially in the Central Plain. The wetter undrainable portions form one of the few sanctuaries of an undisturbed flora, since cattle cannot readily penetrate, and they often yield a number of rare plants. The substratum may be acid or alkaline, but it is the marshes on the limestone which yield the more varied and more interesting flora. The presence of lime in wet grassland is mostly made conspicuous at once by the replacement of *Juncus effusus* by *J. inflexus* and *J. subnodulosus*; with them are often much *Parnassia, Cnicus pratensis, Orchis maculata, Gymnadenia conopsea, Platanthera bifolia, P. chlorantha, Selaginella*; and when the meadow degenerates into marsh, limy or boggy, a taller, coarser vegetation comes in—much *Schœnus, Carex lasiocarpa, C. acutiformis*, with colonies of *Thalictrum flavum, Lysimachia vulgaris, Epipactis palustris*, and sometimes the rare *Lathyrus palustris.* The edges of pools, which gleam white in summer with limy incrustation, are tenanted by tall groves of *Cladium* and *Phragmites,* while in the water *Ranunculus circinatus, Potamogeton coloratus, Chara aculeolata,* are characteristic species, often heavily incrusted with lime. To give an extreme example, Lough Cloughballymore, 2 miles N of Kinvarra on Galway Bay, consists, in summer at least, of a flat expanse of pinkish limy mud set with stools of *Schœnus,* patches of thinly spread *Cladium* and *Phragmites,* and occasional clumps of *Carex Hudsonii,* and scarcely any other vegetation.

By some small lakes on the limestone in S Clare, the change from ordinary pasture to water some four ft. in depth was indicated by the following succession of index plants :—

> Juncus inflexus.
> Iris Pseudacorus.
> Juncus subnodulosus.
> Schœnus nigricans.
> Carex lasiocarpa.

—————————————————————————— Water Level.

> Carex inflata.
> Phragmites communis.
> Sparganium ramosum.
> Typha latifolia.
> Equisetum limosum.
> Nuphar lutea.

In some places in the Central Plain, vegetation has not yet succeeded in colonizing the stretches of marl left bare by drainage, and the latter still extends white and bare, covered along its margin by black peat. *Phragmites* is usually the first plant to invade it, growing dwarf and sparse.

Around Lough Neagh, marshy land mostly flooded in winter by alkaline or neutral water produces well-marked fen, the flora of which has been lately the subject of careful study.[21]

[21] J. SMALL: ''The Fenlands of Lough Neagh.'' Journ. Ecol. **19** 383–388, 1931. J. M. WHITE: ''The Fenlands of North Armagh.'' Proc. Roy. Irish Acad., **40 B** 233–283, Pl. **VI. 1932.**

Lakes.

58. Ireland is essentially a country of lakes, varying from lowland pools or mountain tarns to Lough Neagh, the most extensive area of fresh water in Ireland, and larger than any in Britain. Their number is to be reckoned in thousands rather than hundreds, and in some areas— SW Connemara, the Rosses in Donegal, and the valley of the Erne—there are bewildering networks of land and water. The flora of most of the larger lakes is dealt with in the Topographical Part. In general, it may be said that the pH factor is usually of much importance in relation to the flora. Certain aquatic species are essentially calcicole : *Ranunculus circinatus, Potamogeton coloratus, Chara aculeolata, C. tomentosa*, are seldom or never seen save in alkaline water. On the other hand, *Elatine hexandra, Callitriche intermedia, Lobelia Dortmanna, Juncus bulbosus, Potamogeton polygonifolius, P. obtusifolius, Isoetes lacustris, Nitella translucens*, show a marked preference for acid waters. As a consequence, the former group shows in Ireland a decidedly "Central" range, and the latter a decidedly "Marginal" one (**41**).

The largest lake, Lough Neagh, very open, with shores often sandy, and fed by both acid and alkaline rivers, has a flora peculiar in several respects (**463**). Most of the major lakes—Loughs Corrib, Mask, Conn, Derg, Ree, Erne, as well as the Lower Lake at Killarney and the lakes of Westmeath, lie mostly or wholly on the limestone ; Lough Allen is a notable exception. But many hundreds of smaller lakes, as in Kerry, W Galway, W Mayo, Donegal, Down, Cavan, lie on non-calcareous rocks. They produce a flora which is not encrusted with lime, unlike that of the limestone lakes, and which is much richer in Charophyta, often beautifully clean. Very little dredging has so far been done in any of the Irish lakes, and no doubt the discovery of some new plants and the extension of the range of others will ensue when this has been carried out.

Of special interest is the flora of the turloughs, the lakelets lying on the driftless fissured limestones of the west, which fill in rain and are empty in dry weather. A note on their vegetation is given later (**360**).

The Calcicole Flora.

59. The calcicole flora finds its best expression on the limy gravels of the eskers of the Central Plain (**240–3**) and on the limestone crag-lands of the west (**346, 352,** etc.). Some of the characteristic plants re-appear on the basalts of the NE (**455**), and limy sea-sands often enable these or others to extend their range far over the non-calcareous rocks, as in S Connemara (**386**). The ice of Pleistocene times has sometimes pushed calcareous material far over the boundary of the acid rocks, as on the northern slopes of the Dublin mountains (**235, 268**); and the converse phenomenon also occurs.

On the eskers and limestone pavements of the centre and west the most characteristic and abundant species include *Arabis hirsuta, Poterium Sanguisorba, Galium sylvestre, Asperula cynanchica, Antennaria dioica, Carlina, Blackstonia, Gentiana Amarella, Ophrys apifera, O. muscifera, Gymnadenia conopsea, Sesleria.* The limy marshes and pools, often lying on thick soft calcareous white marl, have also a flora which includes may calcicole plants— *Ranunculus circinatus, Thalicrum flavum, Stellaria glauca, Lathyrus palustris, Myriophyllum verticillatum, Galium uliginosum, Epipactis palustris, Juncus inflexus, J. subnodulosus, Potamogeton coloratus, Scirpus pauciflorus, Carex acutiformis, C. Pseudo-Cyperus, Chara aculeolata.* The following are also characteristic of the limestones of the Central Plain :—*Hypericum perforatum, Geranium lucidum, Rhamnus catharticus, Euonymus, Anthyllis, Sorbus Aria* segregates, *Erigeron acre, Leontodon hispidus, Centaurea Scabiosa, Primula veris, Origanum, Anacamptis pyramidalis, Orchis morio, Trisetum, Avena pubescens ;* many of these are also widely if thinly spread over much of Ireland, finding a sufficiency of alkalies in local sources, on sea sands, and so on.

It may be noted that while distinctly calcicole in Ireland, many of these plants would not enter into this category in countries on the Continent, and some not even in England. This is a well-known phenomenon, due presumably to the fact that with increase of precipitation neutral soils tend to become acid owing to the accumulation of humus compounds, driving basiphilous and neutrophilous plants to the limestone in order to escape from inimical edaphic conditions. (See also **61**.)

The penetration into outlying limestones or limy-soil areas by the characteristic Central Plain flora offers some points of interest. Two of the most marked cases are the Killarney lakes and E Down. Killarney (**320**) and the band of limestone that zigzags northward from it is cut off from the broken-up edge of the Central Plain limestones by a minimum of 20 miles of hilly Coal-measures and Devonian slates: E Down (**477**) has 40 miles of Silurian slates between it and the nearest tongue of the main area of Carboniferous limestone, and only a quite small and local outcrop of this rock; yet both possess a number of the most characteristic calcicoles of the Central Plain flora (see **322, 477**). The converse is seen in the much commoner phenomenon of the presence of a characteristic calcifuge flora on even quite small outcrops, from under the limestone, of slates or sandstones, as is well exemplified in the minor foldings in the southern portion of the Central Plain, where inliers of this nature are frequent. These cases merely illustrate the dictum of A. R. Wallace, that plants have existed long enough to have been carried to all suitable localities, and the determining factor in their distribution is to be found in their power of adaptation to their new conditions.

The continuity of the calcicole flora in the Central Plain is much obscured by the abundance of peat-bog on the limestone, with its essentially calcifuge flora. Distribution-maps of species, say by counties, may thus be very misleading, just as the presence of limy shell-sands in areas of acid rocks upsets the continuity of the calcifuge flora when mapped except in a very detailed manner.

The range of the calcicole flora has been extended, as mentioned below (**75**), by the use of lime mortar in walls, which has allowed some of the calcicole ferns, for instance, to spread into every part of the country.

The most remarkable calcicole flora in Ireland is that which colonizes the bare limestone "crags," "clints," or "pavements" which stretch with little interruption from Lough Carra in Mayo south to Askeaton in Limerick, and attain their main development in the Burren region of Clare, on the S side of Galway Bay (Plates 19, 23) They yield a number of plants unknown elsewhere in Ireland, and an amazing profusion of some others rare in the country. They offer a surprising mixture of types, from Mediterranean (*e.g.*, *Neotinea*) to Arctic-alpine (*e.g.*, *Dryas*). Their flora is dealt with especially in **346, 352**, and also in **338, 344, 359–363, 365**, etc.

The lowlying parts of this area are characterized by turloughs — depressions in the limestone which fill with water in wet weather by subterranean passages, and empty by the same means. They yield a peculiar flora (characterized especially by *Viola stagnina*), which is referred to in **360**.

The Calcifuge Flora.

60. The calcifuge flora occupies a much wider range of geological formations than the calcicole. While the latter attains its main development on the Carboniferous limestone and soils derived from it (the Chalk of NE Ireland being too limited in area to have more than a quite slight effect), calcifuge plants find a home equally on the wide areas of metamorphic rocks (NW and W Ireland), granites (E especially), Silurian slates (NE), Devonian sandstones and slates (S), and Coal-measure shales (SW, etc.). The Central Plain is in fact ringed round with a calcifuge flora. This is of two types—the familiar peat flora, found equally in the Central Plain and elsewhere, and the flora of non-calcareous soils, wide-spread outside the Central Plain wherever peat is absent, and occupying in characteristic form the lowlands such as those of N Kerry (Coal-measures), Cork (Devonian), Wexford and Down (Silurian), Donegal (metamorphic). Here we find abundance of :—

Ranunculus hederaceus	Senecio sylvaticus
Lepidium heterophyllum	Jasione montana
Raphanus Raphanistrum	Vaccinium Myrtillus
Polygala serpyllacea	Digitalis purpurea
Spergula vulgaris	Stachys arvensis
Montia fontana	Teucrium Scorodonia
Hypericum humifusum	Polygonum Hydropiper
Radiola linoides	Rumex Acetosella
Cytisus scoparius	Juncus squarrosus
Ulex Gallii	bulbosus
Lotus uliginosus	Potamogeton
Lathyrus montanus	polygonifolius
Cotyledon Umbilicus-Veneris	Scirpus fluitans
Sedum anglicum	Deschampsia flexuosa
Peplis Portula	Nardus stricta
Galium saxatile	Blechnum Spicant
Gnaphalium uliginosum	Athyrium Filix-fœmina
Chrysanthemum segetum	Lastrea aristata
	Equisetum sylvaticum

It is to be noted that old woods on the limestone, with a humus soil, tend like bogland to introduce a calcifuge flora as islands. This applies for instance to the occurrence of *Scilla non-scripta* and some of the ferns.

The general change of vegetation observed as one passes from non-calcareous to calcareous soils is illustrated in **472** by contrasting the flora of the NE Silurian area with that of the Carboniferous limestone to the south of it.

Where the drift is absent at the junction of the alkaline and acid rocks, as about Lough Corrib and Lough Mask, the abrupt change of flora may form a very remarkable and striking feature (**371**). The same phenomenon is seen on a grand scale in the contrasting floras of the N and S shores of the deep indentation of Galway Bay—on one side the brown heath vegetation of the vast bogs of Connemara with its rare Ericaceæ, etc. (**385, 386**); on the other the gaunt grey limestone hills of Burren with *Sesleria* dominating a gramineous formation, also full of rare and interesting plants (**346–350**).

61. Sometimes there are found puzzling transgressions between the calcicole and calcifuge floras in areas markedly acid or alkaline as regards their rocks. The presence of a large group of calcicole plants on the metamorphic area of SW Connemara, referred to elsewhere (**388**), is sufficiently accounted for by the highly calcareous nature of the sea-sands in the vicinity; but the explanation of the presence of *Erica cinerea* and *Calluna* on bare limestone tracts here and there is not so obvious, and, like the other cases quoted below, calls for a series of close observations and chemical experiments not yet carried out. A few other exceptional occurrences of calcifuge plants on limestone, or *vice versa*, may be quoted :—

. At the E base of Keshcorran in Sligo, on drumlins formed of limestone drift lying on the Limestone Plain, *Cytisus, Digitalis, Blechnum, Athyrium*, and other characteristic calcifuge species flourish ; here, probably, the lime has been leached out of the soil in which the plants grow (**382**).

On scarps of Yoredale sandstone in the Carrick district of Fermanagh, *Sesleria, Arabis hirsuta, Asplenium Ruta-muraria*, all conspicuous calcicoles, grow mixed with such calcifuge species as *Vaccinium Vitis-Idæa, V. Myrtillus, Calluna, Erica, Digitalis*, and *Blechnum* (**438**). About New Ross the calcicole *Origanum, Sedum acre, Ceterach,* and the calcifuge *Erica cinerea, Sedum anglicum*, etc., grow together on Gotlandian slates (**259**). The strongly calcicole *Euphrasia salisburgensis* furnishes other examples (**344**). Of the Galway-Clare calcicoles, *Neotinea* has been found at Mount Gable near Cong on the metamorphic rocks, and also on the Coal-measures near Ennistymon 6 to 8 miles from the nearest limestone; *Gentiana verna* flourishes on peaty banks on similar shales between Bally-vaughan and Lisdoonvarna, amid a calcifuge vegetation.

Near Corco Gap in Maam Turk *Saxifraga spathularis, Athyrium*, and *Blechnum* grow with *Asplenium Ruta-muraria* in dry almost soil-less chinks of primitive lime-stone. On L. Corrib and L. Cullin the last-named flourishes in chinks of wave-washed slates and gneisses.

In these and similar cases actual soil-analysis alone can throw light on the relations between plant and sub-stratum, and this has not yet been done. But it is clear that the question is seldom simple. To calcicolous plants the attraction of the limestone may be its chemical qualities, or its physical: and the upper layers of a limy soil may be neutral or even acid, owing to leaching or to the accumulation of humus. See for instance E. J. SALISBURY: The ''Significance of the Calcicolous Habit'' (Journ. Ecology, **8** 202–215, 1921—with good bibliography).

Coasts.

62. There is a rather marked contrast betweeen the sea-side flora of the E and W sides of Ireland. That of the E coast is much the richer, attaining its optimum in Wicklow and Wexford (**272**), and including among its rarer members *Matthiola sinuata, Trifolium glomeratum, Diotis, Asparagus, Glyceria Borreri*, and the endemic *Equisetum Moorei*. The cause of the difference between E and W is partly edaphic, in the larger amount of gravelly and sandy beaches which prevail in the east, and partly climatic, in the very great exposure prevailing on the western sea-board. Quite a number of E coast plants —for instance, *Thalictrum dunense, Trigonella, Trifolium glomeratum, T. scabrum, T. subterraneum, Inula crith-moides, Diotis, Atriplex maritima, Asparagus, Scilla verna, Juncus acutus*, some of them widely spread there—do not venture into the west, or at most make a rare and tentative appearance in Donegal or Kerry.

On the other hand, very few maritime plants prefer the W coast. *Cochlearia grœnlandica,* which ranges from Antrim to W Cork, appears to be the only well-marked example.

The poverty of maritime species in the west is illustrated by the sand-dune plants much more than by those inhabiting rocks or salt-marshes. While in some favoured spots in the west, as on the Mullet, at Strandhill in Sligo, and Killala in West Mayo, we find well-developed dunes, backed by undulating stable moss-grown sandy ground, in other parts the sands are reduced by exposure to mere deserts. The great tracts beyond Bunowen in SW Connemara, at Keel strand in Achill, and on North Inishkea will serve as examples. These large level areas are quite bare of vegetation, or support a miserably poor and starved flora. But even where the sands are well colonized, the characteristic flora is much reduced as compared with the east. Among the dunes we miss the blue spires of *Echium,* the grey leaves of *Cynoglossum*; *Thalictrum dunense, Lychnis alba, Trifolium arvense* are likewise absent from the sands. *Lycopsis, Euphorbia portlandica, E. Paralias* are quite rare; and on the beaches, *Cakile, Eryngium maritimum, Salsola, Polygonum Raii* are but seldom found. The sand is often blown up adjoining hill-sides to a height of several hundred feet; elsewhere, as at Keel Lough in Achill, its advance has dammed the natural drainage and flooded a considerable extent of country, or has buried houses and demesne lands, as at Dunfanaghy and Rosapenna in Donegal.

Semi-exposure to the Atlantic sometimes produces the phenomenon of a salt-marsh flora at the foot of a storm-beach of boulders, as below Mountcharles on Donegal Bay, where a sward of *Glyceria maritima* with *Limonium humile* and *Atriplex* occupies such a position.

The flora of the western sea-rocks and cliffs, as contrasted with the east coast, compares favourably with it, yielding, in addition to all the common species, such plants as *Spergularia rupicola* in abundance, *Lavatera, Limonium binervosum,* and other local species.

On the cliff-ranges (and down to sea-level on beaches) *Sedum roseum* joins the maritime group, and is often profuse, with quantities of *Angelica, Eupatorium, Matricaria inodora, Beta,* and other rank-growing species on the damp ledges.   The salt-marsh flora is likewise similar to that of the E coast, but rather poorer.

63. **Plantago Sward**.—One of the most distinct formations of the extreme west is a very close dwarf maritime sward, occupying wind-swept cliff-tops and exposed slopes by the sea, composed mainly, often almost wholly, of *Plantago maritima* and *P. Coronopus*, growing extremely small.   As noted on Clare Island,[22] this association consists mainly of the two species named, with *P. lanceolata, Thymus Serpyllum*, and *Euphrasia*, and in small quantity nearly thirty other species, all forming a dense sward about ½-inch in height, with flower-stems (even of the taller species like *Scabiosa Succisa* and *Hypochœris*), rising about 2 inches).   *Radiola* grows ½-inch high, usually unbranched ; *Ophioglossum* the same height, and barren.   The flowers of the majority of the plants rise level with those of *Angallis tenella* and *Radiola*.   At the west end of Inishkea. *Plantago* sward, consisting almost entirely of the two species first named, growing extremely minute, covers a considerable area ; it forms a shining green carpet, so close that no lawn-mower, however close-set, could cut a leaf off it, and as smooth to the hand as a newly-ironed table-cloth. *Plantago* sward is characteristic of areas of particularly great exposure close to the sea.   It occurs all along the W coast, and ranges from sea-level to about 400 feet.

64. The occurrence of maritime plants in alpine or inland situations in Ireland is a feature almost confined to the western mountains, from W Cork to Donegal, the species involved there being *Cochlearia officinalis (alpina), Silene maritima, Armeria maritima* (not *planifolia), Plantago maritima.   Cochlearia* extends to the Galtees, in the southern midlands ; in Derry the fine cliff of Benevenagh shelters all four species (as also *Cerastium semidecandrum*), and *Silene* continues along the basaltic scarp to Cave Hill overlooking Belfast.   In the west, *P. maritima* has in addition a wide inland range over the low limestones and along the shores of Loughs Derg (up to 25 miles from the sea), Corrib, Mask, Carra, Conn, etc.   Rather unexpected also is the occurrence of *Carex distans* in two stations near Tuam in NE Galway.   The Lakes of Killarney harbour a curious outlier of the maritime flora,

[22] R. Ll. Praeger: ''The Flora of Clare Island.''   Irish Nat., 1903, 281.

embracing *Silene maritima, Cerastium semidecandrum, Armeria maritima,* and *Asplenium marinum.* The most interesting inland outlier of the maritime flora is on L. Neagh, where are found *Viola Curtisii, Spergularia rupicola, Cerastium semidecandrum, Erodium cicutarium, Trifolium arvense, Plantago maritima, Scirpus maritimus, S. Tabernæmontani,* and *Carex extensa* (see **464**). Most of these are elsewhere in Ireland exclusively maritime.

Islands.

65. The islands off the west coast—*e.g.,* Inishbofin, Inishturk, Clare I.—offer some points of interest. They have larger floras than most equal areas of the adjoining mainland, insularity being more than compensated by the variety of habitat which they afford. None of them possesses any limestone. They are less smothered in bog than the mainland, and tend to yield drier soils—and I believe have a smaller rainfall, though this has not been fully tested.[23] Frost and snow are practically unknown. The mean annual range of temperature is only about $15\frac{1}{2}°$ F. The following points regarding their flora may be mentioned :—

1. They furnish, owing to limy sands or lime-built walls, a home for some calcicole plants, *e.g., Arabis Brownii, Centaurea Scabiosa, Ophrys apifera* (all on Bofin), *Ceterach* (on Bofin and Clare I.).

2. They possess a few representatives of the Lusitanian and American groups—*Saxifraga spathularis* is on all three, *S. Geum* (in a hybridized state) and *Erica mediterranea* on Clare I., *Euphorbia hiberna* on Turk, *Eriocaulon* on Bofin.

3. They offer refuges for some plants very sparsely distributed in Ireland, some of them probably relict—*Helianthemum guttatum* on Bofin and Turk (elsewhere only in Cork), *Orobanche rubra* and *Cephalanthera ensifolia* on Clare I., *Calamagrostis epigejos* and *Lycopodium inundatum* on Bofin. These would appear to be at least local relicts, swamped out of their parent stations on the mainland by the spread of the overmastering peat.

[23] See W. J. Lyons: ''Climatology,'' In Biological Survey of Clare Island, Proc. Roy. Irish Acad., **31**, part 6. 1914.

4. Their flora shows an interesting mixture of plants of great exposure and others of damp shady places. Thus all three islands possess *Cochlearia grœnlandica, Sedum roseum, Juniperus sibirica, Lastrea œmula, Athyrium Filix-fœmina*, all growing close to sea-level.

Achill is not included in the above comparison, as it is scarcely separated from the mainland.

The Great Blasket, a desperately exposed high narrow ridge of slate, has *Cochlearia grœnlandica, Saxifraga spathularis, Lastrea œmula, Athyrium, Hymenophyllum peltatum*, and as an index to the degree of oceanic influence, *Asplenium marinum* on walls at 750 ft. above the sea (**329**).

The Aran Islands, large reefs mostly of bare fissured limestone, yield a very different flora, of high interest (**352**).

Mountains.

66. The mountainous areas in Ireland are situated mainly in the maritime counties, the only exception of importance being the fine range of the Galtees (3015 ft.) rising from the rich limestone plain of southern Tipperary. A number of other hill-groups—mostly NE and SW ridges—have been formed, like these, by folding, which has brought up the underlying Devonian and Silurian slates and sandstones; this results in wide heathery moorlands, but the hills do not generally rise high enough to effect much other change in the vegetation, which is usually of a conventional calcifuge type. Wicklow possesses the largest continuous area of high land (over 200 sq. m. are above 1000 ft.), with Lugnaquilla rising to 3039 ft., but like that other fine eastern range, the Mourne Mountains in Down (2796 ft.), it is poorer in alpine plants than many lower and smaller areas in the west. Thus, Slieve League in Donegal (1972 ft.) yields on a quite restricted area more plants of alpine type than Wicklow and Down taken together. The Irish mountain-groups are formed of granites, schists, slates, sandstones, and shales, mostly of pre-Permian age (the Eocene Mournes being a marked exception). The only limestone mountains are the interesting Ben Bulben group—a plateau standing up some 2000 ft. (fig. 24), with cliff-walls rich in alpine plants. An upland somewhat similar to the last in being cliff-walled and yielding a rich flora of calcicole proclivities, is formed by the Eocene basalts of Antrim and Londonderry. In view of the perennial interest of "alpines," the mountain plants are dealt with rather fully in the Topographical Section below, and only a few general considerations need detain us here.

In the first place, the Irish alpine flora is poor. But the general scarcity of alpine plants need not diminish the interest of mountain botanizing. "It may sound like a paradox" writes one of the most critical of Irish botanists[24] "to say that the botanical survey of an Irish mountain region derives a peculiar zest from the very poverty of our flora in alpine species. Yet the assertion may be made with perfect truthfulness. That the rapture of discovery varies directly with the rarity of the object sought for, that the value of the thing attained is measured by the labour of attainment—these are time-honoured truisms in every system of proverbial philosophy; and their essential truth is daily borne in upon the mind of the botanist who devotes himself to the exploration of any of the mountain-groups of Ireland. The fans of the Alpine Club-moss, which he spurns with callous feet on the slopes of Snowdon, he half worships when they meet his longing eyes in the Wicklow or Kerry highlands; and so with many others of our alpine species—unconsidered trifles abroad, they become for him objects of enthusiasm at home."

67. Out of 67 species comprising Watson's "Highland" type in Britain, 42 occur in Ireland.[25] This is a not unduly small number when we consider especially the area and height of mountainous country in Scotland and Wales; but in Ireland the plants are mostly thinly distributed and rare. Seven of the rarest—*Arabis petræa, Alchemilla alpina, Rubus Chamæmorus, Saxifraga nivalis, Epilobium alsinefolium, Carex pauciflora, Poa alpina,* have only eleven stations between them all. Against this, the following are relatively much more widespread in Ireland than in Britain (but it is to be noted that in Ireland all of them but *Salix herbacea* descend to sea-level, and attain their increase of range not by their abundance on the

---

[24] N. COLGAN: "Botanical Notes on the Galway and Mayo Highlands." Irish Nat., 1900, 111–118.

[25] In addition, there is in Ireland *Arenaria ciliata* on the Ben Bulben range, 1000–1950 ft., unknown in Britain; and *Euphrasia salisburgensis,* recently found in Yorkshire, which Watson would have included in his Highland type: it is in western Ireland widespread and mainly quite lowland (to sea-level) though ascending to 1000 ft. on Ben Bulben.

mountains, but by widespread occurrence on the low-lands) :—

Dryas octopetala	Carex aquatilis
Hieracium anglicum	Sesleria cœrulea
iricum	Selaginella selaginoides
Salix herbacea	Isoetes lacustris
Juniperus sibirica	

The "Highland" plants of Watson as a group do not in Ireland fit well with his definition, "Species chiefly seen about the mountains." Much less do they constitute the alpine flora in this country. Taking the Highland Type flora as set down in "Cybele Hibernica," p. xliv (the *Hieracia* are a difficulty, and one may for convenience accept the compromise there used), we find that 16 of the 42 occur down to sea-level, often not in the proximity of mountains. These are :—

Draba incana	Hieracium gothicum
Dryas octopetala	Arctostaphylos Uva-ursi
Saxifraga oppositifolia	Vaccinium Vitis-Idæa
aizoides	Juniperus sibirica
Sedum roseum	Carex aquatilis
Galium boreale	Sesleria cœrulea
Hieracium anglicum	Selaginella selaginoides
iricum	Isoetes lacustris

Sixteen more come in before the 1000 ft. contour is exceeded :—

Thalictrum alpinum	Hieracium strictum
Subularia aquatica	Polygonum viviparum
Silene acaulis	Oxyria digyna
Saxifraga stellaris	Salix herbacea
Epilobium alsinefolium	Carex rigida
Saussurea alpina	pauciflora
Hieracium prenanthoides	Cryptogramme crispa
senescens	Asplenium viride

and three die out below 1000 ft.

Between 1000 and 2000 ft. seven more have appeared :—

Arabis petræa	Poa alpina
Saxifraga nivalis	Polystichum Lonchitis
Alchemilla alpina	Lycopodium alpinum
Rubus Chamæmorus	

and no less than twelve more have died out.

No new species appears above 2000 ft. except *Deschampsia alpina,* and of the 21 plants which continue upwards all but eight—

Saxifraga stellaris,	3400	Carex rigida,	3300
Sedum roseum,	3150	Deschampsia alpina,	3370
Oxyria digyna,	3150	Poa alpina,	3100
Salix herbacea,	3050	Asplenium viride,	3150

are gone when 3000 ft. is reached. (The figure attached to each shows its extreme limit.) These last are all on Macgillicuddy's Reeks. It is to be remembered that the Reeks (Carrantual 3414 ft.) is the only range that exceeds 3000 by more than at most 127 ft. The accompanying graph (fig. 9) exhibits the increase in the number of "Highland" plants on the Irish hills up to a maximum at 1000–1200 ft., and their steady diminution with greater elevation.

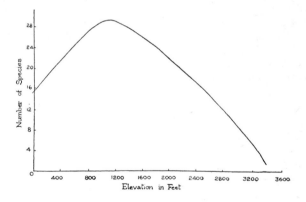

FIG. 9.—IRISH ALPINE PLANTS IN RELATION TO ELEVATION.

68. As regards their general distribution in Ireland, the "Highland" plants tend, like the mountains, to be marginal. With a maximum in the N, they range down either coast—much more plentifully over Ben Bulben and the Mayo-Galway highlands into Kerry than over the Mournes into Wicklow. The height at which they occur increases from N to S, and is greater in the E than in the W. Comparing the Donegal-Derry area with Kerry-Cork (data for 17 species are available), we find that the "Highland" plants common to both have a mean lower limit in the N of 550 ft., which rises in the S to about double that amount. This is one of the few definite effects of latitude that one can find in the Irish flora.

A similar comparison of the behaviour of the "Highland" plants of west and east—of the Mayo-Galway mountains with those of Down-Wicklow—shows a corresponding decrease in elevation towards the west. Of 25 species in the western area and 18 in the eastern, 13 are common to both, of which data are available for 12. These 12 show a mean lower limit of 1070 ft. in Down-Wicklow, and of 720 ft. in Mayo-Galway. This effect is climatic, and illustrates the manner in which the whole flora is depressed in altitude on account of the conditions prevailing on the Atlantic sea-board.

69. None of our mountains is high enough to reach the region of an alpine-plateau vegetation. Indeed, the summits, commonly covered with peat, are usually very poor in alpines compared with rocky scarps on the flanks of the hills, especially where the latter face northward. Taking the highest points in Donegal (2466 ft.), Down (2796), Sligo (2115), Mayo (2688), Wicklow (3039), Tipperary (3015), and Kerry (3414), we find that the average summit-flora numbers about 16 species, of which not more than two are usually alpines. These figures remain very uniform all over Ireland. The prevailing summit plants are species of wide vertical range, all descending to sea-level :—

Potentilla erecta	Luzula sylvatica
Galium saxatile	Eriophorum vaginatum
Vaccinium Myrtillus	Deschampsia cæspitosa
Calluna vulgaris	Anthoxanthum odoratum
Rumex Acetosella	Festuca ovina
Empetrum nigrum	Lycopodium Selago
Juncus squarrosus	

The alpines which may occur with these on the summits named are *Saxifraga stellaris, Vaccinium Vitis-Idæa, Salix herbacea, Carex rigida,* and *Lycopodium alpinum.*

The nearest approaches to an alpine sward are found here and there in exposed ground in the west—on Curraun Achill, at about 1200 ft., where *Arctostaphylos Uva-ursi, Empetrum,* and *Juniperus sibirica* form a dense carpet;

and on the Meenawn range on Achill Island, where a very close sward formed mainly of *Calluna* and the three species before mentioned, with a little *Salix herbacea*, occupies the ground at 1300–1500 ft. But indeed an almost equally alpine facies is supplied by the sheets of *Arctostaphylos Uva-ursi*, *Dryas*, *Gentiana verna*, *Sesleria*, that in Burren range from over 1000 ft. down to sea-level, though mixed with such southern types as *Neotinea* and *Adiantum* (see Plate 2).

For a brief account of the alpine flora of each mountain-range, see the Topographical Part below.

Human Influence on the Flora.

70. The natural climax vegetation of Ireland—forest, now chiefly of *Quercus* and *Betula*—has long since passed away save in places unsuitable for grazing, and has been replaced by grassland, its boundaries further extended by drainage and the removal of peat-bog. Tillage furnishes a further step in the destruction of the native flora, but, fortunately for the botanist, Ireland is not essentially a tillage country. Potatoes, oats, turnips furnish the principal crops: all crops put together do not occupy half the area which is covered by grass (see **76**). The amount of tillage increases from west to east, and reaches a maximum in certain NE counties, mostly in the Silurian area (fig. 10). Tillage has brought with it the introduction of a great number of extraneous species, largely annual plants of Central European and Mediterranean origin. These reach their maximum around Dublin and in the SE, which is the region of maximum summer temperature. The indigenous vegetation of the lowland rich soils has been almost obliterated; the flora becomes more and more aboriginal as the soil becomes unsuitable for agricultural operations, as on bogs, marshes, and mountains. The native plants are in general very persistent, clinging to any remnant of primitive ground, whence, if allowed, they tend to spread back into their old territory. But many others, more sensitive, have shrunk back or disappeared before the changes induced by man. Thus, for instance, the poverty of the woodland flora of Ireland is conspicuous as compared with England or Scotland—no *Helleborus*, *Linnæa*, *Trientalis*, *Daphne*, *Corallorhiza*, *Goodyera*,

Cephalanthera grandiflora, Maianthemum, Ruscus, Convallaria, Paris. This is one of the prices Ireland has paid for being a cattle country, for probably at least a few of these were here formerly.

The plants of the seashore seem especially sensitive. *Euphorbia Peplis* is gone, and *Crambe, Matthiola, Diotis, Asparagus, Mertensia* and others are decreasing.

FIG. 10.—TILLAGE IN IRELAND.

White = under 25 per cent.
Black = over 35 per cent.

71. The medley of introduced plants includes some vigorous species which spread and colonize; many of these are mentioned in the succeeding paragraphs. One of the most difficult tasks, therefore, of the botanical geographer, endeavouring to trace the routes and dates of the immigrations of the flora, and its reactions to soil, climate, and so on, is set by this destruction of the broad features of especially the lowland vegetation, and the intricate mixture of native and introduced plants which now prevails. Various definitions of a NATIVE PLANT have been proposed: I use that employed by Dunn[26] :—*A species is only held to be native in a natural locality to which it has spread by natural means.* This excludes all interference by man,

---

[26] S. T. DUNN: ''Alien Flora of Britain,'' p. x. 1905.

direct or indirect, as regards either the origin of the seed or plant, its transport, or its subsequent development. But these tests are not easy to apply, and cannot mostly be applied directly. We have to rely rather on the absence of any evidence of introduction, making "native," in its application, a purely negative term, as H. C. Watson has pointed out. By most botanists, "native" is not used in so strict a sense as that quoted above, and generally includes individuals of species truly native in the district or country, which are growing by natural spreading *either in natural or in artificial habitats*; and in view of the fact that these two categories comprise the limits which the native plants have achieved by natural dissemination, there is a good deal to be said for the practice, which is adopted in the pages which follow.

Dunn's definition involves three points: (1) the origin of the plant; (2) the means by which it reached its present habitat; (3) the nature of the habitat. I have elsewhere[27] proposed a formula for indicating the standing of plants according to these tests. If we use N as meaning uncontaminated conditions, whether of source, dispersal, or habitat, and * for contaminated conditions, then NNN, and only that combination, fulfils Dunn's formula for a fully native plant. To this one adds NN*—native plants which have spread by natural means to a non-natural habitat, like *Lonicera* in a roadside hedge, *Ceterach* on a wall, or the many native plants in drained land or planted woods. All other combinations involve either a contaminated source or artificial introduction, and while plants thus defined may be fully naturalized, they cannot be reckoned as native. Thus, the American garden *Mimulus* by a mountain stream is *NN, having spread by natural means to a natural habitat, and *Pinguicula grandiflora* brought from Kerry into wild ground in Wexford is N*N. The plants of our gardens are mostly ***. This matter will be found more fully discussed in the place indicated.[27]

There has been much loose thinking and loose practice as regards this admittedly difficult matter, and I think it will be found that appeal to this formula tends to clarify one's ideas.

72. The most obviously introduced plants in our flora

[27] R. LL. PRAEGER: "Phanerogamia." (Clare Island Survey.) Proc. Roy. Irish Acad. **31**, part 10. 1911.

(excluding exotic species deliberately planted) are the many casuals which occur especially about centres of industry—docks, railway yards, factories, etc. While the majority of these are mere fleeting waifs, a few of them establish themselves, and have become permanent members of our flora. Examples of such in Ireland are :—

Diplotaxis muralis
Coronopus didymus
Lepidium latifolium
Arenaria tenuifolia
Malva rotundifolia

Valerianella carinata
Matricaria suaveolens
Senecio squalidus
Linaria minor
Poa compressa

But though these may spread widely, only seldom do they succeed in penetrating into the native vegetation on wild ground. Most of them continue to be denizens of waysides, railways, tilled land, walls, and other artificial habitats, and remain *N*. Their tenure is dependent on the continuance of the artificial conditions under which they live, which tend to restrain the native vegetation from ousting them. Along with the crops of the farmland and the flowers of our gardens, they are quite alien to the native flora; but being often unfamiliar, and offering interesting problems regarding their mode of introduction and possible acclimatization, they are well worthy of attention. As an example of the number of plants which are unsuccessfully introduced in this way into our country, see KNOWLES in Irish Nat. 1906, 143–150.

A large section of the introduced flora, in Ireland as elsewhere, owes its presence to agricultural and garden seed, and to arrival in soil on the roots of planted shrubs and herbs. To this category belong all or most of the species of *Papaver, Fumaria, Brassica, Valerianella, Lamium, Galeopsis,* some species of *Veronica* and *Euphorbia,* and so on—at least 60 or 70 in all. Some of these are at present spreading rapidly, like *Draba muralis, Crepis biennis, Orobanche minor*; but like the last group they usually do not mix with the undisturbed native flora, and a large number are dependent for their continuance on the annual turning over of the soil, an operation which holds the indigenous flora in check. Almost without exception they remain *N*.

73. Another group consists of ornamental plants, or species formerly used as pot-herbs or for medicinal pur-

poses. Many of these are aggressive or at least persistent perennials, with great staying qualities, but again with little power of competing on equal terms with the native flora; as a consequence, they are still mainly found about houses and villages, on old walls, or on disturbed ground. Examples of successful colonizers among the kitchen-garden plants are :—

Chelidonium majus
Cochlearia Armoracia
Saponaria officinalis
Smyrnium Olusatrum
Ægopodium Podagraria
Myrrhis odorata
Fœniculum vulgare
Sambucus Ebulus
Inula Helenium

Tanacetum vulgare
Lactuca muralis
Verbena officinalis
Lycium chinense
Mentha *spp.*
Ballota nigra
Chenopodium Bonus-
            Henricus
Allium *spp.*

and 20 or more others. The garden plants in this category include :—

Clematis Vitalba
Corydalis lutea
Cheiranthus Cheiri
Hypericum hircinum
Sedum *spp.*
Kentranthus ruber
Aster *spp.*

Petasites fragrans
Campanula rapunculoides
Mimulus Langsdorffii
Elodea canadensis
Stratiotes Aloides
Narcissus biflorus

and a number of others. One or two of them—*Kentranthus, Aster, Mimulus,* and *Elodea,* for example—have spread into quite wild ground, and, mixing on equal terms with the native flora, merit *NN. But most remain *** or *N*. The naturalization of alien plants goes on slowly but steadily. Among those which appear to be at present establishing themselves as members of the permanent flora of wild ground are the Himalayan *Impatiens glandulifera* (river-sides in Dublin, Sligo, etc.), the Himalayan *Cotoneaster microphyllus* (mountain rocks in Mayo, river-gravels in Dublin, railway banks W of Athenry, etc.), the New Zealand *Epilobium nummularifolium* (Blackstairs at 1200 ft., bank by the sea S of Killybegs, railway ballast for ½ mile at Helen's Bay (Down), roadside at Lougha-veema and Ramore Head (Antrim), Portsalon (Donegal), etc.), the Australian *Acœna Sanguisorbœ* (woods in

Cork, Dublin, Down), the S. American *Fuchsia gracilis* (Berehaven district, Co. Cork), the West Asiatic *Lactuca tatarica* (shore near Galway), the South European *Calystegia sylvestris* (L. Gill), the N. American *Mimulus moschatus* (wet ground in Wicklow, Armagh, Down, Antrim), and the S. African *Montbretia Pottsii* (rocky shores of L. Gill in Sligo and many other places).

74. As to many other aliens, the means by which they reached their present stations is not clear, and they can only be classed as "followers of man." For instance :—

Sisymbrum Irio	Rumex pulcher
Reseda lutea	Mercurialis annua
Galium Cruciata	Urtica urens
Picris hieracioides	dioica
Hieracium, *several spp.*	Hordeum murinum

and 20 or 30 others. They mostly remain as dependents on human activities, but some of them spread into the native flora.

75. Another aspect of the question of human influence is the obvious fact that man's operations have often encouraged a material extension of the range of native plants. Mortar-built walls, for instance, arising during the last thousand years all over the country, have allowed lime-loving species to extend far beyond their natural boundaries. There seems little doubt, for instance, that *Ceterach* was formerly confined to Clare and the adjoining limestone tracts; now it is found, frequently in abundance, in all the 40 Irish vice-counties, and often in every part of them. *Asplenium Ruta-muraria,* with an original range not so restricted (it occurs, for instance, on basalt in Antrim, and on metamorphic and slate rocks (on lake-shores) in Mayo) is now much more universally distributed even than *Ceterach.* The ferns, indeed, on account of the abundance and lightness of their spores, are especially enterprising colonists. Many of the extreme outposts of Ireland—Inishbofin, Inishturk, the W end of Achill—now support one or both of those just mentioned. At least six species have penetrated to the centre of Dublin City (see Irish Nat. 1920, 108).

Spores, possibly from a local garden, possibly native and from a long distance, are responsible for the fleeting appearance of *Asplenium septentrionale* in Down (Irish

Nat. 1912, 154) and *Lastrea rigida* near Drogheda (Cyb. Hib. ed. I, 371). Similar cases are *Asplenium viride* on old walls at Convoy (about 200 ft. elevation), and *Polystichum Lonchitis* close to sea-level at Killybegs, both in Donegal (Hart Fl. Don. 286, 288), and (if the records may be trusted) the same fern at Edgeworthstown and Dungannon (Cyb. Hib. ed. I, 372). *Cryptogramme* has in Down been twice found at quite low elevations on walls or stone-heaps (Irish Nat. 1903, 36, and Irish Nat. Journ. **1**, 242).

Roads and road traffic have spread native plants such as *Juncus macer*. Canals have extended widely the range in the Central Plain of plants like *Butomus* and *Sagittaria*, as also of *Potamogeton* spp., *Charophyta,* etc. They have brought *Apium Moorei* and *Potamogeton coloratus,* otherwise unknown in the county, inside the Dublin boundary, and *Lycopus, Potamogeton densus, Glyceria aquatica,* etc., almost into the heart of the city.

Railways in Ireland furnish a classic example of their power of promoting dispersal, by the rapidity with which in recent years *Diplotaxis muralis, Arenaria tenuifolia,* and *Linaria minor*—none of them native—have extended their range through the country. *Senecio squalidus* has travelled by rail from Cork to Dublin, and *Tragopogon pratense* from Leinster (presumably) to Belfast.

76. The maximum change of vegetation due to human operations is found in cities, in well-tilled land, and in gardens, where the change amounts to a full 100 per cent. as regards NNN plants, and may be nearly equally high for NN*. Even leaving out of account the fact that much of the present grassland was originally wood, the influence of grazing animals is still profound in the richer soils, tending to produce a uniform and limited flora, of little interest to the botanist. On poorer soils, especially on limestone, the change is less; it is less still on the mountains, and there affects the relative abundance of the constituent plants rather than the question of their presence or absence. The surface of Ireland has about one-fifth under crops and a little over one-half in grass, with about 11 per cent. mountain land and 8 per cent. peat bog and marsh, and 4 per cent. water; so that over at least three-quarters of the surface the indigenous vegetation has been destroyed or greatly altered. The farmland runs up the hills (mostly following the drift deposits) to heights which vary from

200 ft. in the exposed areas of the west to about 1200 ft. in sheltered spots in the east. From these figures, the very small amount of ground which remains in anything like its pristine condition may be judged.

The net result of human operations comprises two categories—(1) extermination or reduction or redistribution of the native plants, and (2) introduction of alien plants.

These changes are difficult to estimate, since they involve comparison with the past, which can only vaguely be reconstructed. They are greatest in those parts of the country which are richest, were earliest settled, and support the largest human population. In the County of Dublin, which comes foremost in these respects, the alien flora is estimated (qualitatively) at 22 per cent. of the total.[28] In Kerry it is estimated at 14 per cent.[29] Quantitatively the difference is much greater: in Kerry, with its large areas of almost undisturbed mountain land, there is, relatively to area, probably not one-tenth the *amount* of introduced plants which exists in the small County of Dublin.

In the light of evidence now available, there can be little doubt that certain plants, set down in former works such as "Cybele Hibernica" as introduced or probably introduced, are indigenous in Ireland in at least some of their stations. Such are *Viola odorata, Trifolium glomeratum, T. subterraneum, Prunus Avium, Rosa stylosa, R. rubiginosa, R. micrantha, Pyrus Malus (acerba), Lysimachia Nummularia, Cuscuta Epithymum, C. Trifolii, Sisyrinchium angustifolium, Leucojum æstivum, Allium Babingtonii, Juncus macer, Brachypodium primatum*—see Journ. Bot., 1934, 68–75.

THE BOTANICAL SUBDIVISION OF IRELAND.

77. Several plans have been put forward, at different times, to allow the distribution of plants within the country to be shown by numbers or symbols representing subdivisions, large or small, of the whole area—as Watson did for Britain in his "Topographical Botany," and as indeed it is necessary to do if one attempts to express in detail the range of plants within any but a small area.

---

[28] COLGAN: Fl. Dublin, xxxvi.   [29] SCULLY: Fl. Kerry, xxxiii.

No such scheme is required or used in the following pages, save in the "Census List of the Irish Flora" at end, on account of the geographical arrangement of the subject-matter; but it may be well to indicate the plans which have been proposed or used.

1. Division of Ireland into 12 DISTRICTS (based mostly on county boundaries) and numbered XIX to XXX from S to N, corresponding to and continuing the 18 PROVINCES used by H. C. Watson in his " Cybele Britannica " (**3**, 1852). Proposed by Prof. C. C. Babington [30] in 1859 (and used, re-numbered 1–12, by Moore and More in "Cybele Hibernica," 1866, and by Colgan and Scully (who preferred Roman numerals) in the second edition of the same work (1898)); also division into 37 subordinate VICE-COUNTIES (a few of the larger of the 32 counties being bisected), corresponding to the 112 VICE-COUNTIES used by Watson in his "Topographical Botany" (1873–4).

2. Partition into 40 DIVISIONS, numbered 1–40 from S to N (derived from the 32 counties of Ireland by partition of the larger ones), corresponding to the 112 VICE-COUNTIES used by H. C. Watson in his " Topographical Botany " (1873–4). Proposed by R. Ll. Praeger in 1896,[31] and used in " Irish Topographical Botany " (1901) and in a number of subsequent papers by various authors dealing with other groups of the flora and fauna. See fig. 29.

3. Partition of Ireland into 12 SUB-PROVINCES, derived by dividing each of the four Provinces of Ireland into three (and designating them U[1], U[2], U[3], etc., U = Ulster); each of these contains two to five of Praeger's VICE-COUNTIES. Proposed by J. Adams in 1908.[32] This suggestion has distinct advantages over the corresponding scheme of Babington, inasmuch as the coast-line and the mountain areas are somewhat evenly distributed among the 12 areas, which tends to emphasize differences of biological distribution not dependent on habitat only.

[30] BABINGTON (C. C.): "Hints towards a *Cybele Hibernica.*" Proc. Dublin University Zool. and Bot. Assoc. 1 246–250, 1859, and Nat. Hist. Review 6 (Proc.) 633–637, 1859.

[31] R. LL. PRAEGER: "On the Botanical Subdivision of Ireland." Journ. Bot. 1896, 57–66, and Irish Nat. 1896, 28–38, map.

[32] J. ADAMS: "On the Division of Ireland into Biological Sub-Provinces." Irish Nat. 1908, 145–151, map.

4. Proposal by R. Ll. Praeger[33] to employ his subdivision or that of Babington graphically, when desired,
by printing, in two kinds of type (representing presence
or absence) the Division-numbers in positions approximating
to those which the Divisions occupy on a map.

Amended by A. W. Stelfox,[34] to allow the use, instead
of numbers, of two-letter symbols to represent the
Divisions, these symbols being obtained by contracting
the names of the Divisions.   Used in several subsequent
botanical and zoological papers.

The forty divisions into which Ireland is partitioned
in ''Irish Topographical Botany'' (fig. 29) have an
average area of 813 square miles (maximum 1336 (W
Mayo), minimum 316 (Louth))—which approximates
closely to the average of Watson's 112 vice-counties,
which is 804 sq. miles.   This allows comparisons to be
instituted between the two, which show clearly the reduction in the flora on the western side of the Irish Sea.   The
actual figures vary according to the standard adopted for
that flexible term *species*.   If we use it in a conservative
sense (*e.g.*, that adopted in '' Irish Topographical Botany,''
and shown there by heavy type), a standard which approximates to that of Babington's ''Manual'' (excluding suppl. 2
of the 10th edition), the maximum for any Irish division
stands at present at 800 (Antrim), the minimum at 520
(Monaghan), the average at 640.   If we include the plants
treated as sub-species in '' Irish Topographical Botany ''
(still a conservative estimate), the maximum for any Irish
division is 942 (Down) and the minimum 554 (Longford).
(The increase in the difference between the extremes is
largely due to the unequal working out of segregates : for
instance, only one fruticose *Rubus* is on record from
Longford.)   If estimated on the standard of the '' London
Catalogue,'' the numbers would be larger.   These figures
may be compared with such English county totals as have
been published.   The latter are greater by some 50 per
cent.

[33] R. LL. PRAEGER : ''A Simple Method of representing Geographical
Distribution.''   Irish Nat. 1906, 88–94, and Journ. Bot. 1906, 128–
130.

[34] A. W. STELFOX : ''A List of the Land and Freshwater Mollusks
of Ireland.''   Proc. Roy. Irish Acad. **29** B 65–164, Pl. VII.   1911.

WORKERS AT THE IRISH FLORA.

78. In this book it is not possible in most cases to give the names of the finders of the rarer plants; and all that can be attempted in the present section is to indicate briefly the principal workers whose investigations have made possible a compilation such as this, and to mention in a few words the scope of their researches. For convenience of reference the names are arranged alphabetically, though a chronological arrangement would be more appropriate. Details of their published communications, often merely indicated below, up to 1900, will be found in the Bibliography of " Irish Topographical Botany." The more important subsequent papers or notes are indicated, like the earlier ones, in the latter portion of this book, where the plant or area dealt with is mentioned. In the following notes "first finder" means of course first finder of the plant in Ireland. Only those writings are referred to which deal especially with the Irish flora, and only the rarer plants which were found are mentioned.

Apart from the books noticed, the bulk of material relating to Irish plants has appeared in journals as follows :—

For the earlier period, say 1830–50—" Magazine of Natural History " and its successor the well-known "Annals and Mag."

1841–1863—" The Phytologist."

1863 to date—"Journal of Botany."

1892–1924—" Irish Naturalist."

1926 to date—" Irish Naturalists' Journal."

Also Proceedings of the Dublin Natural History Society, 1849–71, and of other Irish societies.

A few contractions are used below :—B. & B. signifies Britten and Boulger's " Biographical Index of British and Irish Botanists," 2nd ed. by A. B. Rendle, 1931; D.N.B., the " Dictionary of National Biography "; Fl. Dublin, Colgan's well-known Flora, 1904; Fl. NE.I., Stewart and Corry's similar work for Down, Antrim, and Derry; Lett, H. W. Lett's " Botanists of the North of Ireland," in Proc. Belfast Nat. Field Club, 1912–3, 615–628 (reprinted in Irish Nat. 1913, 21–33; the page-references below are to the original paper).

ADAMS (John): "A Student's Illustrated Irish Flora," 8vo, London, 1931. Proposed a useful scheme for the botanical subdivision of Ireland (**77**). Papers and notes in Irish Naturalist.

ALLIN (*Rev.* Thomas): "The Flowering Plants and Ferns of the County Cork," 8vo, Weston-super-Mare, 1883. And many notes on Cork plants in Journ. Bot. 1871–4.

ALLMAN (*Prof.* George James): First finder of *Diotis maritima* (1845). Biography—D.N.B. **1** 335; B. & B. 5.

ANDREWS (William): Notes and papers especially on ferns, mostly in Proc. Dublin Nat. Hist. Soc. 1841–71. Found *Cerastium arvense* var. *Andrewsii, Trichomanes radicans* var. *Andrewsii.* Claimed to have found *Saxifraga Andrewsii* (*S.* ? *Aizoon* × *spathularis*). Biography— D.N.B. **1** 409; B. & B. 8.

BABINGTON (*Prof.* Charles Cardale): Many notes and papers, mostly in Ann. & Mag. Nat. Hist., 1836–72. First proposer of a scheme for the botanical subdivision of Ireland (**77**). Biography—"Memorials of Charles Cardale Babington," 1897 (portrait); Journ. Bot. 1895, 257–266 (portrait); etc.

BAILY (Katherine Sophia, afterwards *Lady* Kane): "The Irish Flora," 8vo, Dublin, 1833. Biography—Lett 623.

BALFOUR (*Prof.* John Hutton): Papers (mostly Connemara) in Phytologist and Trans. Bot. Soc. Edinb., 1853–76. Biography—D.N.B. **2** 56; B. & B. 17.

BALL (John): Finder of *Potamogeton Babingtonii* (1835). Biography—Journ. Bot. 1889, 365–370; D.N.B. Suppl. 1, 115–8.

BALL (Robert): First finder (with William THOMPSON) of *Astragalus danicus* (1834), and *Allium Babingtonii* (1834). Biography—Nat. Hist. Review 1858, 134; D.N.B. **2** 77–8.

BARRETT-HAMILTON (Gerald E. H.): Papers (some with Miss L. S. GLASCOTT, or with C. B. MOFFAT) on Wexford plants, in Journ. Bot. and Irish Nat., 1887–1908. Biography — Irish Nat. 1914, 83–93 (portrait, list of papers).

BARRINGTON (Richard Manliffe): Reports on Tory I., the Blaskets, L. Erne, and (with R. P. VOWELL) Ben Bulben and L. Ree, and other papers, mostly in Journ. Bot. and

Proc. Roy. Ir. Acad., 1872–1915. First finder of *Caltha radicans* and (with R. P. VOWELL) of *Epilobium alsine-folium* (1884), and (with H. and J. GROVES) of *Nitella gracilis* (1892). Biography (with portrait and list of papers) Irish Nat. 1915, 193–206; Journ. Bot. 1915, 364–7 (portrait); B. & B. 22.

BENNETT (Arthur) : Papers and notes, in Journ. Bot. and Irish Nat., 1881–1919. Biography—Journ. Bot. 1929, 217–221 (portrait).

BLASHFORD (J.) : First finder of *Microcala filiformis* (before 1804), and *Colchicum autumnale* (1799).

BRENAN (*Rev.* Samuel Arthur) : Ulster notes, in Journ. Bot. and Irish Nat., 1884–1901. Biography—Irish Nat. 1908, 43; Lett 625.

BRITTEN (James) : Papers and notes in Journ. Bot., 1872–1919. Portrait—Journ. Bot. 1912, frontispiece. Biography (with portrait)—Journ. Bot. 1924, 327–343.

BRUNKER (James Ponsonby) : Papers and notes on the Wicklow flora, 1918——.

BULLOCK-WEBSTER (*Rev.* George Russell) ; Papers (some with J. GROVES) on Charophyta, in Irish Nat. and Journ. Bot., 1917–20. First finder of *Nitella mucronata* (1901), *N. spanioclema* n.sp. (1916), *C. muscosa* n.sp. (1917).

CARROLL (Isaac) : A few short papers in Phytol., Journ. Bot., etc., 1854–75. Biography—Journ. Bot. 1881, 128.

COLGAN (Nathaniel) : " Flora of the County Dublin," 8vo, Dublin, 1904. Editor (with R. W. SCULLY) of " Contributions towards a Cybele Hibernica," ed. 2, 1898. Many papers and notes in Irish Nat. (mainly) and Journ. Bot., 1885–1918. First finder of *Scrophularia alata* (1894),[35] and (with F. W. BURBIDGE) of *Senecio* × *albescens* (1902). Biography (with portrait and list of papers)—Irish Nat. 1919, 121–6; B. &. B. 69.

CORRY (Thomas Hughes) : Joint author with S. A. STEWART of "Flora of the North-east of Ireland," 8vo, Belfast, 1888. Papers and notes, mainly in Journ. Bot., 1880–84. First finder of *Hieracium hypochœroides* (1879). Biography—Fl. NE.I. v–vi; Journ. Bot. 1883, 313–4.

[35] Isaac Carroll collected it at an earlier date, but did not recognise it. See 334.

D'ARCY (Elinor): First finder of *Carex magellanica* (1901).

DAVIES (John Henry): Many papers and notes (Antrim and Down) in Irish Nat., 1892–1907. Biography—Irish Nat. 1909, 235–6; Lett 625–6; B. & B. 86.

DICKIE (*Prof.* George): "A Flora of Ulster and Botanist's Guide to the North of Ireland," 8vo, Belfast, 1864. A few other contributions. Biography—Fl. NE.I., xx–xxi; D.N.B. **15** 32; B. & B. 90.

DRUCE (George Claridge): Papers and notes, mostly in Journ. Bot., Bot. Exchange Club Reports, and Irish Nat., 1890–1929. First finder of *Utricularia ochroleuca* (1875), *U. Bremii* (1875), and many critical plants. Biography—Journ. Bot. 1932, 141–4.

DRUMMOND (James): First finder of *Pinguicula grandiflora* (1809) and *Spiranthes gemmipara* (1810). Papers in Munster Farmer's Mag., 1818–20. Biography—D.N.B. **16** 33; B. & B. 95.

FOOT (Frederick James): Papers and notes, chiefly on ferns, mostly in Proc. Dublin Nat. Hist. Soc., 1860–71. Biography—Geol. Mag. 1867, 95.

GROVES (Henry) and James GROVES: Papers, mostly on Charophyta, in Journ. Bot. (mainly) and Irish Nat., 1880–98. First finders of *Nitella tenuissima* (1892), and *N. gracilis* (with R. M. BARRINGTON, 1892).

GROVES (James) and *Rev.* G. R. BULLOCK-WEBSTER: Papers on Charophyta, in Journ. Bot. and Irish Nat., 1917–24.

HART (Henry Chichester): "Flora of the County Donegal," 8vo, Dublin, 1898. "Flora of Howth," 8vo, Dublin, 1887. Many reports and papers on flora of Irish mountains, rivers, islands, in Journ. Bot. (mainly), Proc. R. Irish Acad., etc., 1873–1908. First finder of *Cochlearia grœnlandica* (before 1896), *Helianthemum Chamæcistus* (1893), *Hieracium hibernicum* (1883), *Carex Bœnninghauseniana* (1883–4). Biography —Irish Nat. 1908, 249–254 (portrait, list of papers); Journ. Bot. 1911, 121–2 (portrait); B. & B. 141–2.

HEATON (*Rev.* Richard): First finder of *Dryas octopetala*, *Gentiana verna*, *Scilla verna*, and very possibly *Saxifraga spathularis* (*umbrosa* auct.) and *Euphorbia*

hiberna, published as Irish in How's " Phytologia,"
1650. Biography—Colgan Fl. Dublin xix; B. & B. 143.

HIND (*Rev.* William Marsden): Papers, mostly in Phytologist, 1851–71. Biography—Lett 619.

K'EOGH (John): " Botanica Universalis Hibernica," 4to,
Cork, 1735. Biography—D.N.B. **31** 33; B. & B. 172.

KINAHAN (George Henry): Papers on ferns (western),
mostly in Proc. Dublin Nat. Hist. Soc., 1860–71. Biography (with portrait)—Irish Nat. 1909, 29–31.

KINAHAN (*Prof.* John Robert): Papers on ferns, mostly in
Proc. Dublin Nat. Hist. Soc., 1854–63. Biography—
B. & B. 173.

KIRK (Thomas): First finder of *Potamogeton sparganii-folius* (before 1856).

KNOWLES (Matilda Cullen): Papers and notes (Tyrone,
Limerick, etc.) in Irish Nat., Irish Nat. Journ., Proc.
R. Irish Acad., 1897–1932.

LEEBODY (Mary Isabella): Notes (northern) in Journ. Bot.
and Irish Nat., 1893–1911. First finder of *Teesdalia
nudicaulis* (1896). Biography—Irish Nat. 1911, 218.

LETT (*Rev.* Henry William): Notes (NE area) in Journ. Bot.
and Irish Nat., 1884–1914. First finder of *Rubus Lettii*
(1901), *Hypochœris glabra* (with C. H. WADDELL, 1900),
Carex pauciflora (1889). Biography—Irish Nat. 1921,
41–3; Journ. Bot. 1921, 75–6; B. & B. 186.

LEVINGE (Harry Corbyn): Papers (mostly on Westmeath
plants) in Journ. Bot. and Irish Nat., 1891–6. First
finder of *Chara denudata* (1892), and (with H. and J.
GROVES) of *Nitella tenuissima* (1892). Biography—
Irish Nat. 1906, 107; B. & B. 186–7.

LHWYD (Edward): First finder of *Arenaria ciliata* (1699),
Saxifraga Geum (1699), *Potentilla fruticosa* (1699),
Dabeocia polifolia (about 1699), *Adiantum Capillus-Veneris* (before 1700). Paper in Phil. Trans., 1712.
Biography—B. & B. 187.

LINTON (*Rev.* Edward Francis): Notes and papers in
Journ. Bot. and Irish Nat., 1886–1909. Finder of *Carex
trinervis* (1885).

LYNAM (John): First finder of *Sisyrinchium angustifolium*
(1845). Biography—B. & B. 186.

MACALLA (William), phycologist: First finder of *Erica Mackaii* (before 1835). Biography—B. & B. 197.

MACKAY (James Townsend): '' Flora Hibernica,'' 8vo, Dublin, 1836. Catalogues of Irish plants, and notes, 1806–60. First finder of *Erica mediterranea* (1830), *Arabis Brownii* (1805), *Bartsia viscosa* (before 1806), *Sibthorpia europaea* (1805), *Helianthemum canum* (before 1806), *Poa alpina* (1804), and other plants. Biography—Fl. Dublin xxvi–xxviii; B. & B. 199.

MARSHALL (*Rev.* Edward Shearburn): Papers (one with W. A. SHOOLBRED) and notes (Wexford, Mayo), in Journ. Bot. (mostly) and Irish Nat., 1892–1912. First finder of *Ranunculus scoticus* (1899), *Leucojum aestivum* (1897), *Sisyrinchium californicum* (1896), *Tolypella nidifica* (1896), *Chara connivens* (1896). Biography— Journ. Bot. 1920, 1–11 (portrait).

MOFFAT (Charles Bethune): ''Life and letters of Alexander Goodman More,'' 8vo, Dublin, 1898. Papers and notes (mostly Wexford) in Irish Nat. (chiefly) and Journ. Bot., 1889——.

MOORE (David): Joint author with A. G. MORE of ''Contributions towards a Cybele Hibernica,'' 8vo, Dublin, 1866. Many papers and notes in Phytol., Journ. Bot., etc., 1843–78. First finder of *Trifolium glomeratum* (1869), *Rosa Moorei* (c. 1835), *Apium Moorei* (before 1866), *Inula salicina* (1843), *Pyrola secunda* (1835) *P. rotundifolia* (1870), *Ajuga pyramidalis* (1854), *Carex fusca* (1835), *C. divisa* (1866), *C. elongata* (1837), *C. paradoxa*, *Calamagrostis neglecta* var. *Hookeri* (1836), *Glyceria Borreri* (before 1866), *Equisetum Moorei* (with J. MELVILLE, 1851), *Chara tomentosa* (1841), *Tolypella prolifera* and *T. intricata* (before 1860), and other plants. Biography—Gard. Chron. 1879, 739 (portrait); Fl. NE.I. xviii–xx; B. & B. 219; Fl. Dublin xxvii–xxix.

MORE (Alexander Goodman): Joint author with David Moore of ''Contributions towards a Cybele Hibernica,'' 1866. Many papers and notes in Journ. Bot., Trans. Bot. Soc. Edinb., etc., 1855–93. First finder of *Viola stagnina* (1851), *Trifolium subterraneum* (1867), *Scirpus nanus* (1868), *Deschampsia alpina* (before 1872), *D. setacea* (1869). Biography—'' Life and Letters '' by C. B. MOFFAT, 1898 (portrait); Irish Nat. 1895, 109–116 (portrait, list of papers); Journ. Bot. 1895, 225–7 (portrait); Fl. Dublin xxvii–xxx; B. & B. 220.

MORE (Frances M.) : First finder of *Neotinea intacta*, 1864. Biography—Irish Nat. 1909, 132; Fl. Dublin xxvii–xxix.

MURPHY (*Prof.* Edmund) : Paper in Mag. Nat. Hist. 1829, etc. First finder of *Trollius europæus* (before 1829), *Rubus Chamæmorus* (1826), *Polystichum Lonchitis* (1826). Biography—B. & B. 224.

NEWMAN (Edward) : Papers and notes, chiefly on ferns, in Phytol. (mainly), 1839–54. Biography—Journ. Bot. 1876, 223–4; Memoir, by his son (portrait), 1876.

O'BRIEN (Robert Donough) : First finder of *Scirpus triqueter* (1900). Biography—Irish Nat. 1917, 113.

O'KELLY (Patrick B.) : First finder of *Potamogeton perpygmæus* (1891), *Limosella aquatica* (1893).

OLIVER (Daniel), jun. : Papers in Phytologist, 1851–3. First finder of *Naias flexilis* (1850) and *Euphrasia salisburgensis* (1852). Biography—Journ. Bot. 1917, 89–95 (portrait); B. & B. 233.

O'MAHONY (*Rev.* Thaddeus) : Paper in Proc. Dublin Nat. Hist. Soc., 1860. First finder of *Simethis planifolia* (1848) and *Epipactis atropurpurea* (1851). Biography —B. & B. 233.

PETHYBRIDGE (George Herbert) : Joint author (with R. Ll. PRAEGER) of "Vegetation of the District lying south of Dublin," Proc. Roy. Irish Acad. **25** B 124–180, 5 plates, coloured map, 1905.

PHILLIPS (Robert Albert) : Notes and papers (S and SE Ireland) in Irish Nat. 1893——. First finder of *Ranunculus lutarius* (1894), *R. tripartitus* (1896), *Sorbus latifolia* (1924), *Œnanthe pimpinelloides* (1896), *Serratula tinctoria* (1925), *Brachypodium pinnatum* (1898).

POWER (Thomas) : "The Botanist's Guide for the County of Cork" (in "Contributions towards a Fauna and Flora of the County of Cork," by J. R. HARVEY, J. D. HUMPHREYS, and T. POWER), 8vo, London, Cork, 1845.

PRAEGER (Robert Lloyd) : " Irish Topographical Botany" (1901), and Supplements for 1901–5, 1906–28, and 1929–34, all in Proc. R. Irish Acad. "A Tourist's Flora of the West of Ireland," 8vo, Dublin, 1909, etc Also many papers and notes in Journ. Bot., Irish Nat.

(mainly), etc., 1890——. First finder of *Medicago sylvestris* (1894), *Sorbus anglica* (1933), *Arctium majus* (1893), *Polygonum laxiflorum* (1896), *Spiranthes stricta* (1892), *Poa palustris* (1896), *Glyceria Foucaudii* (1903), *Lastrea remota* (1898), *Phegopteris Robertiana* (1932), *Equisetum litorale* (1917).

ROGERS (*Rev.* William Moyle): "Handbook of British Rubi," 8vo, London, 1900. Papers and notes on *Rubi* in Journ. Bot. and Irish Nat., 1894–1918. Described several Irish *Rubi* and determined very many. Biography—Journ. Bot. 1920, 161–4 (portrait).

SCULLY (Reginald William): "Flora of County Kerry," 8vo, Dublin, 1916. Joint editor with N. COLGAN of "Contributions towards a Cybele Hibernica," 2nd ed., 8vo, Dublin, 1898. Papers and notes, mainly on Kerry plants, Journ. Bot. and Irish Nat., 1888——. First finder of *Hieracium Scullyi* (1894), *Polygonum sagittatum* (1889), *Juncus macer* (1889), *Carex hibernica* (1889), *Nitella batrachosperma* (1889), *Chara canescens* (1894).

SHERARD (William): First finder of *Mertensia maritima* (c. 1691). Biography—Journ. Bot. 1874, 129–138; B. & B. 274.

STELFOX (Arthur Wilson): First finder of *Carex Pairaei* (1920). Papers and notes in "Irish Naturalist," etc., 1914——.

STEWART (Samuel Alexander): Joint author with Thomas Hughes CORRY of "A Flora of the North-east of Ireland," 8vo, Belfast, 1888; and (with R. Ll. PRAEGER) of "A Supplement to the Flora of the North-east of Ireland," 1895. (See also under WEAR, S.) Reports, papers, and notes (NE (mostly), Fermanagh, L. Allen, Lower Shannon) in Proc. R. Irish Acad., Journ. Bot., and Irish Nat., 1865–1904. First finder of *Ranunculus fluitans* (1865), *Hieracium Stewartii* (1891), *Carex aquatilis* (1884). Biography—Irish Nat. 1910, 201–9 (with portrait and list of papers); Journ. Bot. 1911, 122–3 (portrait); Lett 623–4; B. & B. 289.

STOKES (Whitley): First finder of *Trichomanes radicans* (before 1804). Biography—B. & B. 290.

TATE (*Prof.* Ralph): "Flora Belfastiensis," 16mo, Belfast, 1863. A few notes in Journ. Bot., etc. Biography—Fl. NE.I. xx-xxi; Irish Nat. 1902, 36–9; B. & B. 296.

TEMPLETON (John): MS. " Flora Hibernica" and other MSS. in Library R. I. Academy. Letter (on *Rosa hibernica*) in Trans. R. Dublin Soc. **3** 162, 1802. First finder of *Rosa hibernica* (1795), *Sisymbrium Irio, Ligusticum scoticum* (1793), *Adoxa Moschatellina* (with J. L. DRUMMOND, 1820), *Orobanche rubra* (before 1793), and many other plants. Biography—Fl. NE.I. xvi–xviii; B. & B. 298; Fl. Dublin xxiv–xxv; Lett 616–7.

THOMPSON (William), author of " The Natural History of Ireland," 4 vols. Notes in Ann. Nat. Hist. and Phytologist, 1842–3. First finder of *Elatine Hydropiper* (1836), *Astragalus danicus* (with R. BALL, 1834), *Allium Babingtonii* (with R. BALL, 1834). Biography— Memoir, by R. Patterson, in Thompson : Nat. Hist. of I., **4** (portrait).

THRELKELD (Caleb): "Synopsis Stirpium Hibernicarum," 12mo, Dublin, 1726. Biography—Fl. Dublin xix–xx; B. & B. 301.

TOMLINSON (W. J. C.): Paper and notes in Irish Nat., 1903–16. Biography—Irish Nat., 1921, 108.

TOWNSEND (Miss H.): Finder of *Helianthemum guttatum* (before 1843).

TRENCH (Miss): Finder of *Euphorbia Peplis* (1839). Biography—Lett 620–1.

VOWELL (Richard Prendergast): Joint author with R. M. BARRINGTON (*vide*) of reports in Proc. R. Irish Acad., 1885–8. First finder (with R. M. BARRINGTON) of *Epilobium alsinefolium* (1884). Biography—Irish Nat. 1911, 218.

WADDELL (*Rev.* Cosslett Herbert): Notes (mostly NE area) in Irish Nat., 1892–1917. First finder of *Hieracium senescens* (1900), *Hypochœris glabra* (with H. W. LETT, 1900), *Erythrœa litoralis* (with J. E. SAUL, 1913), *Rhinanthus major* (1908), and some critical plants. Biography—Irish Nat. 1919, 108; Journ. Bot. 1919, 358–9; B. & B. 311.

WADE (Walter): "Catalogus systematicus plantarum indigenarum in comitatu Dublinensi inventarum," Pars prima. 8vo, Dublin, 1794. Papers in Trans. R. Dublin Soc., 1802–4. First finder of *Matthiola sinuata* (1804), *Eriocaulon septangulare* (1801). Biography—Fl. Dublin, xxiii–xxiv; B. & B. 312.

WEAR (Sylvanus): "A Second Supplement and Summary of Stewart and Corry's Flora of the North-east of Ireland," Belfast Nat. Field Club, 1923. Biography—Irish Nat. 1921, 23.

WHITE (J.): "An Essay on the indigenous Grasses of Ireland." Dublin, 1808. Contributor to "The Irish Flora," 1833. Biography—Fl. Dublin, xxvii; B. & B. 180.

WHITLA (William): First finder of *Equisetum trachyodon* (1830). Biography—B. & B. 325.

WYNNE (*Rt. Hon.* John): First finder of *Arabis petræa* and *Saxifraga nivalis* (1837). Biography—B. & B. 336.

SOME RARE OR INTERESTING PLANTS.

79. For convenience of reference it may be well to list the more interesting members of the Irish flora in systematic order, with brief notes as to their distribution, etc. This list deals mostly with such well-marked and well-known species (Linneons) as are of rare occurrence in Ireland or Britain or both, but some interesting critical plants and hybrids have been included. A fair amount of work has been done at the more critical plants in Ireland, but a great deal remains. Many of the finer splits, such as those of *Capsella, Viola, Taraxacum, Thymus, Salicornia,* have hardly been touched as yet. In so far as segregates have been awarded specific rank in the eleventh edition of the "London Catalogue," their Irish range is shown in brief in the "Census List" which appears at the end of this volume. For further information, the obvious Irish works of reference should be consulted, as well as papers by E. S. Marshall, G. C. Druce, C. H. Waddell, etc., referred to in the Bibliography in "Irish Topographical Botany"; also my own papers, "Gleanings in Irish Topographical Botany," in Proc. Roy. Irish Acad., **24**, B, 1902, and "A Contribution to the Flora of Ireland" (including 3rd Suppl. to "Irish Topographical Botany") in Proc. Roy. Irish Acad., **42**, B, 1934. Druce's "Comital Flora of the British Isles" contains some additional records (*Viola, Alchemilla, Thymus, Rhinanthus,* etc.).[36] As regards hybrids, work is required in *Epilobium, Euphrasia, Rumex, Salix,*

[36] See **492.**

and some other genera. The few hybrids mentioned below are ones in which interesting species take part, or which are known to be of very infrequent occurrence.

80. Ranunculus fluitans Lamk.—Known only from the Sixmilewater in Antrim, where it grows abundantly for some miles (**461**). First found by S. A. Stewart in 1865. Widespread in England and southern Scotland, and in Europe generally.

81. R. tripartitus DC.—Found by R. A. Phillips in 1896 growing plentifully in a small lake S of Baltimore in W Cork (see Journ. Bot., 1896, 277, and Irish Nat., 1896, 166), and the closely allied *R. lutarius* in a pool near Adrigole, 20 miles to the NW in 1894. Only Irish stations. In Britain *R. tripartitus* is exceedingly rare, and in England in SW only. In Europe western. See also **297, 306**.

82. R. scoticus E. S. Marshall (*R. petiolaris* E. S. Marshall, 1892, non H. B. & K., 1821. *R. Flammula* var. *petiolaris* Lange).—Only recently separated from *R. Flammula*, and distribution imperfectly known. First found in Ireland by Marshall on Achill I. in 1899; since by many lakes in W Mayo, and by Loughs Corrib, Mask, Conn, Achree, Gill, Key, Macnean, Erne, Neagh (Londonderry), and Dan (Wicklow). Also by L. Bofin in Leitrim, whence there was a doubtful old record (Journ. Bot., 1892, 377). From 50 ft. (L. Corrib) to 1000 ft. (lakelets S of Lower L. Erne). It occurs both on and off the limestone. Figured in Journ. Bot., 1892, tab. 328. Known from 14 vice-counties in Scotland.

82a. Caltha radicans Forst.—Long known only from Devenish Island near Enniskillen, it proves to be widespread in Ireland, though not in such an extreme form. It is lowland, favouring lake-shores especially in the Central Plain and elsewhere, and becomes rare only in the E and SE. See Irish Nat. Journ., **5**, 98–102, 1934, for particulars of its distribution.

83. Meconopsis cambrica Vig.—Isolated stations on the hills in many counties, 50–1500 ft., but absent from some likely ground, such as Kerry, Mayo, Donegal. First found by John Templeton at Fair Head in Antrim before 1804. Southern in Britain, being not native in Scotland. Southwestern in Europe (Ireland to Spain and the Jura).

84. Matthiola sinuata Br.—A rare and decreasing species in Ireland, but still grows on the Wexford coast (see Irish Nat. Journ., **1**, 96), and in Clare (Irish Nat., 1912, 153, and Irish Nat. Journ., **5**, 33); apparently extinct in Kerry (Fl. Kerry, 16). In Britain confined to SW England and Wales. Elsewhere on the shores of W Europe and the Mediterranean.

85. Arabis Brownii Jordan Diagn. 123, 1864. (*A. ciliata* auct. angl., non R. Br. *A. hibernica* Willmott, Journ. Bot., 1924, 26). A critical plant, which has been involved in much confusion, on account of its proximity to forms of *A. hirsuta*. As now segregated, it appears to be endemic in Ireland (see Salmon in Journ. Bot., 1924, 236–8). First found by Lhwyd in 1699. In sandy places along the west coast in S Kerry, W Cork, Clare, W Galway, and W Donegal.

86. Cochlearia anglica Linn.—Till lately considered a southern plant in Ireland (Wexford to the Shannon), with an outlying station on the Foyle, but now known to occur in muddy places and by muddy waters all round the Irish coast. It crosses everywhere with the ubiquitous *C. officinalis*, the hybrids being usually much more abundant than pure *anglica*, and including "var." *Hortii* (Praeger in Proc. R. Irish Acad., **41**, B, 96–100, 1932). In Britain southern, but extends to the W coast of Scotland. In Europe along the NW coasts.

87. C. grœnlandica Linn. (see Marshall in Journ. Bot., 1892, 225–6, tab. 326. *C. scotica* Druce).—West and north coasts, in Cork, Kerry, Clare, Galway, Mayo, Sligo, Donegal, Derry, Antrim, frequent on exposed sea-rocks. See Praeger in Proc. R. I. Acad., **41**, B, 101–2, 1932. Frequent in similar situations in Scotland; apparently absent further S in Britain. NW Europe to the Arctic. First found by H. C. Hart before 1896. Crosses in Kerry, Donegal, and Derry with *C. officinalis*.

88. Sisymbrium Irio Linn.—A characteristic Dublin alien, unknown elsewhere in Ireland. Formerly abundant; much rarer now, owing mainly, according to Colgan (Fl. Dublin, 21), to the substitution of concrete for gravel on suburban foot-ways. Occasionally found from Swords to Dun Laoghaire (Kingstown), but mostly near the city, where it was first noted by John Templeton at the beginning of the 19th century. In Britain it is equally rare and local. Widespread in Europe, extending to N Africa, etc.

89. Helianthemum guttatum Mill.—The type and var.

Breweri grow plentifully at Three Castle Head in W Cork, and on Inishbofin (**405**) and (var. *Breweri* only) the neighbouring Inishark in W Galway; the type also sparingly on Inishturk (**406**), a few miles N. First found by Miss Townsend before 1843. The variety grows on Anglesea, but the nearest station for the type is the Channel Islands. The plant has a wide foreign range—Europe, N Africa, W Asia.

90. Helianthemum canum Baumg.—Locally abundant in Burren, Co. Clare, with outliers on Inishmore and the Cliffs of Moher to the W, and Salthill in W Galway to the N. First found by J. T. Mackay before 1806. W England, rare. More widely spread abroad, in Europe, W Asia, N Africa.

90a. Helianthemum Chamæcistus Mill.—Found by H. C. Hart in 1893 on bare limestone south of Ballintra in SW Donegal (Journ. Bot., 1893, 218); disallowed subsequently as a casual escape or introduction. Refound on the same ground in 1933, amid a native flora, and undoubtedly indigenous. (Irish Nat. Journ., 5, 76–77.) The only Irish station for a plant very widespread in Britain.

91. Viola stagnina Kitaibel.—Very local, in damp places on the western limestones in Clare, SE Galway, E Mayo and Fermanagh, and on the banks of the Shannon near Roosky in Longford. It is essentially a plant of the "turloughs"—deep grassy hollows in the limestone which are flooded, often for long periods, especially in winter (**360**). The neighbourhood of Gort seems to be its headquarters. Hybridizes freely with *V. canina*. First found in Ireland by A. G. More in 1851 at Garryland near Gort. In England very rare, mostly on bogs and fenlands in the E. Also in NW and central Europe.

92. Viola lutea Huds.—In only three areas—one in W Cork, one in Clare, and a third about the upper Liffey in Kildare, Wicklow and Dublin, where it is very abundant in places. From sea-level to 1000 ft. In hilly districts almost throughout Britain. Europe.

93. Polygala vulgaris var. **Ballii** (Nyman). (Var. *grandiflora* Bab. *P. grandiflora* Druce. *P. Babingtonii* Druce).—This fine variety has been long known from the Ben Bulben range (**422, 429**) in Sligo and Leitrim (to 1700 ft.) and from Benevenagh (**456**) in Derry. G. C. Druce has recorded it also from Ardrahan in SE Galway and

Inchnadamph in Sutherland (Irish Nat., 1912, 240) and Ostenfeld from the Faeröes (Bot. of the Faeröes, 71, 1901), but the identity of these plants with the Irish one appears somewhat doubtful. See A. Bennett, Journ. Bot., 1912, 228–9, and G. C. Druce, *ibid.*, 1913, 60–61.

94. Silene acaulis Linn.—Common on the Ben Bulben range, with outlying single stations in Mayo, Donegal, and Derry, 400–1600 ft.—a NW distribution characteristic of alpine and "Scottish" species in Ireland. Alpine in Britain. Circumpolar, in Arctic and northern regions.

95. Arenaria ciliata Linn. subsp. **hibernica** Ostenfeld and Dahl (plate 3). In Nyt Magazin for Naturvidenskaberne, 1917, 215–225, these authors distinguish the Ben Bulben plant under the name *hibernica* from subsp. *pseudofrigida* occurring in Arctic Europe, and subsp. *norvegica* (*A. norvegica* Gunnerus and auct. britt.) of Shetland and Arctic Europe and America, the three forming "Kollektivart" *A. ciliata.* The plant is of frequent occurrence on the W part of the Ben Bulben range in Sligo, on limestone rocks, 1200–1950 ft. (**423**). Found by E. Lhwyd in 1699 (Druce Com. Fl. Brit., 54); first recorded by J. T. Mackay in 1806.

96. Elatine Hydropiper Linn.—Was apparently confined to Lough Briclan and Lough Shark, Co. Down, and L. Neagh (**465**) until the making of the canals late in the 18th century, when it spread from L. Neagh along the Lagan Canal as far as Belfast, and (presumably from L. Shark) down the Newry Canal to Newry. Not seen recently in most of its old stations, but occupies 15 miles of the canal from Newry northward (see Irish Nat. Journ., 5, 102–4). First found in Ireland by William Thompson at Newry in 1836. Equally rare in England and Wales, though not quite so local. Widespread in Europe, but local; Siberia, N America.

97. Lavatera arborea Linn.—An aboriginal inhabitant of sea-rocks, with a present range much extended by introduction into cottage gardens. Indigenous in the S (Waterford to Clare), and in Wicklow, Dublin, Antrim, on cliffs and sea-stacks. Of doubtful standing elsewhere. Britain as far N as the Clyde. W and S coasts of Europe; N Africa, Canaries.

98. Erodium moschatum L'Hérit.—A plant of uncertain standing, chiefly about roadsides and towns by the sea. Of 20 divisions (out of 40) in which it grows, only one (Offaly) is wholly inland. More frequent in Cork than elsewhere. Widespread in England and Wales. Widespread in Europe, N Africa, W Asia, America, N. Zealand.

99. Ulex europæus f. **strictus** (Mackay). (*Ulex strictus* Mackay in Trans. Roy. Irish Acad., **14**, 166, 1824–5 (*nomen solum*). *U. europæus* var. *strictus* Mackay Fl. Hib., **1**, 74, 1836. *U. hibernicus* G. Don, Loudon's General System, **2**, 148, 1832).—This curious juvenile form of *U. europæus*, soft, erect, with short branchlets and spines, and few flowers, was found at Mountstewart, Co. Down, by a Mr. Murray of Comber; he brought it into his nursery, where it was seen by John White, gardener at Glasnevin and author of the ''The Grasses of Ireland'' (1808), by whom it was introduced into cultivation (see Fl. NE Ireland, 34).

100. Medicago sylvestris Fr. (? *M. falcata* × *sativa*).— Co. Dublin only—a large patch on dunes at Malahide, and a smaller patch at Portmarnock, 3½ miles S (**228**). First recognised by Praeger in 1896 (Irish Nat., 1896, 249–251), but has grown at Portmarnock for at least 80 years (specimens in Herb. Nat. Mus., Dublin). See also Colgan Fl. Dublin, 55–6.

In Britain a very rare plant of the E counties of England. On the Continent local and rather western. Often set down as *M. falcata* × *sativa*. In Ireland, in the area where the plant occurs the former is a rare casual, the second a naturalized alien; *M. sylvestris* is there a very persistent perennial.

101. Astragalus danicus Retz. (*A. hypoglottis* Linn.).— Known only in Ireland from the Aran Islands in Galway Bay, where it was first found by R. Ball and Wm. Thompson in 1834. It appears to be limited there to Inishmore and Inishmaan, where it is a member of the peculiar calcicole flora of the bare limestones (**352**), spreading to sandy pastures. In Britain local, on gravelly and chalky soils from N Scotland to S England. Arctic-alpine in Europe; also in N Asia.

Clearly a relict species in Ireland from Glacial or early Postglacial times.

102. Trifolium glomeratum Linn.—SE coast only, in Wexford (Rosslare and Arthurstown) and Wicklow (riverside at Wicklow, where it was found for the first time in Ireland by D. Moore in 1869). S and SE England. SW and S parts of the Continent, from W France along the Mediterranean; W Asia, N Africa, Canaries.

103. Lathyrus palustris Linn.—Drainage has greatly reduced this plant on Lough Neagh, where it was formerly frequent. Elsewhere it has a number of stations confined to the Shannon and Erne basins, also outposts in Wicklow. (Compare the similar range of *Sium latifolium*). Middle England mostly. A northern circumpolar plant.

104. Lathyrus maritimus Bigel.—Another northern plant, clearly relict, known only from sand-dunes at Rossbeigh on the S side of the head of Dingle Bay. Recorded as long ago as 1756 (Smith: "History of Kerry," 380) from the sands of Inch on the N shore close by, and seen at intervals, apparently in both stations, until 1845. Rediscovered at Rossbeigh in 1918 (Irish Nat., 1918, 113–115). It grows there in fair quantity over a limited area. (See **318**, also Scully Fl. Kerry, 72–4, and Irish Nat., *l.c.*) In Britain down the E and SE coast from Shetland to Dorset. Elsewhere a northern and Arctic circumpolar plant.

105. Rubus.—Though much collecting has been done in this group, as by S. A. Stewart, H. W. Lett, C. H. Waddell, R. A. Phillips, R. W. Scully, E. S. Marshall, W. M. Rogers, and the writer, Irish brambles are still incompletely known. Their range and abundance seem to differ materially from what obtains in England, and Moyle Rogers has remarked on the fact that few species in Ireland tally exactly with their British analogues. A few noteworthy plants may be mentioned here:—

R. hesperius Rogers.—Chiefly in the W, apparently frequent. Very rare in Britain.

R. iricus Rogers.—In nine S and W divisions, from Cork to Mayo, and also Antrim. Type apparently endemic, but var. *minor* is found in Britain.

R. Lettii Rogers.—In Cavan, Armagh, Down, and half a dozen English counties.

R. adenanthus Boul. and Gill.—S and N Kerry and five British vice-counties.

R. mucronatoides Ley.—W Mayo and three English vice-counties only.

R. regillus Ley.—In N and S Kerry, also Down; and three vice-counties in Britain.

R. morganwgensis Barton and Riddelsdell.—In Down and some eight British vice-counties.

R. ochrodermis Ley.—In SE Galway and eight British vice-counties.

Among other *Rubi* rare in both Ireland and Britain are *R. altiarcuatus, nemoralis, Questieri, Salteri, Wedgwoodiæ, thyrsiger, longithyrsiger, serpens,* etc.

106. Dryas octopetala Linn. (plates 2, 20).—He who has viewed the thousands of acres of this Arctic-alpine plant in full flower on the limestones of the Burren region of Clare, from hill-top down to sea-level, has seen one of the loveliest sights that Ireland has to offer to the botanist. It has a second less continuous headquarters about Ben Bulben and northward, and single stations in Donegal, Derry, and Antrim. It is not certain that in its limited stations off the limestone in W Galway (**391**) and Donegal (**445**) the plant is actually in a non-calcareous soil, for the metamorphic rocks among which it occurs may well yield a sufficiency of calcium by the decay of primitive limestone or of serpentine. In this connection see also **193**. In outlying stations on limestone drift at Gentian Hill and Barna Head, west of Galway, as in Burren, it descends to sea-level. Its highest station in Ireland is only 1300 ft. (Slieve League in Donegal). First found by Rev. Richard Heaton before 1650 (How's "Phytologia"). In Britain from Wales to Orkney, mostly on mountains, and essentially a limestone plant. An Arctic-alpine circumpolar species: also in the Caucasus.

107. Potentilla fruticosa Linn. (plate 21).—On limestone rocks along the western edge of the Central Plain for about 40 miles from N Clare to E Mayo, locally abundant. N England on limestone, very local. A northern and alpine plant with a wide but local circumpolar range, S to the Himalayas and Caucasus.

108. Rosa.—Knowledge of Irish roses has not kept pace with recent work in England. Some existing records stand in need of revision, and further work is required. The only species which is known to have a definite *local* range in the country is *R. micrantha,* which is confined to Kerry and Cork. *R. stylosa* (var. *systyla*) (apparently southern and native—see Journ. Bot., 1934, 69). *R. Afzeliana* (= *glauca*), *R. dumetorum, R. coriifolia, R. Sherardi* (= *omissa*), and *R. tomentella* occur, but their range is not worked out yet. *R. spinosissima, canina, tomentosa, arvensis* are common, *rubiginosa* and *mollis* rather rarer. Among the hybrid forms, *R. gracilescens* and *R. Moorei* (*rubiginosa* × *spino-*

sissima), two very rare plants, are old north-eastern finds,[37] and *R. mayoensis* (*mollis* × *spinosissima*) has been recently described (Wolley-Dod in Journ. Bot., 1924, 202) from Mayo and Sutherland.   *R. pilosa* is Irish only, but the station is unknown.   Two other hybrids are mentioned separately below, the first because of its essentially Irish history, the second because it is unknown elsewhere.

109. Rosa hibernica Templeton (*R. dumetorum* × *spinosissima*).—This rose was first found between Belfast and Holywood in Down (**474**) in 1795 by John Templeton, who described it as a new species in 1802.   Long looked on as *R. canina* × *spinosissima*, that combination is now attributed to the plant known as *R. hibernica* var. *glabra* Baker (× *R. glabra* Wolley-Dod).   The "type" has been found about Tillysburn and Stranmillis near Belfast and (?) Magilligan in Derry, *R. glabra* about Glenarm and Carnlough in Antrim, and Magilligan : also in Co. Limerick (Irish Nat., 1903, 250).   Rare in Britain and on the Continent.   See Templeton in Trans. Dublin Soc., **3**, 162–4; J. Britten in Journ. Bot., 1907, 304–5, and Irish Nat., 1907, 309–10; H. W. Lett in Journ. Bot., 1907, 346–7; Wolley-Dod in Journ. Bot., 1931, Suppl., 14.

110. Rosa Praegeri Wolley-Dod (*R. canina* × *rugosa*).— A natural cross between the native *R. canina* and the Japanese *R. rugosa*, recently found at Cushendun in Antrim (Journ. Bot., 1928, 87–88).   Not known elsewhere.

111. Sorbus Aria group.—The complicated group of forms formerly included in *S. Aria* have recently been found to be well represented in Ireland.   *S. anglica* (*S. Mougeotii* var. *anglica* Hedlund), rare in W England, is native at Killarney.   *S. rupicola* is apparently quite rare, but ranges from Kerry to Antrim.   Unexpectedly, *S. porrigens*, which like *anglica* is confined in Britain to the region of the Severn, proves to be the prevailing Irish plant, ranging from Kerry to Dublin and Sligo.   *S. Aria*, very widespread in Britain, is so far in Ireland determined only from the neighbourhood of Galway.   (See Irish Nat. Journ., **5**, 50–52, also Salmon in Journ. Bot., 1930, 172–7.)   More distinct is *S. latifolia*, first found by R. A. Phillips in 1908, occupying an area along the rivers Barrow and Nore above and below New Ross, and I believe native there (see Irish

---

[37] *R. Moorei* is also doubtfully recorded from Kilkenny (Irish Nat., 1906, 170–1).

Nat., 1924, 129; Irish Nat. Journ., **4**, 194), as the others undoubtedly are in the regions indicated.

112. Saxifraga nivalis Linn.—Among *Arenaria ciliata* and other alpines on the cliffs of Annacoona in the Ben Bulben range (**422**), 1200–1950 ft. Only Irish station, where it was first found by John Wynne in 1837. In Britain in Wales, the Lake District, and the Highlands, in alpine situations. A northern circumpolar plant.

113. Saxifraga Geum Linn.—Confined in Ireland in its typical (pure) form to the SW (Kerry (except the north) and W. Cork (**300**)). Elsewhere in the Pyrenees, N Spain, Portugal; also reported from the Alps and Carpathians, but there is no reason to consider it native there. Observation points to the truth of Scully's suggestion that this is a decreasing species in Ireland, now tending to hybridize itself out of existence by crossing with the stronger *S. spathularis* (*umbrosa* auct.). This view is endorsed by its more recent discovery on Clare Island, Co. Mayo (the only station outside the district named), where it occurs sparingly not in its pure form, but in a hybrid state well known in the SW and representing a plant about two-thirds *S. Geum* and one third *S. spathularis* : and along with this on Clare Island (amid abundance of typical *spathularis*), are other plants about half-way between the two species. *S. Geum sensu stricto* is a plant "foliis reniformibus dentatis" (Linn. Sp. Pl. (ed. 2), 575). This is a well-marked form, varying little except in size and in the dentition of the leaves, which may be crenate or apiculate-crenate or truncate-crenate or quite sharply serrate with teeth about equilateral (*i.e.*, with the base of the teeth equalling the sides). Authors subsequent to Linnæus have often included under *S. Geum* forms in which the leaves are not reniform, only orbicular, the leaf-margins entering the petiole at about a right-angle. Forms of this facies (to which the Clare Island plant belongs) have several times been awarded separate names—*S. polita* Haworth, *S. dentata* Haworth, *S. gracilis* Mackay, *S. elegans* Mackay, but they must be looked on as of hybrid origin—the first step from *Geum* towards *spathularis*. *S. hirsuta* represents a further stage towards *spathularis* among the complicated hybrid progeny, being fairly intermediate between the parents. Thence other forms lead on to pure *spathularis* (see under that species). (See Scully Fl. Kerry, 125–6, also Praeger in Irish Nat., 1912, 205–6.)

First found by E. Lhwyd, in Kerry, in 1699.

114. Saxifraga spathularis Brotero. (*S. umbrosa* auct.
non Linn.[38]) (plates 4, 34).—An abundant and characteristic plant at all elevations in the Kerry-Cork (**300**) and
Galway-Mayo (**384**) mountain areas; more local in Donegal.
Eastward it dies out before Cork city is reached, but re-
appears in the mountains of Waterford and S Tipperary
(Galtees (**250**), Knockmealdown (**284**), Comeraghs (**285**)),
with very interesting outlying stations on Lugnaquilla and
Conavalla in Wicklow (**263**). On the Continent it grows
in the Pyrenees, N Spain, and Portugal. It is stated to
occur (under the name *umbrosa*) in Corsica and the Alps,
but Hegi and others look on it as an escape in those
places, and the plant intended is no doubt the strong-
growing true *S. umbrosa* L., which is a hybrid *Geum* ×
spathularis form, a common plant of gardens.

In the absence of *S. Geum*, with which it crosses very
freely, producing a multitude of puzzling forms, *S. spathularis* is in Ireland a plant of uniform facies. The most widespread form, which Dr. Scully takes as the Irish type, is
"a compact fleshy plant with a rosette of spreading obovate
leaves and short but broad footstalks"; his description
cannot be improved upon.

The only distinguishable variant which appears to belong
to pure *spathularis* is var. *serratifolia* (Mackay) (*S. serrata*
Sternberg) with "long and rather narrow erect leaves and
deep serrations"; in Kerry it is rarer than the other. Some
extreme forms of it are found in gardens under several
names.

Var. *punctata* (*S. punctata* Haworth) which is equally
common, at least in Kerry, and has "roundish or slightly
oblong leaves, more or less erect, with almost flat tapering
footstalks" is considered by Dr. Scully, I think rightly, to
represent the first stage showing the influence of crossing
with *S. Geum*. Thence we pass through forms showing
more *Geum* influence, to *S. hirsuta* Linn., which is fairly
half-way between the two species. The forms nearer *S.
Geum* have been referred to under that species (**113**).

Where *S. Geum* and *S. spathularis* grow together in the
SW, each generally shows every conceivable gradation

---

[38] The change of name is due to H. W. Pugsley, who in a forth-
coming paper points out that *S. umbrosa* L. is a hybrid, and that
S. spathularis Brot. is the earliest name for the Hiberno-Pyrenean
plant.

PLATE 3.

ARENARIA CILIATA.
In the author's garden at Dublin in May. Roots from Annacoona,
R. Welch, Photo.   Co. Sligo.

AJUGA PYRAMIDALIS AT POULSALLAGH, CO. CLARE.
Among its most abundant concomitants here are *Helianthemum vineale*,
Geranium sanguineum, *Saxifraga Sternbergii*, *Gentiana verna*, *Sesleria
cærulea*. The snails are *Helicella itala*, a characteristic xerophile
lowland species. May.
R. Welch, Photo.

PLATE 4.

Leaves of *Saxifraga Geum* (top left), *S. spathularis* (bottom right), and hybrid intermediates, collected along 100 yards of river-bank at Blackwater Bridge, Co. Kerry. ⅘.

towards the other (plate 4); where *spathularis* alone occurs, as in Waterford, S Tipperary, W Galway, Donegal, it shows no variation in the direction of *Geum*; on the contrary, variation, when it occurs, tends to be away from *Geum*— leaves more sharply toothed, blade smaller or narrower or more cuneate; observations converse to the above cannot be made, as there is no Irish area where *Geum* grows unaccompanied by *spathularis.*

While absolutely hardy on the highest mountain-summits in Ireland, the plant rejoices in a maximum of shade and moisture. In the Killarney woods, growing among *Hymenophyllum,* rosettes may be seen a foot across, bearing flowering stems a foot and a half in height.

A full and excellent account of these two species as found in Ireland, their varieties and hybrids, with plates, including an account of breeding experiments confirming views long expressed by botanists regarding hybridity in the group, is given by Dr. Scully in his ''Flora of Kerry,'' 96–106. This convincing exposition is not so widely known as it ought to be: it is not referred to, for instance, in ENGLER and IRMSCHER'S Monograph of Saxifraga in the ''Pflanzenreich'' series, though published two years after Dr. Scully's volume. The authors of the monograph describe many varieties and forms of both species which are without question merely hybrids.[39]

115. According to the foregoing remarks, the Irish Robertsonian Saxifrages may be arranged as under, following closely Scully's interpretation: the series of 86 native leaf-forms which he figures should be studied in this connection :—

1. *S. Geum* L. Leaves reniform.
    > forma *a*. Margin crenate (? *Robertsonia crenata* Haworth). (Scully, pl. I, 1–4.)
    > forma *β*. Margin serrate (var. *serrata* Syme. ? *R. dentata* Haworth). (Scully, pl. I, 5–13.)
2. *S. spathularis* Brot. Leaves obovate.
    > forma *a* (*typica*). Leaves spreading, rather broadly obovate, moderately deeply serrate. (Scully, pl. VI, 14–16.)

---

[39] See also Praeger in ''The Garden,'' 4 July, 1925, for a more general account of the Robertsonian Saxifrages, their forms and hybrids.

forma *β.   serratifolia* (Mackay).   Leaves erect,
long and narrower, deeply serrate.   (Scully, pl.
VI, 19–23.)
3. *S. Geum* × *spathularis*.   Leaves of many inter-
mediate shapes.   (Scully, pl. II, 1–34, III, 1–23,
VI, 1–6.)
Here belong—
  S. elegans Mackay (*S. spathularis* var. *punctata* ×
  Geum var. *serrata* (E. S. Marshall)).
  S. gracilis Mackay.
  S. polita Haworth.
  S. hirsuta L.   (Scully, pl. III, 1–8.)
  S. punctata Haworth.   (Scully, pl. VI, 10–12.)

116. Saxifraga Hirculus Linn.—Rare, in lowland peat-
bogs in four midland counties, also in W Mayo, and at
1000 ft. in Antrim.   In Britain equally rare, and mostly
in Scotland.   Elsewhere alpine-arctic and circumpolar.

117. Saxifraga hypnoides aggr.[40]—Ireland is remarkably rich in hypnoid Saxifrages.   Several are as yet
unknown elsewhere, namely: *S. Drucei* E. S. Marshall
(Brandon in Kerry, Arranmore in W Donegal).   *S. Sternbergii* Willd. (plate 22) (Brandon in Kerry, Ballyvaughan,
Black Head, and Aran in Clare, mostly as var. *gracilis*:
a plant with a puzzling range—Ireland, Norway (one
station), Harz, Bavaria).   *S. hirta* Sm. (Kerry, frequent;
Galtees and Clare (Aran)).   *S. affinis* D. Don (Brandon
in Kerry (Mackay, 1805)—not seen since).   *S. incurvifolia*
D. Don (Kerry (several stations), W Galway (Muckanaght)).
S. hypnoides var. *robusta* E. S. Marshall (Black Head in
Clare).   In addition, *S. rosacea* Mœnch (*decipiens* Ehrh.),
which has only one British locality, occurs in several
stations in Kerry, on the Twelve Bens in W. Galway, and
on Clare I. in W Mayo.   *S. sponhemica* and *S. hypnoides*
are frequent, the former ranging over the Kerry and
Waterford mountains and re-appearing in Antrim and
Derry, the latter ranging from Clare round the west and
north to Antrim and Wicklow.   For further particulars
see papers by E. S. Marshall in Journ. Bot., 1917–18.

118. Apium inundatum × **nodiflorum** (*A. inundatum*
var. *Moorei* Syme Engl. Bot., **4**, 102, 1865, and Suppl., 187.
A. Moorei Druce in Bot. Exch. Club Rep., 1911, 20–21).—

[40] See E. S. MARSHALL in Journ. Bot., 1917, 151–161.

This rather controversial plant is essentially an Irish form, occurring as at present known in 24 of the 40 divisions, from Limerick and Kildare to S Donegal and Antrim. Seems very rare in Britain, being on record from only four English vice-counties. According to Druce (Com. Fl. Brit.) it occurs in W Germany. This seems a strange distribution for a hybrid of which the parents have both a wide range, but Riddelsdell (Irish Nat., 1914, 1–11) in a full discussion makes out a very good case for hybridity. The plant varies towards *nodiflorum* (f. *subnodiflorum*), and towards the other parent (f. *subinundatum*). First found by David Moore near Portmore in Antrim about 1835.

119. Carum verticillatum Koch. — An essentially "Atlantic" species, on the Continent ranging down the W coast from Belgium southward, in Britain found along the W coast in England, Wales, and Scotland; in Ireland confined to the SW, where it is locally abundant, and to a few stations in W Donegal, Londonderry and Antrim.

120. Œnanthe pimpinelloides Linn.—Known only from pasture at Trabolgan in E Cork (**293**), presumably native. First found by R. A. Phillips in 1896. S England. Middle and S Europe, N Africa, Asia Minor.

121. Rubia peregrina Linn.—Up the W coast as far as Mweelrea, and the E coast as far as Meath. Never far from the sea, though not maritime. Flourishes equally on or off the limestone. Wales and S England. S Europe and NW Africa.

122. Inula salicina Linn. (plate 25).—A noteworthy plant, being one of the few, other than the Lusitanian and American groups, which are Irish but not British. It is confined to the limestone shores of Lough Derg (**353–4**), the largest of the expansions of the Shannon, in N Tipperary and SE Galway, where it was first found by David Moore in 1843 (Journ. Bot., 1865, 333–5, and *ibid.*, 1866, 33–36, tab. 43). Rather widespread on the shores and islets, at a little above flood-level, from the head of the lough at Portumna as far S as the vicinity of the Carrikeen Islands, near Dromineer. Not yet recorded from the S portion of the lake. Its habitat is the rough grassy, stony, or boulder-strewn ground that intervenes between flood-level and the arboreal zone (plate 26). There it runs about by stolons, forming small colonies, and flowering rather sparingly in July. In the garden it spreads rapidly, and soon forms a large patch, with a network of slender white underground stems. It is not yet fully known what effect the Shannon Electricity Works, by which the water of the lake is raised to about winter level, may have on the plant, but there is little reason to fear for it. Widespread on the Continent, from Greece to Norway and W France to central Russia and on into Asia; and in view of its presence in Ireland its absence from Britain is difficult to explain.

123. Diotis maritima Cass. (*D. candidissima* Desf.) (plate 7).—SE coast—Waterford (not seen recently) and Wexford (**278**), where it still exists in some quantity. First found by J. G. Allman in 1845. In Britain southern, on coasts from Suffolk to Anglesea, discontinuous and now very nearly extinct. A plant of the Mediterranean, extending to Portugal and the Canaries. For a full account of the plant in Britain and Ireland, with good photographs taken in its Wexford habitat (one of them reproduced here), see C. P. Hurst in ''Manchester Memoirs,'' 1901–2, 1–8, pl. 1–2.

124. Artemisia Stelleriana Bess. — This ornamental Siberian plant, thrown out from the gardens of St. Anne's at Clontarf, Co. Dublin, before 1891, drifted across to the adjoining sand-spit of the North Bull (**228**), and is naturalized there now over a distance of at least a couple of miles (though apparently rare of late). See Moffat in Journ. Bot., 1894, 22 and 104–106; Areschoug, *ibid.*, 70–75; Colgan Fl. Dublin, 110–111; and Journ. Bot., 1900, 317–8. Naturalized in Cornwall, New York, Kamtschatka (Druce).

125. Senecio squalidus × **vulgaris.**—The S European *Senecio squalidus*, naturalized and abundant about Cork (**291**), where it appeared about a century ago (and has since spread to many Cork towns, and recently to Dublin), hybridizes freely at Cork and sparingly at Dublin with *S. vulgaris*, giving rise by continued crossing to a progeny stretching from the one species to the other, complicated at Cork by the frequent presence of *S. vulgaris* var. *radiatus*. Its nomenclature is involved—see Burbidge in Irish Nat., 1897, 300, and Phillips, *ibid.*, 1898, 22. At Maryborough a small colony of *S. squalidus* has produced crosses with *S. Jacobæa*.

126. S. Cineraria × **Jacobæa** (*S. albescens* Burbidge and Colgan).—The Mediterranean *S. Cineraria*, escaped from a garden at Dalkey, Co. Dublin, has colonized the rocky shore in abundance for nearly a mile. Here it crosses freely with *S. Jacobæa,* the hybrids showing every gradation from one species to the other—see **233.** Similar crossing has been recorded where the plant has run wild in SW England.

127. Cnicus palustris × pratensis (*C. Forsteri* Smith).— Not uncommon where the parents grow together : recorded from Clare, Carlow, Leix, Dublin, SE Galway, Westmeath, E Mayo, Londonderry. See Proc. Roy. Irish Acad., **24,** B, 72 (1902).

128. Hieracium.—Irish *Hieracia* have been tolerably well worked, and the older names revised according to modern knowledge. The Hawkweed flora is fairly rich. Three "species"—*H. Scullyi* Linton (S Kerry), *H. Stewartii* F. J. Hanb. (Down), and *H. subintegrum* Stenstr. (N. Kerry) are, so far as at present known, in the British Isles confined to Ireland. Among other rare Irish forms are :—*H. flocculosum* Backh. (Down, Antrim), *proximum* F. J. Hanb. (Kildare, Wicklow, E Donegal), *scoticum* F. J. Hanb. (W. Donegal), *Leyi* F. J. Hanb. (Derry), *repandum* Ley (Sligo), *orimeles* W. R. Linton (S and N Kerry, E and W Donegal), *hypochæroides* Gibson (W. Cork, Clare, W Mayo, Sligo, Louth), *hibernicum* F. J. Hanb. (Down and E Donegal), *lepistoides* Johanss. var. *sublepistoides* Zahn (N. Kerry, Limerick, Down), *grandidens* Dahlst. (Down), *cinderella* Ley (Down), *pachyphylloides* Zahn (Antrim, Derry), *killinense* Zahn (Antrim), *sagittatum* Lindeb. (Down), *maculosum* Dahlst. (Dublin), *crebridens* Dahlst. (Clare), *cymbifolium* Purchas (Clare, Sligo), *cordigerum* Norrl. (Derry), *farrense* F. J. Hanb. (Wicklow, Antrim), *orarium* Lindeb. (W. Mayo, Antrim), *senescens* Backh. (Down), *sparsifolium* Lindeb. (S. Kerry), *strictum* Fr. (Wicklow, Antrim, Derry), *dumosum* Jord. (Wicklow). The most widespread forms in Ireland are *anglicum* (in 20 divisions), *iricum* (16), *triviale* (17), *umbellatum* (18).

129. Arbutus Unedo Linn. (plate 13).—The most striking and handsome of the Hiberno-Lusitanian group, forming in the Killarney woods (**322**) and at Lough Gill (**420**) not a bush, as usual along the Mediterranean, but a tree up to 30 or 35 ft. in height when growing among other trees, or with a rounded form and a height of up to 20 ft. in the open. Scully mentions boles 12–14 ft. in circumference, but the tree mostly branches widely from near the base, and trunks of any length are rare. The *Arbutus* is essentially lowland, the highest station on record being 525 ft. in Kerry. It shows no preference as to soil (provided the soil is dry) growing equally on bare limestone and on bare sandstone or metamorphic rocks, or in woodland humus, or peat. While in the woods and on the islands of the Lakes of Killarney it is still abundant, it has generally been greatly reduced in quantity and in range in the south-west by being cut for iron-smelting in the 18th century and earlier; it still extends sparingly and discontinuously as far as Lough Currane to the SW, and Adrigole and Glengarriff to the S. By place-names which embody the Irish name of the *Arbutus* (*caithne* and *cuinche*) we are enabled to trace its former extension as far N as Mayo, and recently (Proc. R. Irish Acad., **41**, B, 105–113, 1932) I have endeavoured to show that it is undoubtedly indigenous on Lough Gill in Sligo, where many very old trees occupy chinks in limestone rocks and elsewhere in wild situations among a native tree-flora (see **420**). A gap of 160 miles separates the Kerry and Sligo stations. An attempt to find intermediate habitats on the Galway-Mayo lakes was unsuccessful. The islets in L. Mask and L. Corrib, mostly drift-covered, have a different aspect and a different vegetation. The S end of L. Conn seemed much more likely. There the pointer plants *Taxus* and *Sorbus rupicola*, which are its companions at Killarney and L. Gill, grow on the rocky lake-margin, backed by dense oak-wood, with a vegetation closely recalling that of its Kerry and Sligo habitats; but *Arbutus* does not appear to be there now. The earliest mention of *Arbutus* in human records dates back to the eighth century. In the Brehon Laws (IV, 147) it is included among the classified lists of trees interference with which constitutes trespass:—"The Shrub trees are; black thorn, elder, spindle tree, white hazel, aspen, arbutus, test-tree." *Caithne* is the word there used for Arbutus. The first printed record is in Parkinson's "Theatrum Botanicum," 1489–90 (1640).

South of its Irish stations, its next appearance is on the Brittany coast—cliffs of Trieux near Paimpol in Côtes du Nord; it re-appears on the coasts of SW France, and extends across the Pyrenees and along the whole length of the Mediterranean.

An excellent and very full notice of *Arbutus* is given in Scully's "Flora of Kerry," 179–183.

130. Erica Mackaii Hooker. (*E. Mackaiana* Bab.)
(plate 29).—The most limited in range of the Hiberno-
Lusitanian plants. It occurs in W Galway over a couple
of square miles from Urrisbeg to Craigga-more Lough 4
miles W of N of Roundstone, and in a slightly different
form over an undetermined area on the E side of Carna,
6½ miles SE of Roundstone. At Craigga-more (**389**), where
it was discovered by William M'Alla prior to 1835 (see
Companion to Bot. Mag., 1835, 158), it is quite abundant. A
double form, with the stamens converted into petals, giving
the flowers a very obese appearance, was collected at
Craigga-more by A. G. More in 1869 (fine specimens in
Herb. Nat. Mus., Dublin), and again by F. C. Crawford of
Edinburgh in 1901, after whom it has been called *E.
Crawfordii*.[41]

At that place, and southward towards Urrisbeg, it
hybridizes rather freely with *Tetralix*, forming *E. Praegeri*
Ostenfeld in New Phytol., 1912, 120.

E. Stuartii Linton in Ann. Scott. Nat. Hist., 1902, 176–
177 and Irish Nat., 1902, 177–178 is diagnosed as *E.
Mackaii × mediterranea*—I have no doubt correctly. It
was found (one clump) at Craigga-more (**389**) by Charles
Stuart of Chirnside, Edinburgh, in 1920,[41] growing with
Mackaii: *E. mediterranea* has its nearest station on
Urrisbeg 3 miles to the south. The plant is, like *Crawfordii*
and *Praegeri*, in cultivation.

E. Mackaii has been very variously placed by botanists,
most often as a variety of *E. Tetralix*, but it is undoubtedly
entitled to rank as a distinct species. This is to my mind
finally proved by the study of the leaf-anatomy of this
group of *Ericæ* by Miss Margaret Smith, published (in
part) in Trans. Bot. Soc. Edinb., **30**, 198–205 (1930). In
Mackaii, alone of the Irish Heaths, the upper epidermal
cells have a transverse septum, dividing each into two
approximately equal parts. In *E. mediterranea* and *E.
Tetralix* this remarkable character is absent. In *E.
Crawfordii* it is present, showing that this curious plant
has been correctly interpreted as *E. Mackaii flore pleno*.
In *E. Praegeri* and *E. Stuartii* it is present in half of the
epidermal cells, the other half being undivided, pointing
along with other characters to crossing between *Mackaii*
and other species, which others are undoubtedly *Tetralix*

---

[41] J. M. MacFarlane in Trans. Bot. Soc. Edinb., **19**, 58–64, pl. I.
(1891).

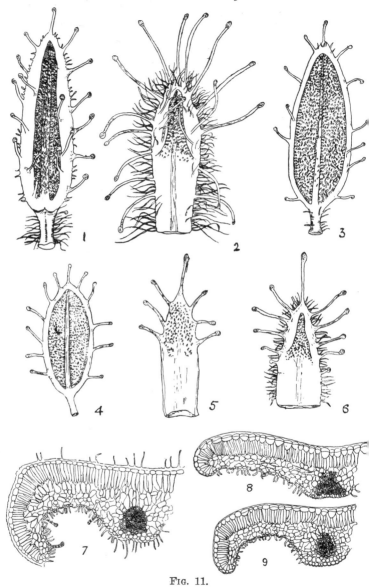

FIG. 11.

1, 3, 4, leaves (× 10), 2, 4, 5, sepals (× 13), 7, 8, 9, transverse sections of leaves (× 56), of *Erica Tetralix* (1, 2, 7), *E. Mackaii* (4, 5, 8), and *E. Praegeri* = *Mackaii* × *Tetralix* (3, 6, 9). By permission of Miss Margaret H. Smith.

and *mediterranea* respectively. By the kindness of Miss Smith I am able to reproduce here (fig. 11) some of the drawings which illustrate her paper. First found by Wm. M'Alla before 1835. Outside Ireland *E. Mackaii* is confined to the Pyrenean region. A useful discussion of it is to be found in Duriæus' "Iter Asturicum Botanicum" (see Lacaita in Journ. Bot., 1929, 256–258).

131. E. mediterranea Linn. (*E. hibernica* Syme. *E. mediterranea* var. *hibernica* Hooker) (plate 35).—Has a fairly wide range in the great Galway-Mayo bog area (**384**), extending from Urrisbeg near Roundstone northward at intervals to the Mullet and Lough Conn, growing best where the peat is well drained—at Mallaranny over six ft. in height.

A plant of wet boggy ground, generally sloping but of no great elevation. Particularly fond of the edges of streamlets and lakes. Usually forms scrubby rounded bushes about 2 ft. in height, easily picked out from among the other heaths by the erect twigs and spreading leaves. Often commences to flower in January. At its best in April; frequently flowers sparingly in autumn. The flower-buds of the following season are already well developed in September. The plant sometimes suffers severely from winter sea-winds in its maritime stations. In 1931 it was much blasted about Mallaranny, so that there was practically no blossom in that spring. In Erris and Achill a dark-flowered dwarf form occurs; but when grown in the garden alongside the typical plant for a few years, very little difference is perceptible. First found by J. T. Mackay in 1830 on Urrisbeg.[42] Not in Great Britain. Elsewhere confined to Portugal, NW Spain, and SW France. Its hybrid with *E. Mackaii* is referred to in the preceding paragraph (**130**).

132. Dabeocia cantabrica K. Koch. (*D. polifolia* D. Don. *Menziesia polifolia* Juss.) (plate 27).—Confined to the Galway-Mayo metamorphic mass (**384**). Widely spread in W Galway; local in Mayo, from near Cong to Killery and Curraun Achill. To 1900 ft. on Maamtrasna and 1800 ft. on the Twelve Bens; on very exposed heaths in W Connemara it grows down to the edge of the Atlantic. Absent from Britain, but occurs in SW France, Spain

---

[42] J. T. MACKAY: "Notice of a New Indigenous Heath, found in Cunnamara." Trans. Roy. Irish Acad., **16**, 127–128. 1830.

(where in the Pyrenees it flourishes on ground under snow for five months of the year (see Irish Nat., 1909, 3)) and the Azores.

Commences to flower in May, and blooms all through the summer into late autumn. In nature a rather straggling plant, growing best when trailing through other shrubby species, such as *Erica, Rubus* or *Ulex Gallii* (as in plate 27). Its limit of altitude in Ireland and Spain (*supra*) shows that, like *Saxifraga spathularis*, it is not a tender species, despite its southern range. The sepals are very variable in the Irish plant, being broad-based or narrow-based, entire or with several acuminate teeth near the apex ending in gland-tipped hairs.

A white-flowered form is long in cultivation, as well as a purple and white form known as *bicolor*; the former is figured in Sweet "British Flower Garden," **6**, t. 276.

133. Limonium humile Mill.—In almost every maritime division in Ireland, often abundant. Rather southern in Britain, being in Scotland confined to the S. The segregate *transwallianum* has been found in Clare (**350**) and *L. paradoxum* in Donegal (**452**).

134. Microcala filiformis Hoffmgg. and Link.—SW only, locally abundant from Skibbereen to Dingle Bay, 0–800 ft. First found by W. H. Harvey in 1845. In England south-western likewise, along the coast from Pembroke to Sussex, with a station in Norfolk. Europe, rather western; Azores. Scully (Fl. Kerry, 194–5) gives a good account of it.

135. Gentiana verna Linn. (plate 24).—A plant of the western limestones, sometimes very abundant, from Clare (Ennis) up to Mayo (Lough Carra), W to the Aran Islands, and E to beyond Athenry. Ranges from 0 to 1000 ft., but mostly quite lowland. First found in Ireland by Richard Heaton, and recorded in How's "Phytologia," 1650. Sub-alpine in England, on limestone in the N. Rather widespread on the mountains of Europe and nearer Asia.

Very rare off the limestone, but grows on peaty road-side banks on the shales, 700–800 ft. elevation, between Ballyvaughan and Lisdoonvarna. Its occurrence in profusion on sea-sands near Ballyvaughan (plate 23) cannot rank as an exception to its calcicole proclivities.

Increases vegetatively by slender underground shoots proceeding from a central rootstock, thus making in open ground patches a foot across with 50 to 100 flowering stems, representing a single individual.

136. Linaria repens × vulgaris (*L. sepium* Allman).— An interesting hybrid, originally claimed as a new species, which grows plentifully by the river at Bandon (**295**), Co. Cork, accompanied by the parents. First found by G. J. Allman before 1843 (Watson in Hooker's Lond. Journ. Bot., 1842, 76–86; Allman in Proc. Roy. Irish Acad., **2**, 404–406 (1843); Babington in Trans. Bot. Soc. Edinb., **5**, 20–22, 64 (1858) and Ann. and Mag. Nat. Hist. (2), **14**, 408–411 (1854), **16**, 449–450 (1855)). Occurs also at Killowen in Down with the parents.

137. Sibthorpia europæa Linn.—South-western : confined to the Dingle Peninsula (**324**) in Kerry, 0–1700 ft. First found by J. T. Mackay in 1805. In Britain also south-western (Sussex to S Wales). South-western in Europe (W France, Spain, Portugal).

138. Veronica peregrina Linn.—Thoroughly naturalized in Donegal, Derry, and Tyrone. Rare and seemingly only sporadic elsewhere, in various counties in the N half of Ireland. An American species which has colonized widely on the Continent, but so far has had little success in Britain.

139. Euphrasia brevipila × salisburgensis.—By Lough Coura in Offaly (**246**). Definitely so named by F. Townsend as a new hybrid (Proc. Roy. Irish Acad., **24**, B, 78, 1902). As *E. salisburgensis* is not known so far E in Ireland (its nearest station lying some 45 miles to the W) further investigation is desirable.

140. E. frigida Pugsley.—Summit of Croaghaun (2192 ft.) in Achill Island (**408**); several other northern forms (*Cochlearia grœnlandica, Hypericum pulchrum* var. *procumbens* Rostrup) grow on the same hill. Collected in 1904–5 by R. Ll. Praeger and named *foulaensis* by F. Townsend (Irish Nat., 1904, 285; 1906, 43). A subarctic plant known in Britain from Yorkshire and Scotland (see Pugsley : "Revision of the British *Euphrasiœ,*" Journ. Linn. Soc. (Bot.), **48**, 490, 1930).

141. E. salisburgensis Funk (plate 28).—A characteristic plant of the low western limestones, extending continuously from Askeaton in Limerick to Lough Mask in Mayo, including the Aran Islands, and reappearing on the limestone hills of Sligo, Leitrim, and Fermanagh, and on low-lying limestones south of Donegal town (fig. 12). Mostly on bare limestone rocks or cliffs, but sometimes on walls, sand-dunes or non-calcareous rocks (as on the Calp at Inchicronan L. in Clare, with such calcifuge species as *Digitalis* and *Athyrium*).   First found by Daniel Oliver on Inishmore in 1852 (F. Townsend in Journ. Bot., 1896, 441–4, tab. 363; *ibid.*, 1897, 471–3, tab. 376, 380. N. Colgan in Irish Nat., 1897, 105–8, and Journ. Bot., 1897, 196–9. H. W. Pugsley in Journ. Linn. Soc. (Bot.), **48**, 532–4, pl. 37).

Fig. 12.—Range of *Euphrasia salisburgensis*.

This plant is easily recognised in the field by its dwarf bushy growth, characteristic colour, and jagged upper leaves. In vigorous plants the lowest branches are themselves mu.h branched, and almost as long as the main stem, so that a dense roundish bush is produced. The leaves in exposed places assume a beautiful coppery brown. The plant has a habit of growing in dense clumps, composed of many individuals, on little bare dry patches of ground. The flowers are of medium size, mainly white, and stand out conspicuously against the dark foliage.

Especially luxuriant bushy plants 2½″ high and 12″ in circumference sometimes grow among *Thymus* on small ant-hills of a minute yellow ant—which one might expect to be a distinctly acid habitat—the "earthing up" done by the ants apparently stimulating growth (Irish Nat., 1906, 259–60). In Aran, *Sedum anglicum* is characteristic of the same habitat—see **352**. I cannot separate var. *hibernica* Pugsley, *l.c.*,—see Proc. Roy. Irish Acad., **42**, B, 75.

Hitherto unknown in Britain, *E. salisburgensis* has recently been found in Yorkshire (Pugsley, *l.c.*). Druce's Devon record (Com. Fl. Brit. I.) is erroneous (Pugsley in Journ. Bot., 1933, 90). On the Continent widespread, usually alpine, not confined to calcareous rocks.

142. Bartsia viscosa Linn.—With a rather wide though local foreign range—W and S Europe, N Africa, W Asia, Canaries—this plant is in Great Britain markedly southern and western (Sussex to Argyll). In Ireland it is also conspicuously southern and western—through Waterford, Cork, Kerry continuously, and then disconnected stations in Galway, Donegal, and Derry. But only in the SW does it become common.

143. Orobanche rubra Smith.—Marginal and mainly northern, though it occurs in Kerry and Cork. More frequent in Burren and on the basaltic area of the NE than elsewhere, but is found on all kinds of rock, 0–1000 ft. First found in the British Isles by John Templeton on the Cave Hill, Belfast, before 1793. In Britain chiefly on the W coast from N Scotland to S England. Middle and S Europe, W Asia. Parasitic on *Thymus* only apparently.

144. Utricularia intermedia Hayne.—A plant which, like its ally *Pinguicula lusitanica*, is much more abundant in Ireland than in Britain, occurring in 23 of the 40 divisions, mostly in the centre and W. First found in the British Isles by Dr. Robert Scott in Fermanagh about 1804. Very seldom flowers. Generally in bog-pools with a soft muddy bottom, in which the colourless bladder-bearing branches burrow deeply. In Britain widely spread, but very rare. Of northern range in Europe, Asia, and America.

145. Pinguicula grandiflora Lam. (plate 9 and fig. 13).—The most attractive of the Hiberno-Lusitanian species, and most botanists who have seen it in its glory in Kerry and Cork will endorse the opinion of Dr. Scully that it is the most beautiful member of the Irish flora. The plant grows in extraordinary profusion over wide areas on bog, wet rocks, and damp pastures, with rosettes of yellow leaves up to $2\frac{1}{2}''$ long by $1\frac{1}{2}''$ broad, and numerous great flowers of imperial purple an inch or more across (up to $1\cdot3$ inch!) on stems $6''$ to $9''$ high.

With a tolerably wide SW Continental range—W France, N Spain, Portugal, Swiss and French Alps—*P. grandiflora* is abundant throughout W Cork and Kerry, thinning out northward to the Shannon, and eastward as far as Mallow. Extends from sea-level to 2250 ft. First definitely found by James Drummond in 1809. "Butterwort" is mentioned in Smith's "History of Kerry," p. 85 (1756), as abundant on islands at the head of Kenmare River, and no doubt this species is intended.

The insectivorous features of the plant are not confined to the leaves. The flowering stem, the calyx, and the outside of the lower corolla-lobe bear glandular hairs among which insects may be found entangled, and are no doubt digested.

Plants with white and with pale lilac flowers have been recorded, and I have thrice found a few plants bearing purplish-pink blossoms.

P. grandiflora appears to be an enterprising species, quite ready to colonize new ground. There is no evidence of any recent extension of its natural range, but half a dozen roots introduced in 1879 to boggy ground at Killanne near the foot of Blackstairs in Wexford had increased in 1913 to 95 plants (Irish Nat., 1896, 212), but it has since disappeared (see **271**). It grows on a dripping cliff at Lisdoonvarna in Clare (**345**), where at first I thought it native (Irish Nat., 1903, 269; 1907, 242); but there is reason to think it has come from a garden just above; it is still increasing. H. C. Hart records (Fl. Don., 214, 1898) that the plant "has established itself thoroughly in peat-bogs at Carrablagh" on Lough Swilly in Donegal, but it has not been seen there recently. In Cornwall it was planted on Tremethick moor by Ralfs, and when it was in danger of extermination by collectors, in spite of very rapid increase, plants were transferred to Trungle moor and the Land's End district, where they are reported to have "multiplied with marvellous speed" (Davey Fl. Cornwall, 345–6 (1909)). These "forgeries of nature's signature" are much to be regretted.

In Kerry *P. grandiflora* hybridizes readily with *P.*

vulgaris (fig. 13). The cross is not common on account of the remarkable rarity in that area of the latter species. One sees ten thousand *P. grandiflora* for one *P. vulgaris*. But wherever *P. vulgaris* occurs among its ally, and only there, hybrid forms may be found—corolla about ¾″ across (intermediate between the ½″ of *vulgaris* and the 1″ or more of *grandiflora*), corolla-lobes much larger and edges less divergent than in *vulgaris*, but not with overlapping undulate margins as in *grandiflora*, and the white patch

Fig 13.

1. *Pinguicula grandiflora.* 2. *P. grandiflora* × *vulgaris.*
3. *P. vulgaris.* All ⅟₁. From near Kenmare, Co. Kerry.

on the corolla-throat intermediate in shape between the long cuneate form found in *grandiflora* and the broad short form of *vulgaris*. With these there may often be found forms nearer one or the other parent, presumably arising from secondary crossing. (Praeger in Journ. Bot., 1930, 249.) Druce (Com. Fl. Brit. I., 229) has called the hybrid *P. Scullyi.*

146. Pinguicula lusitanica Linn.—One of the plants, which, while occurring in both Ireland and Britain, is much more widespread in the former. Recorded from 33 of the 40 Irish divisions, from sea-level to 1000 ft., being an apparent absentee only from some lowland inland counties. In Britain in SW England, Isle of Man, and W Scotland alone. On the Continent south-western, in W France, Spain, Portugal.

147. Galeopsis intermedia Vill.—Has been found in Leix (Emo) and Meath (Bective) but appears very rare. Very rare also in Britain.

148. Teucrium Scordium Linn.—Along the Shannon from Roosky in Leitrim to Doonass above Limerick, often abundant, particularly on L. Ree (**379**) and L. Derg (**253**). Also in outlying stations at Ballyspillane L. in N Tipperary and Glanquin in Clare (**361**). Always on limestone. England—very rare, chiefly in the E. Widespread in Europe; Siberia.

149. Ajuga pyramidalis Linn. (plate 3).—Around Galway Bay only : on limestone in Burren (**346**) and the Aran Islands (**352**) on basalt at Bunowen in Connemara (**388**), from about 50 to 200 ft. Crosses in Burren with *A. reptans* (Journ. Bot., 1926, 31). First found by D. Moore, 1854. In Britain from Westmorland northward. Widespread in Europe, mostly northern and sub-alpine; Caucasus.

150. Oxyria digyna Hill.—On the S and W mountains —the Galtees and those of Cork-Kerry, Galway-Mayo, Ben Bulben, and W Donegal, 550–3150 ft. In Britain on mountains from Wales northward, not common. An arctic-alpine plant elsewhere.

151. Euphorbia Peplis Linn.—Once found at Garraris Cove, near Tramore, Waterford (**273**), by Miss Trench in 1839 (Mackay in Proc. Dublin Univ. Zool. and Bot. Assoc., **1**, 1859). Not seen since, though several times searched for (see Cyb. Hib., ed. 2, 520). Very rare in England, on sandy shores from S Wales to Hants; a decreasing species. From France southward and along the Mediterranean, re-appearing on Asiatic salt-tracts.

152. Euphorbia Paralias Linn.—Whole E coast, but very rare on the W (N Kerry, Clare, W Galway). In Britain rare and southern on the E side, commoner on the W, absent from Scotland. SW and S coasts of Europe.

153. E. portlandica Linn.—In Ireland distributed like the last. W and S coasts of Britain, from S Scotland southward. Has a "Lusitanian" range on the Continent—W France and the Peninsula.

154. E. hiberna Linn. (plate 10).—An Atlantic-Mediterranean plant, its Irish habitats linked with its Continental ones (France and Spain to Switzerland and Italy) by stations in Cornwall, Devon, and Somerset. In Ireland it has an extended S and W range—Waterford, Cork, Kerry, and Limerick continuously, and isolated stations in Galway, Mayo, Donegal. In the SW it grows in profusion in rough pastures, hedgerows and woods, mostly lowland, but rising to 1800 ft. in Kerry. Adare in Limerick (not recently confirmed, but well vouched for) is the only station on the limestone.

E. hiberna has long been used in Ireland by poachers
to kill fish by placing bundles of crushed stems in streams.
A good example of its poisonous properties is given in a
letter from Francis Vaughan to John Ray dated April
24, 1697 (Corresp. of John Ray (Ray Soc., 1848)), and
the results of a full experimental enquiry into its effects on
salmonoid fishes, by H. M. Kyle, are published in Proc. Roy.
Soc., **70** (1902), 48–66.

The appearance of the plant in summer is often marred
by a profusion of *Uromyces tuberculatus*, which results in
spindly growth and gives a golden colour to the leaves and
stems.

155. Spiranthes gemmipara Lindley.—It is over 120
years now since James Drummond,[43] Curator of the Cork
Botanic Garden, added to the European flora one of its
most interesting members. Long known as *S. Romanzoffiana*, it has recently been replaced under its older name
of *gemmipara*,[44] while its companion plant, found much later
in Armagh[45] and other northern counties, and at first
identified as the same, is now separated as *S. stricta*.[44]
Drummond's station was at Berehaven in W Cork (**296**),
where the plant has been seen frequently since. Dunmanway
and Desert, further to the E, were subsequently[46] added
as additional stations; and recently[47] (1921 and after) it has
been found by several observers about West Cove in the
Derrynane area (**314**) of S Kerry, 12 miles NW of
Berehaven. Its general distribution is at present uncertain
pending the working out of the North American segregates.

It has been erroneously recorded from Devon (Journ.
Bot., 1909, 385), and its being in fact gemmiparous has been
questioned—without reason so far as the Irish plant is
concerned.[48]

156. S. stricta Nelson (plate 41).—The northern form
of the Irish "*S. Romanzoffiana*" has been determined (see
under *S. gemmipara*) as *S. stricta* Nelson. It occupies the
basin of the Bann, from above Portadown, all round Lough
Neagh (**462, 465**), and down to Coleraine (**462**), growing in

[43] J. DRUMMOND: Munster Farmers' Mag., 1818–20 (not seen).
[44] A. J. WILLMOTT: Journ. Bot., 1927, 145–9. See also M. C.
KNOWLES, Irish Nat. Journ., 2, 2–6, pl. 1, 1928.
[45] R. LL. PRAEGER: Journ. Bot., 1892, 272.
[46] T. ALLIN: Fl. Plants and Ferns of Co. Cork. 1883.
[47] R. W. SCULLY: Irish Nat., 1921, 79.
[48] M. C. KNOWLES: Irish Nat., 1924, 75–6.

cut-away bog, damp meadows and on wet stony lake-shores. See (*e.g.*) Tomlinson in Irish Nat., 1907, 311–4.

The plant is like some other orchids (*e.g.*, *Ophrys apifera*) a very uncertain flowerer. On Brackagh Bog near Portadown (its original Irish station) in 1931 over 200 blooms were counted. In 1932 three persons searching for an hour saw only three. It is affirmed that in 1930 blossom was much more abundant than in 1931, and that the bog was "white with them."

First found by Praeger above Portadown in 1892 (Journ. Bot., 1892, 272–4). In Britain known only from Colonsay, where it was recently discovered (Journ. Bot., 1930, 346). Unknown in Continental Europe. Widespread in northern N America.

Papers on these two *Spiranthes,* by Col. Godfery, with figures of the northern and southern plants will be found in "Orchid Review," 1922, 261–3, and 1930, 291–5.

157. Epipactis atropurpurea Raf. (*E. atrorubens* Schultz).—On the western limestones from S Clare to Cong, growing in crevices of the bare rock, 0–1000 ft. Similar habitats in Britain, from Devon to N Scotland, very local. Europe, Caucasus, Persia.

158. Neotinea intacta Reichb. fil. (*Habenaria intacta* Benth.) (plate 2).—Essentially a plant of the bare limestone "pavements," but occurs also on calcareous sea-sands and occasionally on non-calcareous rocks (Coal-measure shales 6 to 8 miles off the limestone near Lehinch in Clare (Irish Nat., 1901, 143), and on Silurian or metamorphic rocks at Mount Gable near Cong). The plant is abundant in the Burren region in Clare and SE Galway, and extends along the limestone shores of L. Corrib and L. Mask to L. Carra. There are interesting outliers in SW Connemara, on limy sands or light moory soil overlying the acid rocks (Irish Nat., 1906, 260; 1907, 243) (fig. 14); it is strange that it does not occur on the Aran Islands. First found by Miss F. M. More at Castle Taylor, Galway, in 1864 (D. Moore in Journ. Bot., 1864, 228, and A. G. More in Trans. Bot. Soc. Edinb., **8**, 265 (1865). See also H. G. Reichenbach, Journ. Bot., 1865, 1–5, tab. 25). Unknown in Britain. On the Continent from W France around the Mediterranean to Asia Minor and the Canaries.

Neotinea appears above ground in October, sending up first one and then another short, broad, acute glaucous leaf, which can withstand hard frost. Flowers in the middle of

May. In summer easily recognised by the short fruit with close-ranked capsules, which is dead and dry by the end of June, but persists till autumn if not trampled.

The commoner Irish form, which H. W. Pugsley has recently named var. *straminea,* has unspotted leaves and wholly greenish-white flowers. What must rank as the type is rarer: it is a more robust plant with glaucescent darker green leaves faintly spotted with purple and flowers of which the sepals are veined or shortly striped with dull

Fig. 14.—Range of *Neotina intacta.*

purple and the lip marked towards the base with purplish pink in dots, blotches or short stripes. As the records are founded in many cases on fruiting plants or on dried specimens, the separate range of the two forms cannot be given accurately. The spotted plant is not uncommon in Burren and SE Galway along with var. *straminea*; it was the second which I collected in SW Connemara. In no case does the Irish plant bear flowers of a uniform pink colour, as shown in several well-known figures (*e.g.,* "English Botany," ed. 3; Reichenbach in Journ. Bot., 1865). R. A. Phillips has noted both forms near Menlo in NE Galway. Pugsley's recent paper on the Irish plant (Journ. Bot., **72,** 54–5) should be consulted.

158a. Orchis O'Kellyi Druce.—There is confusion about this plant. The *Orchis* first observed by P. B. O'Kelly in Clare, and pointed out by him to many botanists, which he called *O. maculata* var. *immaculata*, is a handsome white variety of *O. elodes*.   This is the plant which Druce was shown, and which he proposed to name after the finder : but by inadvertence white *Fuchsii* was sent to him by O'Kelly, and his description refers to this.   The other is a handsome and distinct plant, characterized by its slightly glaucous unspotted leaves and its very fragrant flowers with buff anthers.   *O. O'Kellyi* of Druce does not appear to be worth a separate name ; the other is more distinct, and might have a varietal name under *O. elodes,* but it needs further observation.   It is abundant in Burren.   See Irish Nat., 1933, 143 (where, however, it is called *O'Kellyi*).

159. Sisyrinchium angustifolium Mill. (*S. anceps* Cav.) (plate 11).—The status of this plant in Ireland has often been impugned on account of its known rapid increase following introduction in places as far apart as Germany, Norway, and Queensland.   But in Ireland its habitats and distribution afford no suggestion of this kind.   It never occurs about houses, gardens, or places of traffic, but affects old marshy meadows and hill-sides where the flora is strictly native, many of its stations being remote from all human influence.   First found by James Lynam in 1845 near Woodford in SE Galway (Phytol., 1846, 500) and later in S Kerry (Journ. Bot., 1882, 8), it has since been shown to be widespread in Kerry and to occur also in W Cork, Clare, Tipperary, Sligo, Leitrim, Fermanagh, W Donegal, with a suspicious station in Antrim (Irish Nat., 1916, 187)—a wide western range in wild ground, very unlikely for an introduced species.   Dr. Scully, who has studied the plant for a long period in its Kerry head-quarters, considers it native there, and points out that in 20 years it has not increased in stations which have been under observation (Fl. Kerry, 283–4).   My own view coincides with his. Another objection is based on the alleged ease with which it spreads from gardens.   In Ireland evidence of this is absent.   Indeed, R. M. Barrington records (Irish Nat., 1904, 208) that seeds scattered during three different years in suitable moist coarse meadow by the Enniskerry River in Wicklow failed to produce plants.   The suspicious station by L. Neagh in Antrim seems an exception, but need not disturb us ; the same thing occurs in the case of *Saxifraga Geum, Pinguicula grandiflora, Leucojum æstivum* and other rare indigenous plants which are also occasionally grown in gardens. Outside Ireland the plant as a native is confined to temperate N America. It has escaped in a number of places in S England.

160. S. californicum Ait.—The case of this species is different from that of the last. Also an American plant (and unknown elsewhere in Europe), it grows in some abundance over an area of flat marshy rushy meadow-land, one mile N of Rosslare station in Wexford (**278**). First found here by E. S. Marshall, who in spite of all difficulties considered it indigenous—or if not, then deliberately planted, which appears equally improbable (see Journ. Bot., 1896, 366); his later view was that it is a case of survival, not introduction (*ibid.*, 1898, 49). See also A. B. Rendle, *ibid.*, 1896, 494—5, tab. 364. A native of California and Oregon. Its presence in Ireland is difficult to account for, but it is almost certainly an introduction.

161. Leucojum æstivum Linn. (plate 17).—A plant whose standing has been often questioned because it is grown in gardens, but E. S. Marshall, who first found it in Ireland in the marshes of the Slaney,[49] thought it undoubtedly native there, a view also held by M. C. Knowles and R. A. Phillips, who studied its distribution in the Shannon marshes about Limerick and generally, in an exhaustive paper.[50] I think there can be little doubt that it is indigenous in the stations above-mentioned, and probably also by the Little Brosna river in Tipperary and Offaly, and in Waterford, Leix, Tyrone, and Antrim. The authors of the paper referred to pertinently point out that *L. pulchellum,* a near ally, which is also much cultivated, has never been found "wild" in Ireland; they draw the conclusion that *L. æstivum,* instead of being an escape, has been brought into gardens from native stations. Southern England, Europe, Caucasus, Asia Minor.

162. Asparagus maritimus Mill.—A very rare plant of the Wicklow, Wexford, and Waterford coasts (**272**), with only four recent stations; it was evidently more abundant on the SE coast in former times. First recorded by C. Smith ("History of Waterford") from Tramore (where it still grows) in 1746. Some other records refer to escaped plants of the large cultivated form. In Britain south-

---

[49] Journ. Bot., 1898, 49.
[50] "On the Claim of the Snowflake (*Leucojum aestivum*) to be native in Ireland." Proc. Roy. Irish Acad., 28, B, 387—399, pl. xx-xxii. 1910.

western, from Dorset to Wales.   Rather widely spread
along the coasts of Europe, in Russia and the Canaries.

163. Simethis planifolia Gren. and Godr. (*S. bicolor*
Kunth).—One of the most restricted in range of the
Hiberno-Mediterranean plants, occupying a single area of
ground in S Kerry and another much smaller one in
Hampshire.[51]   It grew formerly in Dorset also, and has
been replanted there.   The Irish station consists of some
10 to 12 sq. miles of wild rocky and furzy heath with a
peaty subsoil, extending from Derrynane eastward; it is
fairly abundant there (**314**).   First found by Rev. T.
O'Mahony (Lond. Journ. Bot., 1848, 571).   On the
Continent from Normandy to Spain and on to Sardinia,
Corsica, and Italy.

164. Allium Babingtonii Borrer.—Along the W and
NW coast from Galway Bay to Lough Swilly, favouring
the islands—Aran (especially), Inishbofin, and probably
the Donegal Arranmore.   On the three Aran Islands it
looks thoroughly native, in rock-chinks and sandy places.
There and elsewhere it is often near houses, but it is no
doubt an indigenous plant brought into cultivation (see
Journ. Bot., 1934, 74).   Elsewhere in Cornwall only.

165. Juncus macer S. F. Gray (*J. tenuis* auct. britt.
non Willd.).—This American plant has been under suspicion
of introduction into Ireland, since it is undoubtedly alien
on the European continent and elsewhere, where it has now
spread very widely.   In N America its usual habitat is the
edges of paths or roads, which tends to easy dispersal by
moving traffic.   But while in continental Europe its rapid
spread is notorious, twenty years' observation by Dr. Scully
of the plant in Kerry reveals no such tendency (Fl. Kerry,
292); and as in the case of *Sisyrinchium*, its habitats lie
mostly in wild remote areas, far from all obvious sources
of introduction such as ports or railways.   Outside of
Kerry (where it is locally abundant) and W Cork the plant
is in Ireland known from roadsides in Wicklow, salt
marshes on Inishmore in Galway Bay, and one suspicious
station in Down, now destroyed, in Victoria Park close to
the Belfast docks.   While it will be difficult to find proof,
it seems probable that research will tend to establish the
position of this plant as indigenous.

    M. L. Fernald, to whom we are indebted for the

---

[51] Where it still survived in 1919—Journ. Bot., 1919, 285.

correction of the name of this plant (Journ. Bot., 1930, 364–7), considers that *Juncus macer* may be a native of the Scottish mountains (Clova, G. Don, 1795 *circa*), but a recent introduction in England. Ireland he does not refer to.

The plant has been known as an alien spreading on roadsides in England (since 1883—see Ridley in Journ. Bot., 1885, 1) and elsewhere.

166. Juncus acutus Linn.—This fine rush occurs along the S and E coasts from Cape Clear to Wicklow town, especially in Wexford. In Britain southern, absent from Scotland. Southern on the Continent, from France eastward. Asia, Canaries, America.

167. Potamogeton perpygmæus Hagstr. (*P. coloratus* × *pusillus*, *P. lanceolatus* auct. quoad pl. hibernicam).—There has been much discussion concerning *P. lanceolatus*. This has centred round the English plant (Cambridgeshire and Wales), which has been variously interpreted as *alpinus* × *pusillus*, *coloratus* × *pusillus*, and *gramineus* × *pusillus*. We are here concerned only with the Irish plant. Examination of this in its original Clare station (Cahir River S of Black Head) will I think confirm the view that it is *coloratus* × *pusillus*. The plant grows with both its reputed parents, the only other Pondweed in the stream being *P. natans*. It has moreover precisely the peculiar red tint of the local *P. coloratus*. A large form (Irish Nat., 1896, 243) grows in shallow rippling streams at Clonbrock (**369**) and Barbersfort (**368**) in NE Galway. The name *lanceolatus* being applicable to the English plant (which, on the evidence, would appear to be *alpinus* × *pusillus*), the Irish hybrid has been named by Hagström *P. perpygmæus* (Bot. Exch. Club Report, 1922, 630). This plant, which appears to be endemic in Ireland, was first found (in the Cahir River) by P. B. O'Kelly (Journ. Bot., 1891, 344; 1892, 195).

168. P. sparganiifolius Læst. (*P. Kirkii* Syme. *P. gramineus* × *natans*).—Long known as growing in the Maam River (**394**) in W Galway, where it was first found by T. Kirk in 1853 (Babington Manual, ed. 4, 351, 1856). The station is near the bridge at Maam. It has been gathered also in Lough Corrib by M. Norman in 1858 (Bot. Exch. Club Report, 1930, 394), and there is a doubtful record from Lough Neagh (see Irish Nat., 1909, 83–5). England and the Continent, very rare.

169. P. Babingtonii Ar. Benn. (*P. longifolius* Bab. non Gay. *P. lucens* × *prælongus*).—A single floating stem picked up by J. Ball in Lough Corrib (**367**) in 1835 is all that is known of this plant, which is figured in Engl. Bot. Suppl., 2847. Its rediscovery is very desirable. See A. Bennett in Journ. Bot., 1894, 204–5.

170. Potamogeton filiformis Nolte.—N half of Ireland, in lakes mostly on the limestone, rare. Scotland and Anglesea, local. Northern in Europe, and spread as far as America and Australia.

171. Ruppia maritima Linn. (*R. spiralis* Hartm.).—From Kerry up the S and E coasts to Donegal, discontinuous, more frequent in Down than elsewhere. Throughout Britain, but local and rare. World-wide in range.

172. Naias flexilis Rosk. and Schmidt.—This little water-plant, very wide-spread in N America, has in Europe a marked "Atlantic" distribution—Kerry, W Galway, and W Donegal in Ireland, Perth and Skye in Scotland, Esthwaite Water in Lancashire, and a few stations in NW Europe. Being small and inconspicuous in its submerged habitat, it is probably more widespread along the western Irish coast than would appear. First found by Daniel Oliver at Cregduff Lough, Roundstone, in 1850 (Bot. Gazette, **2**, 278, 1850).

173. Eriocaulon septangulare With. (plate 12).—The most wide-ranging of the Hiberno-American plants, found from Adrigole in W Cork to the Rosses, N of Dungloe, in W Donegal. Locally abundant in S Kerry, W Galway, W Mayo. Avoids the limestone tracts, the only station being Killower Lough in NE Galway, which is also unusually far inland, 18 miles from the head of Galway Bay. Often very abundant where it occurs, covering the peaty bottom in shallow water. It is extremely buoyant, and in rough weather breaks away from its anchorage, sometimes in large mats, and is often washed up in quantity, when its curious soft white jointed worm-like roots form a conspicuous feature. The flower-stem (July–Sept.) varies in length from a couple of inches, when growing on bare wet ground, to over three ft. when the plant grows in deep water.

Differences in water-level during summer are shown by the flower or fruit heads being submerged, or so much raised above the water that the stems slope : the normal height of the flower above water being 4 inches. In July, the previous year's stems are often still undecayed, lying criss-cross across the green rosettes on the bottom, looking like pale straws.

To 1000 ft. in Cork, but usually lowland. First found by Walter Wade in Connemara in 1801. In Britain confined to the Scottish islands of Skye and Coll. Not on the European continent. In N America has a very wide range.

174. Scirpus nanus Spreng. (*S. parvulus* Rœm. & Schult.).—Very rare in Ireland, confined to two stations— the tidal portion of the Cashen River in N Kerry (**330**), where it is abundant for some 3 miles (Scully in Journ. Bot., 1890, 110) and the river-mouth at Arklow (**276**) in Wicklow (A. G. More in Journ. Bot., 1868, 254 and 321–3, tab. 85). In Britain almost equally rare, in Wales and S England. On the Continent rather widely spread; Asia, Africa, America.

175. Scirpus filiformis Savi (*S. Savii* Seb. and Maur.).—Round almost the whole Irish coast. Rare beyond tidal influence, as in Kerry (850 ft.), Wicklow (700 ft.), E Donegal (300–550 ft.). N to S of Britain, but local. Continental distribution "Mediterranean," from W France eastward. Has a nearly world-wide range.

176. S. triqueter Linn. (plate 16).—As at present known, confined to the foreshore on both sides of the Shannon, from above Limerick to Tervoe lower light about 5 miles below the city, and sparingly in Cratloe Creek a mile further down on the Clare side (**334**). In the tidal river below the town it grows in great profusion on the muddy foreshore. Above the town, it occurs sparingly above the "head of the tide" as far as the old canal entrance 1 mile ENE of Limerick, growing small and starved in running fresh water.

In Britain only on the tidal reaches of rivers in S (especially SE) England. On the Continent widespread, both along the coast and by rivers in central Europe. Asia, Africa, America.

177. Rhynchospora fusca Ait.—On peat-bogs, widespread in the W and centre, often very abundant, its area including half of Ireland. Its E boundary runs roughly from W Cork to Westmeath and on to Donegal Bay. First found by J. T. Mackay near Killarney in 1805. In Britain much rarer, from S England to mid Scotland.   A plant widespread in Europe.

178. Carex pauciflora Lightf.—Widespread and often abundant over an area of some 10 sq. miles of bog, on the Garron Plateau, at about 1000 ft. elevation from Parkmore station to the Trosks above Carnlough in Antrim (**468**), often accompanied by *C. magellanica,* also here in its only Irish station (H. W. Lett, Journ. Bot., 1895, 216–7; R. Ll. Praeger, Irish Nat., 1920, 97).   Alpine and northern in Britain, and arctic-alpine on the Continent and N America.

179. Carex paradoxa Willd.—Very rare, in S Clare (**342**) and central Westmeath (**487**) only.   Very rare also in England, though occurring from N to SE.   A plant of N Europe and Siberia.

180. C. fusca All. (*C. Buxbaumii* Wahl.).—Found by David Moore on Harbour I. in Lough Neagh (**464**) in 1835 (Companion to Bot. Mag., 1835, 307), and seen there occasionally till 1886; apparently extinct now (see Stewart & Corry Fl. NE Ireland 161, and Irish Nat., 1920, 104). For 60 years this remained the only station in the British Isles, but in 1895 it was found by a loch in Arisaig, West Inverness (A. Bennett in Ann. Scott. Nat. Hist., 1895, 247–9).   The plant has a very wide range through Europe, Asia, N and S Africa.

181. C. aquatilis Wahl.—Rather frequent in Ireland, being native in 13 divisions, mostly in the typical form (var. *elatior* Bab.), and chiefly towards the N in low stations near the hills, but extends to the extreme south.   A recent arrival in Dublin (Irish Nat., 1901, 49).   First found by S. A. Stewart at Lough Allen (Journ. Bot., 1885, 49).   In Britain northern.   Elsewhere in Arctic Europe.   See also A. Bennett in Irish Nat., 1892, 48–50.

182. C. hibernica Ar. Benn. (*C. aquatilis* × ? *Goodenowii. C. aquatilis* × *Hudsonii?* Ar. Benn.).—This plant has its only known station by the river half a mile above Galway's Bridge on the old Killarney-Kenmare road in S Kerry, where it was found by R. W. Scully in 1889—see A. Bennett in Journ. Bot., 1897, 250, and Scully Fl. Kerry, 329; also **323**.   Not known elsewhere.   It looks more like a *Goodenowii* than a *Hudsonii* cross, and the former species is present, the latter absent.

183. Carex magellanica Lam. (*C. irrigua* Smith).—

Frequent on the high moorland of the Garron Plateau in Antrim (**468**) at about 1000 ft., with *C. pauciflora*, which also has here its only Irish station. From mid-Scotland to N Wales, rare. A N European plant.

184. Carex punctata Gaud.—Kerry and Cork only, by the sea, from Kerry Head to Ballycottin. SW and W coasts of Britain as far as S Scotland, rare and discontinuous. W Europe (not always maritime), Asia Minor, N Africa.

185. Calamagrostis neglecta P. Beauv. var. **Hookeri** Syme.—This variety was long known only from Lough Neagh (**464**), where it was first found by David Moore in 1836, and has since proved to be of frequent occurrence around that lake and the adjoining L. Beg, though now extinct in some of its former stations. See Fl. NE Ireland and Cyb. Hib. (2), 417. Recently found in a fen at Stow Bedon in Norfolk (A. Bennett in Journ. Bot., 1915, 281, and Irish Nat., 1915, 170), which is its only known station in Britain. The type is absent from Ireland.

186. Sesleria cœrulea Arduin.—The most abundant and characteristic grass of the western limestones, occurring wherever bare limestone appears, save in Limerick; it apparently does not cross the Shannon in the S. Confined in Ireland to the district W of the Shannon save where it colonizes the E bank in Tipperary and Westmeath, and the shores of the Shannon lakes. It follows the western limestones to their limit in S Donegal (**444**). From the limestone rocks it extends to esker-ridges, dry pastures, and sand-dunes; and in SW Connemara (**388**) to rocky heaths where no trace of limy soil is evident. In Fermanagh, also, it grows on ledges of Yoredale sandstone (**438**). Occurs abundantly at sea-level, and thence up to 1000 ft. in Clare and 1100 ft. on the Ben Bulben range.

Rarer and much scattered in Britain, chiefly on limestone hills, from N Scotland to N England and Wales. Rather southern on the continent, widespread. Iceland.

187. Poa alpina Linn.—On Brandon (**325**) in S Kerry, and Ben Bulben and Annacoona in Sligo (**423**), 1500–3100 ft. In Britain from N Wales to N Scotland, 3000–4000 ft. Of alpine-arctic circumpolar range; Himalayas, N Africa.

188. Glyceria Foucaudii Hackel (*G. festucæformis* auct. britt. *non* Heynhold, *G. maritima* var. *hibernica* Druce) (plate 18). Locally plentiful in the Shannon estuary (**331**,

342) (River Fergus,[52] also Foynes[53]) and in Down[54] (especially Strangford Lough (**475**), mostly on islets with a gravelly shore underlaid by clay). It forms a fringe at or slightly below highwater mark, and lower down than the plants (*Aster Tripolium, Atriplex* spp., etc.) which accompany it. It is very distinct when growing, and recognised at a distance by its large, erect tufts, two ft. or more in height. Two plants of different appearance are included in the Shannon records for this species—the one erect, growing on open shores: the other on mud, with prostrate shoots—see the papers quoted in footnotes 52 and 53; further work on these forms is desirable.

In Britain southern, confined to the S English coast. On the Continent a Mediterranean species.

189. Glyceria Borreri Bab.—Dungarvan Harbour (**286**) in Waterford (Irish Nat. Jl., **1**, 96 (1926)), Wexford Harbour (**277**), and Dublin Bay (**236**) only, on muddy ground by the sea. Southern in Britain (Bristol Channel to the Wash, with an outlying station in Forfar). Elsewhere by the North Sea in Holland and Germany.

190. Juniperus.—In Ireland both the Junipers are found chiefly along the W coast. *J. communis* is mainly lowland and calcicole, spread over the low-lying limestones form Killarney to Ballina and on by Sligo and Fermanagh to colonize the metamorphic rocks of Mayo and Donegal. *J. sibirica* (= *nana*) on the other hand is spread over the non-calcareous rocks of the W from Kerry to Donegal (on limestone (? or peat) in Sligo), and on across the Antrim basalts to the Mourne granites from sea-level (such as sand-dunes in Kerry) to 2000 ft. (in Down), being mostly upland. *J. communis* is rarer and *J. sibirica* commoner than in Britain. See Irish Nat. Journ., **5**, 58–61 (1934).

191. Taxus baccata Linn. f. **fastigiata** (Lindl.).—The Irish or Florencecourt Yew, *T. fastigiata* Lindl., now so common in cultivation, had its origin in two seedlings found wild by Mr. George Willis about 1767 on a limestone rock at Carrick-na-Madadh above Florencecourt (**430**), Co. Fermanagh; one of the original specimens still grows in

---

[52] KNOWLES and O'BRIEN, Irish Nat., 1909, 57–68.
[53] KNOWLES, Irish Nat., 1905, 51–3, 1906, 279–80.
[54] PRAEGER in Journ. Bot., 1903, 255; 1904, 79, 310, 352; 1905, 245. Also RENDLE, *ibid.*, 1903, 353–6, tab. 455, and E. S. MARSHALL, *ibid.*, 1918, 56–7.

Florencecourt demesne. These plants were females, and consequently most of the specimens in cultivation (which have been produced by cuttings) are females also. Recently, however, male plants have been found to be in cultivation in England, and these have been traced to the Barnham Nurseries near Bognor, where they have been grown for at least 45 years.[55] How they originated there is not known. Possibly a seedling of the usual *typica* × *fastigiata* cross came true to *fastigiata* (which has apparently happened once before) and was a male: possibly it was a case of sex reversal in *fastigiata,* a phenomenon not unknown among other plants.

Fruit is usually borne in abundance by natural crossing with the Common Yew, but almost invariably the seeds reproduce the typical form, not the variety. It would be interesting to learn whether *fastigiata* ♂ crossed with *fastigiata* ♀ produces *fastigiata.* Dr. Maxwell Masters considered the Florencecourt Yew to be a juvenile form of the species, in which the characters of the seedling (the radial disposition of the leaves and the upright habit) are preserved throughout the life of the plant.[56]

192. Trichomanes radicans Sw.—A characteristic Irish plant, demonstrating the "Atlantic" conditions which prevail everywhere over the island by its occurrence in N, S. E, and W; but it is much more wide-spread in the S, and now extinct in its most easterly stations (Wicklow), where it was found for the first time in Ireland [at Powerscourt] (**268**) by Whitley Stokes before 1804. The raids of collectors have greatly diminished its quantity everywhere. Occurs from sea-level to 1000 ft. See excellent account by Scully in Fl. Kerry, 359–61.

In Britain extremely rare, in Wales, N England, W Scotland. Elsewhere almost world-wide, warm-temperate or tropical.

193. Adiantum Capillus-Veneris Linn. (plate 23).— Along the W coast from Clare to Donegal, but very rare except in the Aran Islands (**246**) and Burren (**252**), where it is often abundant in fissures of the limestone, ascending to 800 ft. Some very limited stations are far from the Carboniferous limestone, and have been quoted as examples of the plant's occurrence in lime-free habitats—such as

[55] W. J. BEAN in Kew Bulletin, 1927, 254–5.

[56] See Gard. Chronicle, 3rd S., **10**, 68, 1891; ELWES and HENRY: Trees of Great Britain and Ireland, **1**, 110–111, 1906; Irish Gardening, **3**, 172–173, 1908.

Lough Bollard in Connemara, Achill Head, and Slieve League. This point can be settled only by a chemical test of the soil in which the plant is actually growing, as in all these cases rocks occur locally which may well yield a soil quite rich in lime, such as epidiorite, bands of primitive limestone, etc. First found by E. Lhwyd in Aran not later than 1699 (Phil. Trans., **27**, 524). SW Britain, Westmorland, Isle of Man. Elsewhere very widely spread in temperate and tropical regions.

194. Asplenium lanceolatum Huds. — South-eastern, from Wicklow to Kerry, rare. Western and southern in Britain, from Westmorland to Cornwall and Kent. Widespread over middle and S Europe to N Africa and the Atlantic Islands.

195. Asplenium Adiantum-nigrum Linn. var. **acutum** (Bory) (plate 4).—In its extreme form, as found in shady places in the SW, this is an exceedingly beautiful plant. It is spread along the S and W coasts from Cork to Mayo, and has outlying stations in Kilkenny, Wicklow, and Down. It is variable, with intermediates connecting it with the type, but it has nevertheless a well-marked and characteristic distribution—Ireland, W England, Spain and Portugal, the Mediterranean region, the Atlantic Islands, and as far as Hawaii. I have discussed the range of variation of the Irish plant (especially) elsewhere (Irish Nat., 1919, 13–19, pl. 2; also 53–4), pointing out that the more developed forms of the plant tend to two extremes, f. *lineare* and f. *ovatum*; the figures there given are reproduced here (plate 5). The former, in Ireland confined to shady rocks in the SW, is a very lovely lacy glossy plant. The latter form, not quite so extreme, occurs also in Cork, Kilkenny, and Down. Near Glendalough in Wicklow a rather less extreme form occurs. The variety is also on record from W Galway and E and W Mayo (see Cyb. Hib. (2), 444). Abroad, the extreme forms mentioned above both occur, and have been recognised (LUERSSEN in Rabenhorst Kryptogamenflora **3**, 260–282), the *lineare* form being set down as rare, but it occurs near Smyrna and in the Canaries; also in Madeira (Irish Nat., 1919, 53–4).

196. A. Ruta-muraria × Trichomanes (*A. Clermontæ* Syme, *A. Petrarchæ* Newman).—One plant, found on the outside of the garden wall of Ravensdale Park at Flurry Bridge, Louth, was robbed by a fern collector—see F. Naylor in Trans. Bot. Soc. Edinb., **8**, 365–6 (1866).

ASPLENIUM ADIANTUM-NIGRUM L.

1. Var. *acutum* (Bory) f. *lineare* Praeger.   2. ditto. f. *ovatum* Praeger.   3. Var. *obtusum* Kit.
4. Type. All ¾.

E. Barnes del.

PLATE 6

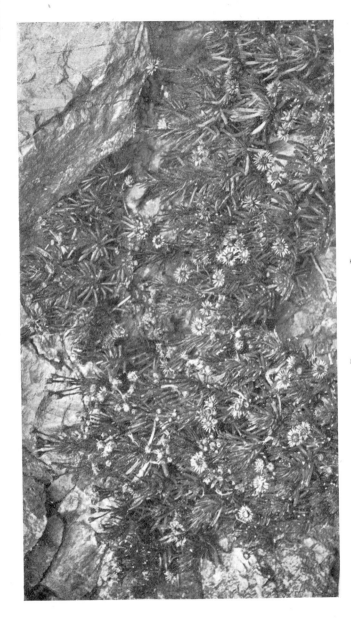

INULA CRITHMOIDES L.
Cliffs at Howth, Co. Dublin, August.

197. Ceterach officinarum Willd. f. **crenatum** (Milde).—This is the prevailing form in Ireland, and is often strikingly developed, with overlapping lobed pinnæ up to ·4 inch broad, giving it a very distinct appearance (fig. 15). The form with pinnæ quite entire is seldom seen.

198. Lastrea remota Moore (*L. Filix-mas* × *spinulosa*).—One clump by the stream in the woods at Dalystown, SE Galway (**357**): see Irish Nat., 1909, 151–3. Here *L. aristata* is also present, but the appearance of the hybrid does not suggest that species as one of the parents: it is quite intermediate between *Filix-mas* and *spinulosa*. The plant is fertile, and in my garden produced self-sown seedlings, which proved practically identical with the parent. A very rare hybrid, known in Britain apparently only from Brathay Woods in the Lake District and Ben Lomond. It occurs occasionally in Central Europe.

199. Phegopteris Robertiana Braun (*Polypodium Robertianum* Hoffm. *P. calcareum* Smith).—In Ireland only on a low limestone hill in Cloghmoyne townland, 3 miles NW of Headford in E Mayo—see **367a**. This fern was planted on Carlingford Mountain, Co. Louth, in 1878, and was still there in 1889 (Irish Nat., 1893, 22). Occurs in 30 English and 1 Scottish vice-county, and has a wide range elsewhere.

200. Ophioglossum vulgatum Linn. var. **polyphyllum** Braun (var. *ambi-*

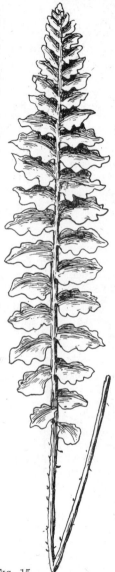

Fig. 15.
Ceterach officinarum f. *crenatum*. ¾.
From Black Head, Co. Clare.

guum C. & G.).—A remarkable dwarf gregarious form grows on the Great Blasket, Brandon Point, and Three Sisters Head in S Kerry, and Slieve League and Horn Head in W Donegal, in all cases in places of great exposure. It has at different times been referred to *O. lusitanicum* of the Channel Islands, or to the var. *polyphyllum* of Orkney and Scilly, but is not the first and more likely the second. As seen on the Great Blasket, it forms a carpet one inch high, with at least 100 plants to the sq. foot (Irish Nat., 1912, 161).

201. Equisetum litorale Kühlew. (*E. arvense* × *limosum*).—Since its discovery in Down (Irish Nat., 1917, 141–7, pl. VI–VII) in 1917, this interesting hybrid has been found in no less than 19 of the 40 county-divisions. Its apparent greater abundance in Ireland than in Britain (where it is recorded from very few stations in England and Scotland) probably signifies want of recognition in Britain rather than actual rarity there. Elsewhere it has been recorded from many places in both Europe and N America. The Irish plant in most cases belongs to the var. *elatius* Milde.

203. E. Moorei Newman.—This Horsetail has a quite definite range along the E coast from Ardmore Point in Wicklow to Wexford Harbour (**272**), on sand-dunes and rocky or clayey banks by the sea, in dry or wet ground. Its appearance is that of a slender *hyemale*, and it has been sometimes assigned varietal rank under that species, sometimes specific rank. One character held to favour the latter view was its alleged deciduous habit, which was later denied. Having collected it in several stations, and grown the plants alongside *hyemale* from different places, I find the stems to be much less persistent than in the other. In a normal winter all the stems except a few produced in autumn die back to within a short distance of the base, while *hyemale* usually remains green throughout, losing at most a few of its ultimate joints. No intermediate forms occur in the *E. Moorei* area, and the plant appears to be strictly maritime, in contrast to *hyemale*. I think it deserves specific rank, in the modern use of that term. First found by D. Moore and J. Melville at Rockfield in 1851 (Phytol., 1854, 17–18). Unknown in Britain.

Schaffner, whose full discussion of the plant should be noted (American Fern Journal, **21**, 95–98, 1931), from a study of herbarium material, identifies it with *E. hyemale* var. *Schleicheri* Milde and *E. occidentale* Hy. He maintains it as a distinct species.

204. E. trachyodon Braun (*E. Mackaii* Newman).—The frequency of this northern plant in Ireland is noteworthy, in view of its extreme rarity in Britain (Kincardine only, doubtful). It grows in no less than 15 divisions, from Kerry to Antrim and from Mayo to Dublin, being rather northern and western. First found by F. Whitla in Colin Glen near Belfast before 1830. A rare northern species on the Continent and in America.

205. Equisetum variegatum Schleich. — There are several forms of this in Ireland. What ranks as the type (var. *majus* Syme) is not infrequent, especially along sandy river-banks—a plant of medium size, with spreading or suberect stems of a lightish green colour. Var. *arenarium*, a smaller (often minute) prostrate very slender form growing in damp places in sand-dunes, looks very distinct from the last, but when grown side by side it becomes only slightly smaller and more prostrate, and has just the same colour. More distinct is a form characteristic of canal banks in the Central Plain, but found in other places also— strong, with very erect very numerous stems 2 to 3 feet high, of a very dark green colour. This plant, as seen in Glen Cahir, Clare, was noted by G. C. Druce as intermediate between *variegatum* and *trachyodon,* but the sheath characters are all those of *variegatum*. Lastly there is *E. Wilsoni* of Newman, as tall as the last, but much stouter and of a lighter green, known only from Kerry— the Lower Lake of Killarney (Muckross Bay and Ross Bay) and Caragh Lake. Although I have searched for this plant, I know it only from dried specimens, and cannot express an opinion as to its distinctness: but it does not seem sufficiently removed from *variegatum* to rank as a species.

206. Isoetes lacustris Linn. f. **maxima** Blytt (var. *Morei* Syme., *I. Morei* Moore).—A remarkable long-leaved form of *I. lacustris* known from Upper Lough Bray in Wicklow (**268**), where it was first found by A. G. More (see D. MOORE: "On a new species of Isoetes from Ireland." Journ. Bot., 1878, 353–5, tab. 199). The leaves may attain a length of over 2 ft. A form approaching this has been found by E. S. Marshall in Lough Camelaun in the Dingle

Peninsula, S Kerry; and leaves 18–20 inches were collected by Scully at Lough Slat in the same neighbourhood (Fl. Kerry, 383).

207. Charophyta.—The general range of the Charophyta is often extraordinarily wide, resembling that of many of the minute-spored terrestrial cryptogams rather than the comparatively heavy-seeded phanerogams. The Irish Charophyte flora is fully dealt with in the recent monograph by Groves and Bullock-Webster (Ray Society, 1917, 1924). The following, so far as at present known, appear to be the Irish species which are rarest in Ireland and Britain :—

Tolypella nidifica Leonh.—Wexford Harbour; two Scottish stations; N and W Europe; Kerguelen.

Nitella batrachosperma Braun (*N. Nordstedtiana* H. and J. Groves. *N. confervacea* Braun).—Killarney and Caragh L., Achill Island, Kindrum in W Donegal. In Britain in the Outer Hebrides only. Wide-spread over the world.

N. tenuissima Kuetz.—Ballindooly in NE Galway, Scraw Bog and L. Owel in Westmeath. Britain—Norfolk, Cambridge, Anglesea. World-wide.

N. spanioclema Groves and Bullock-Webster.—Described from L. Shannagh and L. Kindrum in W Donegal; also in Perth.

Chara muscosa Groves and Bullock-Webster.—Described from L. Mullaghderg in W Donegal; also in Orkney.

C. denudata Braun.—Brittas L. in Westmeath. Unknown in Britain. Switzerland, Italy, Cape Colony.

C. tomentosa Linn.—Abundant in the Shannon lakes of Ree and Derg, and in the Westmeath lakes. Unknown in Britain. Widespread in Europe; also known from Asia and N. Africa.

UNVERIFIED RECORDS AND MISSING PLANTS.

208. As in most countries, the list of species recorded on mistaken determination or careless localization is in Ireland a long one (see Cyb. Hib. (2) 471–520). But there are some interesting records which may still prove to be correct, as has been the case with *Limosella*, recorded from Connemara in 1804, set down as not Irish by subsequent botanists, and refound in Clare in 1893; *Helianthemum Chamæcistus,* reported from S Donegal in 1893, set down subsequently as planted or escaped, and refound undoubtedly native in the same locality in 1933; and *Lathyrus maritimus,* known in Kerry from 1756 to 1845, given up as extinct, and refound in its old station in 1918. A few of the more interesting of these are mentioned in order to draw to them the attention of visiting botanists.

209. Thalictrum alpinum Linn.—At about 2000 ft. on Brandon, S Kerry—A. Ley in Journ. Bot. 1887, 374. Never refound on this well-worked mountain, but a mistake seems unlikely. Unknown elsewhere in the S half of Ireland.

Matthiola sinuata R.Br.—Old records from Beal Castle, 1756, and Banna, 1878 (both in N Kerry, **330**), and Straw Island on Aran, 1835 (**352**). Some of the strand plants are notoriously irregular in their appearance, and *Matthiola* may still re-appear.

210. Saxifraga Geum Linn.—Reported from Clifden in Connemara—see 390.

S. spathularis Brot. (*umbrosa* auct.).—Reported from Malin Head in Donegal—see 452.

Erica ciliaris Linn.—Stated by Mackay (Nat. Hist. Rev. 1859, 537) to have been found by Bergin at Craiggamore (**389**), W Galway (the station for *E. Mackaii*), in 1846; and recorded by J. B. Balfour (Phytol. 1853, 1007) as found in the same place in 1852. The Bergin record is backed by a specimen (*E. ciliaris!*) in Herb. Trin. Coll. Dublin, certified by Bergin as portion of the original finding (see Cyb. Hib. (1) 183 and (2) 498). Searched for by many botanists since without success. Quite likely to occur, as this Pyrenean heath extends as far north as SW England (**35**), and others of the same geographical group grow at Craigga-more.

211. Erica vagans Linn.—Cliffs in Islandicane townland, W of Tramore, Waterford (**383**); found by Dr. Burkitt (before 1866) and named by Dr. Robert Ball (Cyb. Hib. ed. I, 184). Searched for in vain since. General range similar to the last, so its occurrence in Ireland is quite possible, but it is certainly not now in the station named. See Proc. R.I.A., **42**, B, 72.

212. Erica stricta Donn.—North of Ireland, 1834 (Dr. Lloyd), spec. in Herb. Hooker (see Cyb. Hib. (2) 499, and

Journ. Bot. 1872, 25). Reported from Sallagh Braes, Co. Antrim, and Downhill (or rather Magilligan **456**), Co. Derry (Irish Nat. 1923, 32). The Magilligan plant has been refound, and was an escape (see Proc. R. Irish Acad. **41** B 114–5, 1932). The other two records are not definitely localized, and are unsatisfactory. A S European plant, to be expected (if anywhere in Ireland) in the SW rather than in the NE. But the Mediterranean *Glyceria Foucaudii* is abundant on Strangford Lough close by.

213. Arctostaphylos alpina Spreng.—Included without comment in the Irish flora by Druce (Com. Fl. Brit. I) on the strength of a specimen in his herbarium labelled as collected by Bishop Mitchison near Kilmacrenan, W Donegal, in 1865. The area has been searched by Druce and by myself. The station is quite unsuitable, and no alpine plant is found in the vicinity (see Proc. R. Irish Acad. **41** B 113–4, 1932). I have no doubt that a mistake in labelling occurred.

Scrophularia alata Gilib.—Near Limerick, 1846 (Isaac Carroll); spec. in Herb. Brit. Mus.! Known to grow by the Liffey (**236**), where it is abundant, and also sparingly by the Bann in Londonderry (**362**), so quite likely to occur also by the Shannon, but cannot be found. See Britten, also Praeger, in Irish Nat. 1909, 222.

214. Limosella aquatica Linn.—"Frequently occurring where the water has stood during the winter, county Galway, near Ballynahinch, Connemara"—Wade Plantæ Rariores, 1804. Now known to occur locally in the adjoining county of Clare, to which as at present known it is confined. With a wide range in Britain, it is very possibly not so restricted in Ireland as would appear. The habitat of the species is quite correctly described by Wade.

Euphrasia salisburgensis Funk.—Reported from Lough Neagh—see **466**. Also from a rock-ledge at 1100 ft. on Benevenagh, Co. Derry — B.E.C. Report, 1924, 588. Doubtful.

215. Euphorbia Peplis Linn. (**151**). — Once found in Ireland—at Garraris Cove near Tramore (**283**), Waterford, 1839 (see Cyb. Hib. (1), 258). In Britain a rare and decreasing species of S England and Wales. Probably extinct, but should be watched for along the Irish E and S coasts.

216. Elisma natans Buchenau.—Druce recorded this as Irish (from Killarney and Clare) in Irish Nat. 1910, 237, on the authority of Glück. But Glück informed me subsequently (see Irish Nat. 1913, 105) that he considered the evidence for the inclusion of this plant in the Irish flora was insufficient. In the "Comital Flora of the British Isles" Druce sets it down without comment as an Irish plant.

217. Potamogeton Babingtonii Ar. Benn. (*lucens* × *prælongus*).—One fragment found by J. Ball in 1835 floating in Lough Corrib; see Cyb. Hib. (2), 378. An extremely rare hybrid, the refinding of which would be desirable.

P. sparganiifolius Læst.—Reported from Lough Neagh —see **466**.

218. Carex fusca All. (*C. Buxbaumii*).—Known from 1835, when it was found by D. Moore (Companion to Bot. Mag. 1835, 307), till 1886 (S. A. Stewart) on Harbour Island, in Lough Neagh, 3 miles S by E of Toome: apparently now extinct through grazing. This is the only Irish station, and it is equally rare in Britain, having a single station in Scotland—by a small loch in Arisaig, W Inverness (A. Bennett in Ann. Scott. Nat. Hist. 1895, 247–9). Still quite likely to occur at other places on the extensive and incompletely worked shores of L. Neagh. The original station, which had been reduced to a bare pasturage (Fl. NE.I. 161), is now again covered with scrub (Irish Nat. 1920, 103). The L. Neagh flora suffered from the lowering of the level of the lake in 1855, and since that date a number of plants very rare in the north have not been seen there—see **465**. The plant has a very wide range through Europe, Asia, N. and S. Africa.

C. elongata Linn.—Also not seen on L. Neagh since Dr. Moore's time, is a case similar to that of *C. fusca*; but it has a second Irish station on L. Erne.

219. Phegopteris Dryopteris Fée.—This fern is remarkably rare in Ireland, most of the records are old, and the refinding of it in its recorded stations is very desirable. These lie in Clare ("roadside between Broadford village and the Cliffs of Moher," T. H. Wright—Cyb. Hib. 2); Wicklow ("at Sheenabeg near Aughrim, very sparingly, 1879," G. H. Kinahan, *ibid.*, and "in good quantity on a hillside overlooking Glendalough," 1879 [but reported

from memory after twenty years], E. S. Marshall in
Journ. Bot. 1899, 269); Leitrim (Benbo Mountain, 800 ft.,
J. Wynne—Cyb. Hib. (1) 368); Sligo (Lough Talt,
Ox Mountains, R. Warren—see Irish Nat. 1897, 27); and
Antrim (north side of Knocklayd, sparingly, D. Moore,
Cyb. Hib. (1) 368—see Irish Nat. Journ., **5**, 36, 1934).

Athyrium alpestre Milde.—Glenveigh, W Donegal—
F. R. Browning in B.E.C. Report 1927, 426. Not reported
from Ireland before or since, and the record needs confirmation.

Tolypella prolifera Leonh.—Not refound at Glasnevin
—see **236**.

TOPOGRAPHICAL PART.

ENTERING IRELAND.

220. When one travels from Britain, whether to Dublin or to Belfast, or arrives from the west at Cobh (Queenstown), one gets the impression on nearing Ireland of a bold and diversified landscape. Approaching Dublin, the long undulating ridge of the Wicklow Mountains (**261**), on the southern side, catches the eye from afar. This is a very ancient highland, a great fold of Caledonian times, with its former Silurian covering worn off, leaving broad granite domes, now peat-covered, rising to 2000–3000 ft. In the southern foreground, close by Dun Laoghaire (Kingstown), foot-hills of the same rock jut steeply from the sea, and guarding the entrance of the bay the Cambrian rocks of Bray (**267**) and Howth (**231**) rise as bold headlands. North of the latter the volcanic Devonian island of Lambay (**230**) is seen, and, behind both, the low lands which form the eastern edge of the great Limestone Plain (**240**) which stretches without interruption across Ireland to the Atlantic.

221. Approaching the northern capital (**470**) on the other hand, one traverses sheltered Belfast Lough (**471**)—the drowned valley of the River Lagan—with the high scarp, in some places cliff-bound, of the Eocene basaltic plateau of Antrim (**455**) on the right (N), and on the S the fertile undulating Silurian country of County Down (**472, 473**). The fringes and floor of the lough are formed of the marls and sandstones of the Trias, smothered, in the valley at the head of the lough, under the red sands which were deposited in the glacial Lake Belfast. Between the dark basalt and the red marls, white Chalk and thin deposits of Greensand and Lias intervene. At both Dublin and Belfast a so wide diversity of rocks produces very varied scenery, and has a marked effect upon the flora.

222. If one crosses via **Fishguard,** on the SE coast, one finds at Rosslare (**278**) a quite different type of country and of flora—great gravel beaches (**272**) backed by low Cambrian rocks. At **Cobh** (**291**), again, another type of scenery prevails. Here the Devonian and Carboniferous rocks were strongly folded in Armorican times : the former stand up as bold ridges, while the latter remain as limestone troughs, and one passes up to Cork from the Atlantic through ancient narrow river-gorges cutting through the ridges, with broad expanses of water spreading over the foundered limestone between (**291**). A sharp contrast of calcifuge and calcicole floras is thus introduced.

223. The same contrast is present in a much more marked degree if **Galway** (**364**) is one's port of entry. Here the low indented coast backed by heathery hills which forms the N side of the bay is a region of ancient metamorphic rocks, mostly peat-covered, with a calcifuge flora which includes many of the most interesting plants to be found in Ireland (**384**); while the gaunt grey limestone hills which rise steeply on the S shore of the bay harbour a calcicole flora of extraordinary interest (**346**).

CROSSING IRELAND.

Four main lines of railway, with main roads following mostly the same routes, radiate from Dublin—northward to Belfast and the Province of Ulster; westward to Galway and the Province of Connacht; south-westward to Cork and the Province of Munster; and southward to Wexford and Waterford.

DUBLIN TO GALWAY.

224. The western line runs for its entire distance from sea to sea over the plain of Carboniferous limestone which occupies the centre of the country (**240**). Only for about one mile does it encounter any other rock, where near Moate it crosses a small worn-down anticline where the underlying Carboniferous sandstone and Devonian rocks are exposed.   Leaving Dublin, the line follows for fifty miles (to Mullingar) the course of the pre-existing Royal Canal, now little used and full of an interesting vegetation (**236**).   The country is at first mainly flat, rich pasture (**241**), the limestone being heavily covered with calcareous Boulder-clay.   Southward, the abutment of the Leinster Chain (**261**), the great granite mountain-mass which dominates Wicklow, rises boldly from the plain.   The fields on either side of the line are divided by high unclipped thorn hedges, from which rise many well-grown trees, but even here one notices a slight eastward bending of the tree-tops due to the prevalence of westerly winds.[1]

Soon the well-groomed aspect of the Dublin area begins to pass away.   The hedgerows,

> "hardly hedgerows, little lines
> Of sportive wood run wild"

full of native shrubs and climbers—*Euonymus, Prunus, Viburnum, Ilex, Rosa,* etc.—assume a more picturesque appearance.

Poorer wetter land begins to come in; twenty miles out fragments of dark cut-away bog appear, purple with heather; before Mullingar is reached great undisturbed areas of "red bog" (**53**) are crossed, showing the usual inverted-saucer-shaped form.   Where drainage and an admixture of ballast have dried and lightened the soil, willows and gorse fringe the railway.

The hills of Dublin and Wicklow are now only a distant misty ridge, and in the flat country to the S the little volcanic neck of Croghan Hill (**769** ft.) looks quite an imposing mountain.

At Mullingar we see, S of the railway, Lough Ennell, one of the larger of the lakes of Westmeath (**487**).   The remaining lakes (and the more interesting) lie a few miles to the N, where the little conical hill of Knock Eyon (**707** ft.) breaks the sky-line.   As we pass on, the surface becomes rougher in the drier areas, owing to a thinner covering of drift, and limestone blocks lie about in increasing numbers. Wet poor pasture and limy marsh increase, with much

---

[1] But where shelter is afforded the native trees in Ireland may attain unusual stature.   Thus in the woods at Headfort near Kells in Meath *Crataegus* grows to 60 ft. in height.   *Taxus* to 70 ft., *Prunus avium* to 90 ft.

Juncus inflexus and *J. subnodulosus,* and gorse and willows on the fences. Large treeless bogs increase the width of the view. At Athlone (**378**) the Shannon is crossed, a very broad slow stream, and we see on the right the marshy meadows which fringe its exit from Lough Ree (**379**). Rough pasture land succeeds, with limestone blocks lying on the surface in ever-increasing numbers. At Ballinasloe (**369**) we cross the sluggish Suck, which for most of its course is now little more than a deepened drain, converting much swamp into rushy pasture.

From Ballinasloe westward the limestone rock, lying in grey horizontal beds, breaks more and more frequently through the surface : the trees become dwarfed and gnarled and much twisted to the E ; hedges give way to rough grey stone walls ; the rare railway cuttings are through limestone gravel, outliers of the great series of moraines which mark the last retreat of the Pleistocene ice ; they are gay with *Antennaria, Blackstonia, Carlina, Anacamptis pyramidalis* and so on.

Before reaching Athenry considerable hills are seen to the S—almost the first since we lost sight of the granite mountains of the E coast. These are Slieve Aughty (**357**) an anticline of Silurian slates. A little further on we catch sight of the grey uplands of Burren—the naked limestone hills of N Clare, which harbour one of the loveliest and most interesting floras in western Europe (**346**). At the same time we glimpse the waters of the upper end of Galway Bay merging almost imperceptibly with the low land, and to the W we first sight the straight horizon-line of the Atlantic. The last few miles into Galway (**364**) is through increasing areas of grey naked limestone ; and the train draws up on the very junction between this formation and the ancient metamorphic rocks of Connemara.

DUBLIN TO CORK.

225. The train climbs out of Kingsbridge station through a deep cutting, where a number of interesting aliens have established themselves—*Hieracium Lachenalii, Linaria purpurea, Senecio squalidus*—the last clearly an arrival by rail from Cork, where it has been established for a long time. The line runs W first through rich pasture-lands, to clear the projecting spur of Lyons Hill ;

then it swings into its SW course across Kildare, with the granite chain of Wicklow (**361**) rising boldly on the left. The Liffey (**237**), a pleasant rippling stream, is crossed near Newbridge, and to the N, looking towards the ridge of Silurian hills at the "Chair of Kildare" we see the most easterly extension (**54**) of the great bogs of the Central Plain (**240**). A little further on, a large gravel-pit, with many light-soil aliens, gives good sections of the great Glacial gravel-dump which forms the Curragh of Kildare. Close by, on the N side, an extensive marsh (**238**) offers half a square mile of *Cladium* and other plants of limy swamps. The Central Plain of Ireland is seldom flat—mostly gently undulating, but between Kildare (with its ancient cathedral and round tower with a modern battlemented top, well seen from the railway) and Monasterevan, we pass across one of the flattest areas (**239**) in Ireland, lately dried out by the deepening of the bed of the River Barrow (**257**). Beyond this we are in typical Central Plain country (**240**), with much pasture, mostly wettish, very little tillage, the skyline broken by many trees but no extensive woods. At Maryborough (**348**) the long high ridge of Slieve Bloom (**347**) on the right marks the beginning of a great anti-clinal area that continues parallel with the railway past Devil's Bit (**351**) almost down to Limerick. Here Silurian slates and Devonian sandstones prevail, with a calcifuge flora differing widely from that of the surrounding plain. The infant Nore (**252**) is crossed near Mountrath, and about Thurles we run for some way down the flat valley of the Suir (**281**). This S end of the great limestone area is on the whole more fertile and especially less bog-covered than the more typical Central Plain country further N. The well-drained railway banks provide a habitat for some plants not found in the land on either side—quantities of *Brachypodium pinnatum*, and great colonies of the alien *Picris hieracioides*. The broad landscape becomes increasingly broken up by anticlines exposing the under-lying Devonian and Silurian rocks. A notable example is the noble range of the Galtees (**250**) which forms a dominating feature on the left beyond Limerick Junction. At Lisduff we cross one of the rare bogs on this route, offering a contrast to the grasslands which prevail over most of the area. At Charleville the line swings S round the flank of the heathery Ballyhoura Hills and we run down to cross the lovely Blackwater valley (**287**) at

Mallow—the first of the E–W valleys which are so characteristic of S Ireland (**10**). Here for the first time we come within the influence of the Lusitanian flora. Directly we pass the river, wherever there is rough ground, we see vigorous clumps of *Euphorbia hiberna,* golden in the spring months, scarlet-leaved in autumn, and this plant follows us as we climb over the broad ridge of Old Red Sandstone and drop steeply downward into the deep hollow where Cork (**291**) stands upon the banks of the Lee.

DUBLIN TO BELFAST.

226. This a varied and interesting journey. For the first half-hour the route lies along the coast of Dublin, with a little bit of Meath in the N. When the deep cutting full of *Clematis Vitalba* and other aliens which takes us past Raheny is cleared, we see the promontory of Howth (**231**), dear to every botanist, rising boldly beyond the Neolithic raised beach which joins it to the mainland. South of it that three-mile sand-bank formed during the nineteenth century, the North Bull (**228**), is conspicuous. At Malahide (**228**) and again beyond Donabate we pass extensive shallow muddy inlets almost closed by sand-dunes, which yield an interesting flora; seaward, the island of Lambay (**230**) stands up boldly. About Balbriggan, Silurian slates form a rocky coast, which again gives way to extensive sands on either side of the estuary of the Nanny River (**488**), where the line curves inland to Drogheda (**488**). The high viaduct affords a bird's-eye view down the long estuary of the Boyne to its sandy mouth (**488**) — excellent ground for the botanist. Now we climb across a mass of Silurian slates which continue E to the promontory of Clogher Head (**491**) and we drop down into low ground, passing Braganstown bog, full of *Osmunda,* and on to the flat land about Dundalk (**491**) with the Carlingford Mountains (1919 ft.) rising boldly to the E. The railway, climbing from sea-level, crosses these granite uplands over a broad boggy flat, with Slieve Gullion (1893 ft., **490**) towering on the left, and traverses Wellington Cutting with its fine geological sections. Then a wide view across County Down bursts on us, with Newry lying in the valley below, the Mourne Mountains (**479**) from Rostrevor to Slieve Donard rising boldly, and more to the left a peep of Carlingford Lough (**481**), gleaming

between the Mournes and Carlingford Mountain (**490**).
A curious flat-bottomed trough in the Silurian rocks, an
old river valley, offers by the railside interesting swamps
and pools full of *Cicuta,* etc. Towards Portadown this
gives way to flats of cut-away bog, where *Spiranthes
stricta* (**156**) was found for the first time in Europe. We
cross the Bann at Portadown, and rising to Lurgan, get
glimpses of Lough Neagh (**463**) to the N—home of many
rare plants—and beyond it the ridge of Slieve Gallion,
and on a clear day the Sperrin Mountains in Tyrone (**439**).
More to the E is the back of the basaltic plateau of
Antrim (**457**). At Moira a white gleam under dark rock
shows the beginning of the Chalk, underlying the Eocene
basalts. We are now in the Lagan valley, with the low
undulating hills of Down on the right, while on the left,
from Lisburn on, the scarp of the basaltic plateau rises
higher and higher, till we see in front the spires and mill-
chimneys of Belfast (**470**), and through the smoke the bold
precipitous outline of the Cave Hill.

THE COUNTY OF DUBLIN.

227. The small county of Dublin (354 square miles)
has one of the largest floras in Ireland. The reasons are—
a long and diversified shore-line, yielding a varied vegetation : owing to the central position of the county on the
E coast, plants of northern range, such as *Mertensia,
Scilla verna,* overlap with southern coastal plants like
Inula crithmoides, Artemisia maritima, Trifolium scabrum.
Next, the Carboniferous limestone of the Central Plain
protrudes across County Dublin to the sea, accompanied
by canals and railways from the midlands, the limestone
bringing in calcicole species, and the traffic-ways such
plants as *Ranunculus circinatus, Arenaria tenuifolia, Carex
diandra, Chara aculeolata.* Then the northern end
of the Wicklow Mountains occupies the southern edge of
the county, with a population of upland and bogland
species like *Saxifraga stellaris, Andromeda, Lycopodium
clavatum.* And finally, Dublin is the centre in Ireland
not only of some introduced plants such as *Sisymbrium
Irio* (**88**), *S. Sophia,* and *Mercurialis annua,* but of several
indigenous species — *Senecio erucifolius,* widely spread
within the county and scarcely spreading beyond it;

Scrophularia alata, found in Ireland chiefly along the lower course of the Liffey; *Tolypella prolifera* and *T. intricata,* known only from the Royal Canal; *Medicago sylvestris* **(100),** doubtfully native, on sand-dunes on the coast. The absence of lakes, on the other hand, reduces the amount of the aquatic flora.

For botanical purposes, the area may be divided into a central region, including Dublin and the Liffey valley **(236)** ; a southern area, mostly suburban towards the coast, rising into mountains (Kippure, 2473 feet) along the southern margin **(232)** ; and an extensive fertile northern tract, interesting mainly on account of the diversity of shore conditions **(228).**

Dublin is the driest county in Ireland, the only area in which the average annual rainfall sinks below 30 inches: along the Liffey valley, and on the coast from Malahide to Dalkey, the average precipitation actually stands below 28 inches. Ireland being a country with an excess of rain, the feature just mentioned is wholly to the benefit of the flora, favouring the presence of sun-loving and dry-soil plants. Thus it comes about also that the alien flora is here at a maximum, since most casuals and colonists are species of more southern latitudes, and the long-continued use of Dublin as a port of entry for friends and foes alike has favoured the introduction of plants from across the sea.

The Dublin area is of interest from many points of view in addition to that of botany—for instance, zoology, geology, archæology. Plenty of literature is available, but only the leading works on its botany can be quoted here.

N. COLGAN: "Flora of the County Dublin: Flowering Plants, Higher Cryptogams, and Characeæ." Dublin: Hodges, Figgis & Co., 1904. 12s. 6d. net. G. FLETCHER (editor): "Leinster" (The Provinces of Ireland) (Botany by R. LL. PRAEGER). Cambridge University Press, 1922. 7s. 6d. net. G. A. J. COLE and R. LL. PRAEGER (editors): "Handbook to the City of Dublin and the surrounding District" (prepared for British Association). Dublin, 1908 (Botany by N. COLGAN).

NORTH COUNTY DUBLIN.

228. Interest centres on the coast, which is very varied. There are many expanses of sand, from the mouth of the Liffey northward, and stretches of low rocks—Carboni-

ferous limestone in the S, Silurian slates in the N; and
at **Portmarnock, Malahide,** and **Rogerstown** considerable
shallow inlets almost closed by barriers of dunes. In
contrast to these there are also the bold Cambrian promontory of Howth (**231**) and the hilly volcanic island of
Lambay (**230**), which are dealt with separately. In the
north, off Skerries, are three islets too small or too much
grazed to harbour rare plants. The flora of the coastal
region includes a good deal of

Thalictrum dunense
Papaver Argemone
 hybridum
Raphanus maritimus
Viola Curtisii
Cerastium arvense
Spergularia rupicola
Linum bienne
Erodium moschatum
Trifolium scabrum
 fragiferum
Crithmum maritimum
Dipsacus sylvestris
Carlina vulgaris
Tragopogon pratense
Blackstonia perfoliata

Limonium binervosum
 humile
Chenopodium rubrum
Atriplex littoralis
 maritima
 portulacoides
Euphorbia Paralias
Epipactis palustris
Ophrys apifera
Allium vineale
Juncus subnodulosus
Scirpus filiformis
 rufus
Glyceria distans
Festuca uniglumis
Hordeum nodosum

Of the species in this list which are usually maritime
in Ireland, *Cerastium arvense, Trifolium fragiferum,* and
Hordeum nodosum have a few inland Dublin stations also.
 More rare, and found on the S half of the coast of
North Dublin, are *Crambe* (apparently extinct, as is
Mertensia), *Trifolium striatum* (not entirely littoral),
Vicia lathyroides, Inula crithmoides (Rush, N limit in
Ireland), *Artemisia maritima, A. Stelleriana* (**124**)
(naturalized on the North Bull), *Erythræa pulchella*
(North Bull), *Cuscuta Trifolii* (abundant and ? native
at Rush and Portrane), *Hippophae rhamnoides* (naturalized
and spreading at Rush), *Zostera nana* (Portmarnock inlet),
Ruppia maritima (Malahide inlet), *Euphorbia portlandica,
Equisetum variegatum* var. *arenarium* (sandhills at Port-
rane and Portmarnock), *Elymus arenarius* (? native),
Eleocharis uniglumis, Carex, divisa (mouth of the Liffey,
? extinct).

The **North Bull**, a sand island backed by salt-marsh, over 3 miles long and ½ mile wide, separated from the mainland by a muddy inlet of the same or less width, and connected with the shore by a long wooden bridge, is interesting because it was not in existence 150 years ago, having been formed by currents consequent on harbour construction. The whole of its flora and fauna are therefore immigrants within that period, and well worthy of study as examples of recent dispersal. The flora is now varied and dense and contains some of the rarer local littoral plants (*supra*), also some rare species alien or doubtful here, like *Artemisia Stelleriana* (mentioned above), *Spartina Townsendii* (recently planted), *Elymus*.

229. Many of the other rarer N Dublin plants, though not halophile, are found mostly near the coast : *Clematis Vitalba* (established in abundance on railway banks, etc.), *Viola hirta* (reaching at Rogerstown its N limit in Ireland), *Sisymbrium Sophia, Silene conica* (established about Portmarnock and Portrane), *Medicago sylvestris* (dunes at Portmarnock and Malahide, only Irish locality (**100**)), *Melilotus altissima* (common), *Fœniculum, Valerianella rimosa, Senecio erucifolius* (widespread except in the extreme N, in Ireland almost confined to County Dublin), *Picris echioides, Crepis taraxacifolia, Lactuca muralis* (Dollymount), *Campanula rapunculoides, Rumex pulcher* (Portrane), *Lemna gibba, Mercurialis annua, Lamium molucellifolium, Galeopsis angustifolia, Calamintha Acinos, Carex axillaris* (Malahide), *Bromus erectus, Hordeum murinum, Selaginella* (sand-dunes). It will be noted how large a proportion of these (all but seven or eight) belong to the alien element. Among the plants which show no preference for the coast are *Geranium columbinum, Crepis paludosa, Orobanche Hederæ, Rumex maritimus* (formerly at Garristown, where alone in former times low-level bog occurred in the county), *Mercurialis perennis* (Finglas), *Hydrocharis, Lemna polyrrhiza, Carex strigosa*. The inland region is highly tilled or grazed, offering little opportunity for the occurrence of an interesting flora.

230. Lambay. — The island of Lambay, high and heathery, area **617** acres, formed mainly of volcanic rocks of Devonian age, lies some three miles off the coast N of Howth, and the varied Howth flora is to some extent repeated there. Among the abundant plants are

Cerastium arvense
Spergularia rupicola
Erodium maritimum
Crithmum maritimum
Carlina vulgaris

Inula crithmoides
Myosotis collina
Scilla verna
Juncus subnodulosus
Asplenium marinum

Less abundant are

Viola hirta
Geranium sanguineum
Trifolium striatum

Limonium binervosum
Hyoscyamus niger
Scirpus filiformis;

while the following have only a single station each :—

Geranium pusillum
Vicia lathyroides
Agrimonia odorata
Rubia peregrina

Tragopogon porrifolium
Lamium molucellifolium
Atriplex littoralis
Scirpus rufus

The greater part of the surface being occupied by *Pteris* and *Calluna* (which are increasing owing to protection), the rarer plants, as will be seen, are mainly maritime. While the vegetation of the precipitous rocky slopes consists largely of *Crithmum, Inula crithmoides* (Plate 6), and *Beta,* that of the less steep earthy slopes higher up is greatly influenced by the large colonies of breeding Herring Gulls, with *Silene maritima, Armeria, Matricaria inodora* or *Atriplex* spp. dominant. At the great Puffin villages, vegetation is often exterminated. The *Pteris* areas are brilliant in spring with sheets of *Primula vulgaris, Scilla non-scripta, Ranunculus Ficaria, Nepeta hederacea.* On sea-slopes facing N, *Lastrea Filix-mas* and *Athyrium* grow in great luxuriance and profusion. *Iris fœtidissima* looks native on cliff-edges (see also **231**). Two areas in the SE, one on andesite and the other on Silurian slate, present an almost soil-less and at first sight plantless area of bare crumbling rock, occupied on closer inspection by a profusion of *Sedum anglicum* and starved *Erodium maritimum,* with a little *Cerastium ietrandrum, Aira præcox,* and very stunted *Teucrium Scorodonia.* Many sea-birds besides those mentioned breed on the cliffs, and Great Grey Seals in the caves.

H. C. HART: "Notes on the Flora of Lambay Island, County of Dublin." Proc. Roy. Irish Acad. (2) **3** (Science) 670–693. 1883. R. LL. PRAEGER: "Phanerogams and Vascular Cryptogams," in "Contributions to the Natural History of Lambay, Co. Dublin." Irish Nat. 1907, 90–99, vegetation map and 5 plates.

231. Howth.—The high peninsula of Howth, formed of ancient Cambrian quartzites and slates, commanding Dublin Bay on the N, deserves special mention, since its flora is interesting and very varied, and its cliffs and heathery top make it a favourite place for an excursion. The flat gravelly isthmus (a Postglacial raised beach) which connects it with the mainland is best dealt with along with the sands that lie to N and S (**228**). Between these gravels and the hill, Carboniferous limestone laps up against the older rocks. Together the two latter form an area of some three sq. miles, with some sandy shores in the N, high slopes and cliffs in the E, and a lower rocky coast in the S. The occurrence of shelly Glacial drift plastered against the cliffs accounts for the occurrence of calcicole plants on the Cambrian rocks. The flora includes

	Raphanus maritimus		Ligustrum vulgare
	Viola hirta		(native)
	Sagina ciliata		Blackstonia perfoliata
	Spergularia rupicola		Orobanche Hederæ
	Geranium sanguineum		Atriplex littoralis
	Erodium maritimum	E	maritima
E	Trigonella		portulacoides
	ornithopodioides		Euphorbia portlandica
E	Trifolium striatum	E	Scilla verna
E	Ornithopus perpusillus		Scirpus filiformis
	Crithmum maritimum		Bromus erectus
	Rubia peregrina		Brachypodium
E	Inula crithmoides		pinnatum
	Artemisia maritima		Asplenium marinum
	Crepis teraxacifolia		Osmunda regalis
	Limonium binervosum		Selaginella
			selaginoides

Also many established aliens, including

E	Sisymbrium Sophia	E	Picris echioides
E	Irio		Galeopsis angustifolia
	Erodium moschatum		Ornithogalum
	Fœniculum vulgare		umbellatum
	Valerianella rimosa	E	Hordeum murinum
E	Senecio viscosus		

Most of these, as will be seen, are plants usually found near the sea ; those marked E are in Ireland characteristic of the E coast. The best ground for the botanist is the S side towards the Baily lighthouse. The warm slopes of the S shore tend to get colonized by plants derived from gardens above. *Bryonia* is an old denizen there, and more recent arrivals are *Cerastium grandiflorum, Vinca major, Polygonum sachalinense, Allium triquetrum.*

A curious state of *Teucrium Scorodonia* (due to the attacks of a mite, *Eriophyes*) in which well-developed inflorescences bore, in place of flowers, collections of small bright green leaves, was found in Howth demesne (Journ. Bot. 1925, 376).

On the N side, off the town of Howth, rises the extremely picturesque islet of **Ireland's Eye**. Some of the rarer Howth plants, such as *Spergularia rupicola, Geranium sanguineum, Erodium maritimum, Trifolium striatum, Crithmum, Inula crithmoides, Limonium binervosum, Atriplex maritima,* occur again here; also *Thalictrum dunense, Crambe* (uncertain in its appearances), *Cerastium arvense, Lavatera* (indigenous on high stack), *Euphorbia portlandica, Festuca uniglumis. Scilla non-scripta* is immensely abundant, flowering before dense groves of *Pteris* arise to give it summer shade. *Cotyledon Umbilicus-Veneris* also grows profusely and luxuriantly on deep humus under the Bracken. *Viola Curtisii* occurs chiefly in a fine dark blue form. *Iris foetidissima* is an unusual ingredient, growing on sandy wastes; a long-established colony of *Hyoscyamus* occupies similar ground.

As regards *Iris foetidissima,* its occurrence here and on Lambay on sands and cliffs remote from houses or former cultivation leads one to consider whether it may not be native (see Journ. Bot., 1934, 73)—a suggestion already put forward by both H. C. Hart (**230**) and E. S. Marshall (Journ. Bot. 1899, 271).

H. C. HART: "The Flora of Howth." Dublin: Hodges, Figgis, & Co. 1887. N. COLGAN: Fl. Dublin.

SOUTH COUNTY DUBLIN.

232. Building, from Dublin eight miles down to Dalkey, has driven out a good deal of the native flora here. Along the coast, now much enclosed and built over, *Lavatera* was probably originally native, as *Crambe* certainly was; *Limonium humile* appears also extinct. But *Spergularia*

rupicola is still abundant, and *Trifolium fragiferum,*
Limonium binervosum and *Glyceria distans* are frequent.
The much rarer *G. Borreri* (**189**) occurs from the mouth
of the Liffey to Booterstown, *Allium oleraceum* has
established itself in chinks of the stonework half way along
the West Pier at Dun Laoghaire (Kingstown), and *Sagina
ciliata* grows there. *Lathraea* flourished near by 150 years
ago, and *Geranium columbinum* at Blackrock (see Journ.
Bot. 1905, 219). *Cerastium arvense* survives at Merrion,
and *Crithmum* at Blackrock.

Also chiefly near the coast are *Papaver Argemone,
Linum bienne, Dipsacus sylvestris, Carlina, Tragopogon
pratense, Blackstonia, Chenopodium rubrum, Ophrys
apifera, Allium vineale,* etc.

233. Dalkey and Killiney.—The rocky hills and shores
between Dalkey and Killiney still harbour a number of
interesting plants. Also this is an exceptionally mild place,
which has an influence on the flora. Here (in addition to
some mentioned above) are

Geranium sanguineum	Rubia peregrina
Erodium maritimum	Inula crithmoides
Corydalis claviculata	Artemisia maritima
Trigonella ornithopodioides	Euphorbia portlandica
Trifolium striatum	Paralias
scabrum	Scilla verna
Crithmum maritimum	Asplenium marinum

The Mediterranean *Senecio Cineraria,* introduced to a
garden about 1875, has spread to maritime rocks and banks,
and now grows in great profusion for the better part of a
mile along the coast S of Dalkey, hybridizing freely with
S. Jacobaa (× *S. albescens* Burbidge and Colgan, see
Journ. Bot. 1902, 401–406, tab. 444; Irish Nat. 1902, 311–
317, Pl. 5; and **126**). About Vico bathing-place the hybrid
may be seen in forms extending from one parent to the
other, and near by *S. Cineraria* mingles charmingly with
the native vegetation—*Spergularia rupicola, Inula crithmoides, Aster, Beta,* etc. Among the naturalized aliens
are *Erodium moschatum, Crepis biennis, Campanula
rapunculoides, Elymus,* and abundance of *Kenthranthus.*
Eschscholtzia californica and a New Zealand *Hebe* also
seem to be quite naturalized here.

234. Further W, between the sea and the mountains, the flora includes *Clematis Vitalba, Ranunculus parviflorus* (last seen at Bray common, 1875), *Corydalis claviculata, Sisymbrium Irio, Viola hirta* (Greenhills), *Cotoneaster microphyllus* (naturalized by R. Dodder), *Senecio erucifolius* (frequent), *Picris echioides* (nearly extinct), *Kentranthus, Crepis taraxacifolia, C. biennis, Blackstonia, Mercurialis annua, Lamium Galeobdolon, Galeopsis angustifolia, Calamintha Acinos* (extinct?), *C. ascendens, Orobanche Hederæ, Carex axillaris* (Milltown), *C. strigosa, C. Pairæi* (Stepaside), *Bromus erectus, Hordeum murinum, Equisetum variegatum* (both along the River Dodder), *E. trachyodon* (**204**) (by stream below Ballycorus lead works)—a very mixed assemblage of plants, in Ireland local or widespread, native or alien.

235. The Dublin Mountains.—The granite hills which form the N end of the Leinster Chain nowhere offer cliffs suitable for alpine plants, though exceeding 2000 ft. in height (**Kippure** 2473 ft.), and are bog-covered on their summits; they bring in a number of plants, not necessarily montane species. The best ground is **Glenasmole**, a granite valley with much limy Glacial drift; here grow *Saxifraga stellaris, Galium uliginosum, Cnicus palustris* × *pratensis* (*Forsteri*), *Cephalanthera ensifolia, Eriophorum latifolium, Carex aquatilis* (introduced apparently by natural means to the reservoir (see **265**), and spreading since first observed in 1896), *Littorella* (a similar case), *Hymenophyllum peltatum, Lastrea montana, L. æmula* (not seen recently), *Phegopteris polypodioides.* Here or elsewhere along the hills are also *Crepis paludosa, Epipactis palustris, Juncus subnodulosus, J. diffusus, Lycopodium alpinum* (at 1700 ft.). *Wahlenbergia hederacea* reaches its N limit in E Ireland about Glencullen and Featherbed, and the very local *Viola lutea* (**92**) occurs in delightful abundance in the district, especially in the Brittas area, painting the pastures yellow in May. More widespread are *Ranunculus Lenormandi* (N limit in Ireland), *Andromeda, Vaccinium Vitis-Idæa, Pinguicula lusitanica, Listera cordata, Malaxis, Lycopodium clavatum, Selaginella.*

The plants of the mountain-range are more fully dealt with below (**261**).

The S Dublin district, with the adjoining portion of Wicklow, is one of the few Irish areas which have been studied from the vegetational point of view.[2] The succession, from below upwards, is shown to be (1) arable land; (2) hill-pasture with much *Ulex europœus* (below) and *U. Gallii* (above), *Pteris*, and sometimes *Nardus*; (3) *Calluna* moor, which in the higher and flatter ground gives way to moor dominated by *Scirpus cæspitosus, Eriophorum* spp. or *Racomitrium*.

DUBLIN CITY AND THE LIFFEY VALLEY.

236. From Dublin one may make a number of short trips of considerable interest for the botanist—to Howth (**231**), Killiney (**233**), Glenasmole (**235**), etc.: but even within the city boundaries there are some good plants. Two rare aliens, *Sisymbrium Irio* and *Mercurialis annua,* are well-known Dublin weeds. *Carex divisa* and *Glyceria Borreri* (**189**), notable rarities in Ireland, haunt or haunted the river-mouth; while *Tolypella prolifera* and *T. intricata,* unknown elsewhere in the country, were found in the Royal Canal near Glasnevin by David Moore some 75 years ago. The latter was refound there in 1896 (see Cyb. Hib. (2) 468): both are now in need of re-discovery; while not strictly indigenous here, since their habitat was non-existent a century and a half ago, I am inclined to refer the origin of the *Tolypella* species to migration, perhaps assisted, along the canal from some as yet undiscovered stations in the midlands rather than to introduction from an extra-Irish source. *Apium Moorei, Sagittaria, Butomus* and several pondweeds (*P. coloratus, P. lucens, P. densus, P. pectinatus*) have entered within the city limits by way of the canals ; and fed by canal water, even the lake in Stephen's Green in the heart of the city yields *Tolypella glomerata.* *Carex diandra* has thus migrated from the Midlands as far as Clonsilla.

For twelve miles the River Liffey traverses County Dublin from W to E, passing through a rich district with villages and many demesnes. A railway and a canal N of the river, and another railway and canal on the S, have a similar course. All have combined to increase the flora by immigration from the W. Thus *Arenaria tenuifolia* has entered the county along the railways, and *Senecio squalidus,* abundant at Inchicore and now spreading into the city, has clearly come by rail from Cork. The river or

[2] G. H. PETHYBRIDGE and R. LL. PRAEGER : ''The Vegetation of the District lying south of Dublin.'' Proc. Roy. Irish Acad. 25 B 124–180. 5 plates, col. map. 1905.

canals yield (in addition to plants mentioned in the preceding section) *Ranunculus circinatus* and *Chara aculeolata,* typical Central Plain plants, as well as quantities of *Equisetum variegatum.* But the Liffey has also one or two plants of its own, notably *Scrophularia alata,* abundant in places from Sallins and Kilcock (in Kildare) eastward, and very rare elsewhere in Ireland (see **237, 334, 462**). Other plants of the vicinity of the river are *Linum bienne, Geranium columbinum, Vicia tetrasperma* (long established at Knockmaroon), *Hypericum hirsutum* (very frequent here, strangely rare in Ireland), *Impatiens glandulifera* (Kingsbridge), *Senecio erucifolius, Crepis taraxacifolia, C. paludosa, Blackstonia, Monotropa* (formerly at St. Catherine's), *Calamintha ascendens, Galeopsis angustifolia, Orobanche Hederæ, Lamium Galeobdolon, Allium Scorodoprasum* (Lucan), *Carex strigosa, Hordeum nodosum.* As one approaches the city, characteristic Dublin aliens make their appearance—*Sisymbrium Irio, Fœniculum, Mercurialis annua, Hordeum murinum.* On railway banks between Kingsbridge and Inchicore a group of unusual aliens occupy the ground in abundance— *Geranium pratense, Hieracium Lachenalii, Linaria purpurea, Senecio squalidus* (already mentioned). Similarly at Liffey Junction *Hieracium triviale* is a profuse colonist. *H. scanicum* is recorded by G. C. Druce from "near Dublin" (Bot. Exch. Club, 1920, 132), and *H. maculosum* from "Glasnevin" (*ibid.,* 1923, 195). At Lucan *Acæna Sanguisorbæ* has established itself, as at Rostrevor (**481**) and a few places in Great Britain (*e.g.,* Dartmoor, Melrose).

The whole county is excellently provided with means of communication; and one of the most complete floras published in Ireland is devoted to this area.

Literature: see **227**.

THE RIVER LIFFEY.

237. The Liffey is a curious stream. Its source is only 14 miles from its mouth, but it traverses 72 miles of country between the two points, making three-quarters of a circle—a Postglacial course dictated by the distribution of the drift. It rises at 1800 ft. SW of Kippure, inside the Wicklow boundary, amid deep peat-bogs with a smooth vegetation of dwarf *Calluna, Scirpus cæspitosus* and *Eriophorum angustifolium,* mixed with a good deal of

Erica Tetralix, Empetrum, Narthecium and *Eriophorum vaginatum, Carex canescens,* with *Andromeda* and *Lycopodium Selago* sparingly spread. Thence it drops rapidly to Ballysmuttan bridge, near which blue forms of *Anemone nemorosa* grow, also *Orobanche major* and (by Shankill River only) *Saxifraga stellaris.* Past Blessington (*Eriophorum latifolium*), receiving important tributaries from N and S, it meanders placidly to the fine Post-glacial gorge of **Poulaphuca,** where the river cuts foaming through a barrier of Silurian rocks, falling in a high cascade. Here are several Hawkweeds, including *H. farrense, H. proximum* F. J. Hanb., and *H. strictum*; also *Equisetum variegatum* and several calcicole plants (*Arabis hirsuta, Geranium lucidum*), whose presence is due to percolation from calcareous drift. Beyond Ballymore Eustace (*Hieracium proximum* again) the stream assumes the character which it maintains almost to its mouth, flowing between silty or sandy banks of its own making, and running in a wide curve on the E edge of the Limestone Plain. At Athgarvan a high gravelly bank of drift supplies one of the most satisfactory inland Irish stations for *Cynoglossum officinale. Mimulus Langsdorffii, Potamogeton decipiens,* and other plants appear at intervals. At Victoria Bridge, *Impatiens glandulifera* is naturalized on the marshy river bank. A little above **Sallins** that especially characteristic plant of the Liffey, *Scrophularia alata,* begins, descending two small tributaries from the neighbourhood of Naas, and continuing down the main stream, in great abundance in places, to within a few miles of Dublin. It also occurs along the whole course of the Rye Water, which is the largest tributary which the Liffey receives. See Proc. R. Irish Acad. **41** B 115. The botany of the lower reaches of the river is mentioned above (**236**).

THE COUNTY OF KILDARE.

238. The County of Kildare, lying SW of Dublin, links up the E coast flora with that of the Central Plain. Eastward of N Kildare, the limestone runs on to meet the sea at Dublin; eastward of the centre and S, Ordovician foot-hills give way to the granite domes of the Wicklow Mountains. The E portion of the county is fertile, hilly, well-wooded; the remainder is typical Central Plain country, flattish, with much bog. The S is drained by the Barrow (**257**); in the NE the Liffey (**237**) makes a broad loop through the county.

The Royal Canal, which fringes Kildare on the N, yields *Apium Moorei* (**118**), *Sagittaria, Butomus, Equisetum variegatum, Chara aculeolata*, etc., often in abundance. The Grand Canal, which passes right across the county, is less interesting botanically, presumably on account of the greater amount of traffic which it conveys, and few rare plants are recorded from its banks or waters. The boggy country of the NW has *Rhamnus Frangula, Galium uliginosum, Andromeda, Malaxis, Juncus subnodulosus, Carex limosa, Lastrea Thelypteris, Osmunda*. The great gravel-mass of the **Curragh** yields light-soil plants such as *Blackstonia, Calamintha Acinos*. A remarkable marsh which runs N from the E end of the Curragh supports half a square mile of *Cladium*, with *Galium uliginosum, Juncus subnodulosus, Eriophorum latifolium*, and so on. Demesne woods harbour *Monotropa* (at Moore Abbey) and *Orobanche Hederæ*. *Viola hirta*, a rare plant of strangely discontinuous range in Ireland, is on the limestone at Carton, and *V. lutea* comes in on the hills about Poulaphuca. *Senecio erucifolius*, essentially a Dublin plant, crosses into Kildare about Leixlip and Celbridge. The varied adventive flora includes *Sisymbrium Sophia, Melilotus altissima*, and *Fœniculum* in the north, *Galium erectum, Lactuca muralis, Anchusa sempervirens, Hordeum murinum*. *Picris hieracioides* and *Brachypodium pinnatum*, which are Central Plain colonists on railway-banks, have invaded Kildare from the west. See under Barrow (**257**) and Liffey (**237**) for some other Kildare plants.

The visitor will find accommodation at Athy, Naas, Kildare, Dunlavin, etc.

239. One of the few really flat areas in the Central Plain is that which lies N and E of **Monasterevan**, on the borders of Kildare and Leix, and attains its most characteristic facies between the Barrow and Figile rivers, showing a uniform brown plain with scattered bushes. Previous to recent drainage operations, this area often formed a great lake for a month at a time in wet weather. It represents an old lake deposit, being underlaid by deep grey sandy silt : the lake was due to a bar of limestone at Monasterevan, lately cut down. The flora before the drainage showed little effect of flooding. *Molinia* prevailed, with large patches of *Juncus sylvaticus* and *J. articulatus,* much *Cnicus pratensis* and *Spiræa Ulmaria, Scabiosa Succisa, Carex fulva, Mentha aquatica,* and beds of starved *Phragmites.* The bushes were mainly *Cratægus,* with some *Salix aurita* : *Rosa canina* and *Populus* sp. also endured the submergences.

A flat more marshy area not far away with a more interesting flora is described above (**238**).

In the centre of the county, the Hill of Allen and other low eminences of Silurian rocks look down on great peat bogs resting on the limestone—a lonely and little-known region.

THE CENTRAL PLAIN.

240. Some account of the extent and characteristics of the Central Plain has been given in a previous place (**59**). Its flora changes from E to W, and to a less extent from N to S. The E–W change is due mainly to the decreasing depth of calcareous drift which covers the Carboniferous limestone. Greater exposure and higher precipitation play a minor part; the effect of the former being visible chiefly in decrease of trees, both in number and stature, as one approaches the Atlantic : but the Central Plain is in most parts guarded from the ocean by hilly masses of ancient rocks, and it is only along the Dublin-Galway transect that the change may be studied over low land from sea to sea (**224**). Here also it is unbroken by the NE–SW series of anticlines, exposing the underlying non-calcareous rocks, which, especially towards the S, disturb its continuity (**225**).

241. The E portion, as seen best in the counties of Kildare and Meath (**244**), is a land of rich pastures with comparatively little bog, marsh or lake. The plants characteristic of the Central Plain which show a predilection for this eastern region include

Thalictrum flavum	Tragopogon pratense
Ranunculus circinatus	Andromeda Polifolia
Lathyrus palustris	Blackstonia perfoliata
Poterium Sanguisorba	Orchis morio
Erigeron acre	Ophrys apifera
Carlina vulgaris	Lemna polyrrhiza
Centaurea Scabiosa	Sagittaria sagittifolia
Crepis taraxacifolia	Carex Pseudo-Cyperus

Half of these, it will be noted, are plants of dry places, the rest marsh or water plants. *Andromeda* is the only bog plant characteristic of the Central Plain which prefers the E to the W. In this E portion of the plain the flora is reinforced by members of the E coast group which penetrate to a varying extent westward. Such are

Lepidium heterophyllum *var.* canescens	Calamintha Acinos
	Lamium album
Chærophyllum temulum	Salix triandra
Anthiscus vulgaris	Carex acutiformis
Carduus crispus	Glyceria aquatica
Lithospermum arvense	Hordeum murinum

Two-thirds of these are light-soil plants, and most of them followers of man. The *Lepidium,* which is indigenous, is calcifuge, and favours the peninsulas of non-calcareous rocks which project into the limestone area. *Glyceria aquatica* has been assisted in its extension by the canals which run W from Dublin.

242. If on the other hand we study the flora of the W portion of the area, we note a marked increase of the following species specially characteristic of the plain :—

Rhamnus catharticus	Juncus subnodulosus
Cornus sanguinea	Equisetum variegatum
Teucrium Scordium	Chara desmacantha
Betula alba (verrucosa)	Tolypella glomerata
Ophrys muscifera	

The character of the flora is emphasized by the invasion of species which have their head-quarters in W Ireland, such as

Thalictrum minus	Taxus baccata
Viola stagnina	Sisyrinchium
Geranium sanguineum	angustifolium
Drosera longifolia	Rhynchospora fusca
Galium boreale	Sesleria cœrulea
Euphorbia hiberna	Cystopteris fragilis
Juniperus communis	

For these the rocky limestone shores of the Shannon lakes are the main attraction. The rapid decrease eastward of *Drosera longifolia* and *Rynchospora fusca,* despite the abundance of suitable bogs, is very marked; neither of them makes any headway E of the Shannon.

243. Among other plants which in Ireland attain their maximum in the Central Plain, being distributed generally therein, and have not been mentioned above, are

Stellaria glauca	Gentiana Amarella
Galium uliginosum	Potamogeton coloratus
Sium latifolium	Chara aculeolata

A few of the rarest of Irish plants belong to the Central Plain, where they have their only Irish stations:—

> _Inula salicina_ (**122**)—Lough Derg on the Shannon. Absent from Great Britain.
>
> _Pyrola rotundifolia_—Bogs in Westmeath.
>
> _Chara denudata_ (**207**)—Brittas Lake in Westmeath. Absent from Great Britain.
>
> _Nitella tenuissima_ (**207**)—Westmeath, NE Galway.

244. The heavy mesophytic grasslands characteristic of the richest part of the plain, as in Dublin and Meath, have a rank vegetation of grasses, _Cnicus_ spp., _Senecio Jacobæa_, etc., with little to interest the botanist. Where the soil gets lighter and drier, _Orchis morio, Anacamptis pyramidalis. Gentiana Amarella, Erythræa Centaurium,_ etc., come in. The more xerophytic type of grassland, often overlying glacial gravels, shows a close grassy sward with much _Briza media, Kœleria. Trisetum,_ and _Avena pubescens_, often a great quantity of _Antennaria dioica_, and a sprinkling of such plants as _Blackstonia, Carlina, Poterium Sanguisorba, Ophrys apifera_. Where the ground becomes wet, a marsh flora supervenes such as is referred to in **57**.

245. Several factors connected with human activities have influenced materially the flora of the Central Plain.

1. There has been much drainage of lakes and marshes, resulting in a reduction of the aquatic and paludal flora.

2. There has been much cutting of peat for fuel, with an analogous result; but so great is the area of bog, that it is doubtful if any species has disappeared save quite locally.

3. The making towards the end of the 18th century of two canals from Dublin to the Shannon has extended the range of some plants—_e.g., Sagittaria, Glyceria aquatica,_

Potamogeton spp., *Equisetum variegatum,* several *Charo-phyta.*

4. Gravel-pits in the great terminal moraine which extends across the Central Plain, and railway cuttings through the same, have allowed the spread of light-soil plants such as *Papaver, Fumaria, Crepis taraxacifolia, Picris hieracioides, Bromus erectus, Brachypodium pinnatum.* The last is still spreading rapidly on railway-banks.

5. Railway ballasting and the passage of trains have given a wide distribution to *Diplotaxis muralis, Arenaria tenuifolia, Linaria minor,* etc. But the recently introduced spraying of the tracks with a chlorate of potash preparation to keep down the weeds now militates against this method of dispersal, and has reduced materially the quantity of these plants.

OFFALY (KING'S COUNTY).

246. A typical Central Plain area (**240**) with the broad anticline of Slieve Bloom (**247**) breaking the continuity of the limestone on the S border, and the Shannon bounding the area on the W. The Barrow (**257**) flows on the SE edge of the county and drains that portion; and the head-waters of the Boyne (**488**) are in the NE. The general characters of the Central Plain flora are dealt with elsewhere (**240–245**).

In the E, around Edenderry, vast bogs extend, and there is an old record for *Saxifraga Hirculus* (**116**) from near Portarlington. About **Tullamore**, in the centre, are *Potamogeton densus, Equisetum variegatum.* Clonad Wood near Geashill is a remnant of original forest with a flora including *Geranium sanguineum, Sorbus porrigens,* both species of *Rhamnus, Lastrea æmula, L. montana, Osmunda, Equisetum hyemale, E. variegatum.* In the SW, about **Birr** (= Parsonstown), are *Monotropa, Mercurialis perennis, Butomus, Sesleria* (between Birr and Roscrea— only station E of the Shannon), *Agropyron caninum.* At **Lough Coura,** now an extensive limy marsh, are *Galium uliginosum, Utricularia intermedia, Cladium, Carex limosa, C. lasiocarpa,* etc. A *Euphrasia* found near by was definitely named by Mr. F. Townsend *brevipila × salisburgensis* (**139**); *E. salisburgensis* is not known anywhere so far inland, its nearest stations lying some 40 to 50 miles to the W; investigation is desirable. There is an old unverified record for *Vicia Orobus* (extremely local in Ireland) from the Offaly side of Cloughjordan. The broad **Shannon**, flowing along the W edge of the county past Banagher, has been little explored in this part; *Lathyrus palustris* (near Clonmacnoise), *Sium latifolium*, *Galium uliginosum*, *Utricularia intermedia* (**144**), *Ophrys apifera*, *Chara aculeolata* have been noted, and the extensive bogs are full of *Rhynchospora fusca*, *Carex diandra* and other Central Plain plants. The rarer aliens include *Erodium moschatum* (Killeagh), *Valerianella rimosa* (Edenderry), *Lactuca muralis* (Leap Castle), *Lepidium campestre* and *Calamintha Acinos* (Edenderry). The visitor can stay at Banagher, Birr, Clara, Edenderry, Portarlington, Roscrea, Tullamore, etc.

R. LL. PRAEGER: "The Seagull Bog, Tullamore." Irish Nat., 1894, 173–175.

SLIEVE BLOOM.

247. Slieve Bloom rises as a broad conspicuous ridge in the centre of Ireland, the first break in the continuity of the Limestone Plain as one goes S. It is formed by a simple NE and SW fold, and the limestone, now worn off the ridge, gives way as one ascends first to Old Red, and it in its turn to the underlying Gotlandian rocks. The hills are featureless, deeply covered with peat-bog. Near the highest point (Arderin, 1,733 ft.), at Glendine Gap, there is a little gorge with *Meconopsis*, *Phegopteris polypodioides*, *Hymenophyllum peltatum*, *Lastrea æmula*. *Listera cordata* among the heather on the summit, and *Lycopodium clavatum*, are among the few other plants of interest which have been recorded; also *Ulmus montana* (native in a glen above Kinnitty). *Leucorchis albida*, *Lastrea montana*, *Botrychium*, *Equisetum hyemale*, *E. variegatum*, etc., also occur. Near Clonaslee, at the N base of the hills, *Eriophorum latifolium* grows in abundance.

H. C. HART: "The Botany of the Barrow." Journ. Bot., 1885, 9–18.

LEIX (QUEEN'S COUNTY).

248. Leix lies fairly in the Central Plain, but the continuity of the drift-covered Carboniferous Limestone is broken in the NW by the conspicuous broad heathery ridge of Slieve Bloom (**247**) formed of slates and sandstones, while in the south (S of Maryborough, the county town) there are hills of Yoredale rocks, which flank the plateau of the Leinster coal-field (**255**).

The W part of the county is drained by the infant Nore (**252**), while the Barrow (**257**) fringes the area on the NE and E. About Maryborough are *Sagina ciliata, Sorbus porrigens, Galium uliginosum, Blackstonia, Calamintha Acinos, Galeopsis angustifolia, Carex lasiocarpa*; and *Senecio squalidus*, probably derived from Cork, has formed a small colony and lately crossed with *S. Jacobœa. Campanula Trachelium* (naturalized), *Eriophorum latifolium*, and *Equisetum trachyodon* grow 1 mile S of Maryborough; at Emo *Valerianella rimosa, Erythrœa pulchella* (only Irish inland station), and *Galeopsis intermedia* (very rare in Ireland); at Mountmellick *Rosa stylosa.* Cullenagh Mountain, S of Maryborough, yields *Drosera longifolia, Festuca sylvatica*, etc.

The usual Central Plain species, such as *Drosera anglica, Carlina, Blackstonia, Epipactis palustris, Juncus subnodulosus, Potamogeton coloratus, Carex diandra, Osmunda, Chara aculeolata*, are widespread over the county. For the rest, see under Slieve Bloom (**247**), Barrow (**257**), and Nore (**252**). The explorer of Leix can put up at Maryborough, Portarlington, Athy, Carlow, Roscrea, Abbeyleix.

THE COUNTY OF TIPPERARY.

249. In both "Cybele Hibernica" and "Irish Topographical Botany" the large south-central county of Tipperary is divided into two for purposes of showing plant-distribution, the division being formed by the main line of the Great Southern Railway; but this does not mark or approximate to any definite botanical frontier. On the N side, the continuity of the limestone plain is interrupted by a broad anticline, the SW continuation of the Slieve Bloom axis, formed as usual of Silurian and Devonian slates and sandstones; many other smaller—and in some cases more lofty—anticlines occur further S.

North Tipperary presents in the N a stretch of unbroken Carboniferous limestone country, terminated on the W by the indented shore-line of Lough Derg, which is dealt with separately (**353**). This is a drift-covered area, without much bog. Some rare plants occur here and there, native or naturalized—*Pimpinella major, Valerianella rimosa, Pyrola minor* (Mt. Butler), *Lamium molucellifolium* (near Roscrea), *Carex strigosa* and *Sorbus porrigens* (Templemore); *Calamintha Acinos, Faniculum* at **Nenagh**; *Geranium sanguineum* and *Valerianella rimosa* grow at **Cloughjordan**, whence also there is an old record for *Saxifraga Hirculus*. *Teucrium Scordium* (**148**) at Ballyspillane L. near Borrisokane (Irish Nat., 1924, 131) is an interesting outlier from its main stations along the Shannon. In the NW, *Leucojum æstivum* (**161**) grows by the Little Brosna River among natural surroundings, but there are gardens not far away. The *Pyrola* mentioned above is probably the *P. vulgaris* Lob. of How's Phytologia, 1650— "In a Bogge by Roscre in the King's County. Mr. Heaton"—which has hitherto been attributed to *P. rotundifolia*, for which see **487**.

The E and S part of the area is occupied by Old Red Sandstone hills, which are referred to elsewhere (**251**). Hotels will be found at Birr, Portumna, Nenagh, Killaloe, Roscrea, Thurles, Limerick Junction.

South Tipperary (divided from N Tipperary by the main line of the Great Southern Railway) is mostly flattish limestone country, watered chiefly by the River Suir (**281**), but in the S the plain is broken by the fine ranges of the Galtees (**250**) and Knockmealdown (**284**). These and the picturesque lower course of the Suir are the chief features, botanical and topographical, of the area, and are described where indicated above. Out in the plain rises the Rock of Cashel, a twisted hump of limestone, crowded with the finest group of early Romanesque buildings in Ireland. **Slievenaman** (2,364 ft.) is an imposing isolated heathery dome of Old Red and Gotlandian rocks rising on the N side of the Suir near Clonmel. It has but little botanical interest, *Carex rigida* on the summit being the only rare plant recorded from it.

THE GALTEES.

250. The Galtees form the loftiest of the local folds which, resulting from the crumpling of the earth's crust in post-Carboniferous times, have broken up the S part of the Limestone Plain into a series of mountain-ranges from which the limestone is now denuded; in the Galtees, the Old Red Sandstone and Silurian rocks stand up boldly in a series of fine conical peaks. For nearly eight miles the main ridge maintains an altitude of over 2,500 ft., and the loftiest point, Galtymore, reaches 3,015 ft., a height surpassed in Ireland only in Kerry and Wicklow. The Galtees thus easily dominate Knockmealdown (**284**) and the Comeraghs (**285**), which adjoin on the SE and E. Along the N slope, high up, is a series of deep corries, occupied by tarns, and, as in the Comeraghs, the cliffs which form the cirques provide the best ground for the botanist. The hard slates produce good foot-hold, and the massive conglomerates have ledges which allow easy access to the cliff-ranges. Cultivation is not found above 700–800 ft.

The best plant of the Galtees is *Arabis petræa*, which occurs on one bluff facing NNE above and W of L. Curra, at 2,600 ft. Its only other Irish station is in Glenade in Leitrim. The best bit of ground is the grand cliffs over L. Muskry. Here, mostly in abundance, grow *Thalictrum minus, Cochlearia alpina, Sedum roseum, Saxifraga stellaris, S. spathularis* (here in its most inland Irish station), *S. sponhemica, S. hirta, Vaccinium Vitis-Idæa, Oxyria, Salix herbacea, Asplenium viride.* Most of these occur again on the cliffs around L. Curra and L. Diheen. L. Curra yields *Cystopteris fragilis* and *Hymenophyllum tunbridgense,* and L. Diheen *Saussurea. Meconopsis* grows by the stream in Glancushnabinnian. Other plants of the range are *Pinguicula lusitanica, Crepis paludosa, Hymenophyllum peltatum, Cystopteris fragilis, Lastrea æmula, L. montana. Carex rigida* is frequent on the high bogs. The summit of Galtymore (3,015 ft.), like most Irish mountain summits, has a small flora almost devoid of alpines; the following are given as occurring above 3,000 ft. :—

Stellaria uliginosa	Salix herbacea
Saxifraga stellaris	Luzula sylvatica
Chrysosplenium oppositi-folium	Eriophorum vaginatum
Galium saxatile	Agrostis tenuis
Calluna vulgaris	Deschampsia flexuosa
Vaccinium Myrtillus	Poa annua
Rumex Acetosa	Festuca ovina
Acetosella	

This very fine range of hills may be worked from Cahir,

Tipperary, or Mitchelstown. The best ground lies along the N slope. In the valley of the Funshion, at the S base of the Galtees, are the Mitchelstown Caves, of which the "New Cave" is one of the longest in the country, its passages aggregating nearly two miles, and offering many picturesque features. (See C. A. HILL, H. BRODRICK, and N. RULE: "The Mitchelstown Caves, Co. Tipperary." Proc. Roy. Irish Acad., **27**, B, 235–268, pl. xiv–xvii, 1909.)

H. C. HART: "On the Botany of the Galtee Mountains, Co. Tipperary." Proc. Roy. Irish Acad. (ser. 2), 3 (Science), 392–402, 1881.

DEVIL'S BIT, SLIEVE FELIM, KEEPER, SILVERMINES.

251. The anticline which forms Slieve Bloom (**247**) broadens southwestward, and produces an extensive area of mountainous ground composed of Gotlandian rocks with a fringe of Old Red Sandstone. Towards the N these strata form the flat-topped ridge of Devil's Bit (1,583 ft.), conspicuous from the Dublin-Cork railway near Templemore, with a gap on its straight skyline, from which it derives its name. Southward, the hills (Slieve Felim) are lower and uninteresting, but towards the W Slievekimalta or Keeper Hill towers up to 2,278 ft., with the deep glen of the Mulkear River separating it from the ridge of the Silvermines Mountains (1,607 ft.).

This extensive range of country provides fine walking but poor botanizing, the heathery hills yielding very little. The single alpine plant of the area is *Vaccinum Vitis-Idæa* on Keeper, where *Hymenophyllum tunbridgense, Cystopteris fragilis, Lastrea æmula* also grow. *Lastrea montana* occurs low down on the same hill. *Trichomanes radicans* and *Crepis paludosa* have been found at Glenstal (**336**) on the southern side of the hills.

H. C. HART: "On the Botany of the River Suir." Sci. Proc. R. Dublin Soc., N.S., 4, 326–334, 1885. *Ibid.*: "Notes on Irish Plants." Journ. Bot., 1881, 167–169.

KILKENNY AND THE NORE.

252. The **County of Kilkenny** is attractive to the botanist mainly along the limestone valley of the River Nore, which flows S down the centre of the area. E and W of it is

rising Coal-measure country of small botanical interest
(**255**). In the S, the river cuts deeply through slates and
sandstones to join the Barrow (**257**).

The **Nore** rises on Devil's Bit (**251**) in Tipperary, and
flows ENE to near Mountrath, receiving tributaries from
Slieve Bloom; then it assumes the SSE direction, which it
maintains right across Co. Kilkenny to its junction with
the Barrow. Its length from source to this point is a little
over 100 miles. The marked contrast between its flat upper
course and beautiful gorge-like lower reach is explained
above (**18**). From the city of Kilkenny onwards, and
especially from Inistioge down, the Nore is one of the most
charming rivers in Ireland. It is interesting to the botanist
as furnishing the Irish headquarters of two very local
plants—*Campanula Trachelium* and *Colchicum autumnale*—
which indeed are in Ireland almost confined to its vicinity.
The *Campanula* haunts the wooded banks from Abbeyleix
down to its junction with the Barrow and on to New Ross,
its only outlying stations being high up the Barrow near
Portarlington. *Colchicum* is abundant in many places
along the Nore and its western tributaries from Freshford
and Callan to Inistioge, and again above New Ross in
Wexford below the Nore-Barrow junction. Otherwise
rarities are few.

253. One of the few interesting plants recorded from
the headwaters of the Nore is *Saxifraga Hirculus* (**116**),
which has been found in the neighbourhood of Mountrath.
Near the upper portion of the stream, *Pimpinella major,
Galeopsis angustifolia,* and *Mercurialis perennis* occur
about Abbeyleix; the last, with *Lactuca muralis, Orobanche
Hederæ,* and *Carex strigosa,* is recorded from the Durrow
neighbourhood; and a rarer plant, *Leucojum æstivum*
(**161**), grows in marshes by the Erkina River, ½ mile below
Durrow. In the middle part of the course of the Nore
Nasturtium sylvestre and *Orobanche Hederæ* are frequent,
and a N American *Aster* is naturalized in several spots.
Sorbus latifolia, a rare south-eastern species, grows
sparingly on the river-bank a couple of miles above
Kilkenny.

Kilkenny.—The more interesting plants which are
found about this historic town include *Ranunculus parviflorus,* a form of *R. auricomus* with perfect petals and
flowers measuring 1¼ inch across, *Lepidium campestre,
Linum bienne, Archangelica* (naturalized here, very rare
in Ireland), *Crepis taraxacifolia, C. biennis, Campanula*

rapunculoides, Blackstonia, Calamintha ascendens, Salvia horminoides, Sparganium minimum, Mercurialis annua, Lemna polyrrhiza—largely an alien assemblage. The town is beautifully situated on the steep bank of the Nore, and possesses much of archæological interest, especially in the way of ecclesiastical architecture; the ancient buildings provide a home for masses of *Cheiranthus, Kentranthus,* and *Parietaria.* (The visitor to Kilkenny may take the opportunity of visiting the Cave of Dunmore, a few miles distant, one of the most famous, though by no means the largest, of Irish caverns (see Irish Nat., 1918, 148–158).

254. About Thomastown, a less interesting place on a wide bend of the Nore below Kilkenny, are *Sorbus porrigens, Fœniculum, Kentranthus, Valerianella rimosa, Rumex pulcher, Spiranthes spiralis, Allium vineale, Potamogeton angustifolius, P. densus, Hordeum murinum, Osmunda*—like the Kilkenny plants, largely early introductions—and at Inistioge, where the river becomes tidal, *Valerianella rimosa, Allium vineale,* and *Potamogeton nitens.* The pollarded willows here harbour an interesting flora, mostly flood-borne (Irish Nat., 1918, 103–4). The fine demesne of Woodstock near by yields *Pinguicula lusitanica, Festuca sylvatica, Hordeum nodosum, Equisetum litorale* (**201**).

This SE part of Ireland is rich in ecclesiastical remains, and to those interested in early architecture the Cistercian abbey at Graiguenamanagh, the much more beautiful similar building at Jerpoint, the Augustinian priory at Inistioge, all in the Nore valley, will appeal, in addition to the many early buildings in Kilkenny.

The botanical tourist visiting the Nore may stay at Maryborough, Abbeyleix, Durrow, Kilkenny, or Inistioge.

H. C. HART: "Botanical Notes along the Rivers Nore, Black-water, etc." Journ. Bot., 1885, 228–233. R. LL. PRAEGER: "Botanical Notes from Inistioge." Irish Nat., 1918, 103–105.

THE LEINSTER COAL-FIELD.

255. An outlier of Upper Carboniferous rocks, resting on the limestone, forms a conspicuous plateau some 300 sq. miles in area between the rivers Nore and Barrow NE of the city of Kilkenny. Some coal is mined here, chiefly about Castlecomer. As in other areas of these rocks—for instance in N Kerry, Clare, etc.—a poor and monotonous flora of calcifuge type prevails, characterized especially by rushy pasture. An old record by John Templeton of *Malaxis* from Old Leighlin is one of the very few notes of plants of any interest.

THE COUNTY OF CARLOW.

256. A small area formed mainly of low undulating granite country, the S end of the great Leinster Chain (**261**); but along the River Barrow (**257**), which flows S down the W boundary, a limestone valley provides a richer flora, and the river continues interesting as it passes S through the gorge which it has cut in the Devonian rocks (**18**).

THE RIVER BARROW.

257. The Barrow rises on Slieve Bloom (**247**), which lies a little SE of the central point of Ireland, and flows at first N, then, when it clears the hills, E, and then S for the greater part of its course. Along with several other Irish rivers, notably its companion the Nore (**252**), it possesses the peculiarity of flowing (save near its source) *towards* instead of away from the higher ground. This is explained by the fact that the Limestone Plain, through which its upper course lies, has been much lowered by denudation, and now extends little higher than the level of the river; but further S, non-soluble granite and Ordovician and Devonian rocks have resisted denudation and stand up high on either side of the valley which the Barrow has cut (**18**). The landscape in the upper part therefore is dull—indeed the stream from the base of Slieve Bloom to Monasterevan flows through one of the flattest areas in Ireland, all marshy meadow and bog (**239**); in the lower part there are high steep banks, often wooded, with lovely scenery. About the place where the Nore and Barrow join the river flows in a grand wooded gorge several miles in length; here it becomes tidal, and steadily broadening, is joined by the tidal Suir (**281**), and the three streams debouch together in the long and broad sea-inlet of Waterford Harbour. Its course from source to open sea is about 130 miles.

258. The botany of the Barrow is not in any way peculiar, and offers little of special interest in its upper course, where the prevailing flora is that of the Central Plain. Lower down, there are more woodland and water-

side plants, and finally a calcifuge flora; the lower reaches
are the more interesting botanically. Among the plants
which occur on the upper waters are *Cystopteris fragilis*
and *Carlina* near the source; *Spergularia rubra, Epipactis
palustris, Ophrys apifera,* and *Equisetum variegatum* (the
last at Tinnehinch Bridge near Slieve Bloom), *Carex
axillaris* (a little lower down), *C. Pseudo-Cyperus* and
Verbena (Mountmellick), *Campanula Trachelium* on the
left bank near Portarlington (an interesting outlier of its
main Irish station along the Nore (**252**)), *Crepis paludosa,
Potamogeton nitens* and *P. densus, Carex diandra,* with
Rhamnus catharticus and *Cystopteris fragilis* at Monasterevan. There is an old record of *Rhamnus Frangula* from
Mountmellick.

Along the middle course of the Barrow from Athy to
where it becomes tidal below Graiguenamanagh, the plants
include *Nasturtium sylvestre, Sagina ciliata, Linum bienne,
Trifolium fragiferum* (very rare inland in Ireland, as at
Leighlinbridge), *Galium uliginosum, Crepis taraxacifolia,
Blackstonia, Verbena, Pinguicula lusitanica, Calamintha
ascendens, Epipactis palustris, Colchicum autumnale* (Carlow, not seen recently), *Ophrys muscifera, O. apifera,
Sagittaria, Potamogeton densus, Lemna polyrrhiza, Scirpus
filiformis, Carex Hudsonii, C. Pseudo-Cyperus, Lastrea
œmula, Equisetum variegatum, Chara aculeolata.* At
Bagenalstown *Arenaria tenuifolia* grows on old walls, the
only instance I know of its forsaking its usual habitat of
railway ballast. The rare *Sorbus latifolia* (**111**) is in a
rocky wood by the river below Graiguenamanagh, I believe
indigenous. About the towns and in tilled land or on
railway banks some local aliens are established, such as
*Lepidium campestre, Fœniculum, Valerianella rimosa,
Picris hieracioides, Brachypodium pinnatum.* At Graiguenamanagh the ground on the W side of the river rises into
heathery hills; Brandon Mountain, a southern outlier of
the Wicklow granite chain (not to be confused with the
Kerry Brandon), rises over the Barrow in a graceful cone
to 1,694 ft. Hereabouts *Hymenophyllum tunbridgense*
occurs, with *Hypericum elodes, Scutellaria minor, Lastrea
montana, Lycopodium clavatum.* The tidal part of the
river that lies between the Barrow-Nore and Barrow-Suir
junctions has yielded some good plants, and will repay
further study. Here there are steep wooded banks alternating with flat alluvial stretches.

259. New Ross forms an excellent centre for this area. *Sorbus latifolia* (**111**) has been found on both sides of the river here, and abundantly at Pilltown, a few miles lower down (where the rare *Carex divisa* also grows). As seen three miles above New Ross on a steep scarp, the concomitants of the *Sorbus* are *Quercus, Ilex, Prunus Avium, Fraxinus, Ulex, Ulmus montana, Sorbus Aucuparia, Prunus spinosa, Corylus,* and one *Acer Pseudo-platanus*— a purely native assemblage save for the last, which is widely spread as a wind-borne alien. *Serratula tinctoria* has its only Irish station a mile above New Ross. Its habitat here is unusual. Just above high water-mark on the rocky steep wooded bank of the tidal Barrow a narrow platform of silt extends with a rank vegetation of *Centaurea nigra, Spiræa Ulmaria, Mentha aquatica, Senecio aquaticus, Leontodon autumnalis, Festuca arundinacea,* and a number of other species (see Irish Nat. Journ., **1**, 188). Among these and in adjoining rock-chinks *Serratula* grows, 2 to 4 ft. high, at intervals for ¼ mile, while the moss *Cinclidotus fontinaloides,* an indication of frequent flooding, extends to 1 ft. above it. About New Ross, *Cochlearia anglica, Geranium rotundifolium, Trifolium fragiferum, Campanula Trachelium, Orobanche Hederæ, Ophrys apifera, Colchicum autumnale, Allium oleraceum,* and *Hordeum nodosum* are also to be seen. *Orobanche major* occurs here and at other places. *Origanum,* usually found on limestone, is abundant over this tract of Ordovician slates, and there is generally a queer mixture of calcicole and calcifuge plants—*Erica cinerea, Ceterach, Sedum anglicum,* etc. The river itself is too muddy for hydrophytes, and *Callitriche,* a little below high water, is the only species seen. Lower down, about Ballinlaw Ferry, are *Spergularia rupicola, Crithmum, Limonium humile, Scirpus filiformis, Hordeum nodosum*; and at Snowhill *Asplenium Adiantum-nigrum* var. *acutum* grows on rocks on the edge of the tidal stream, and *Corydalis claviculata* on White Horse Rock near by. Two other rare ferns, *Asplenium lanceolatum* (**194**) and *Trichomanes radicans* (**192**), occur within the river valley—the former along a lane near Gowlin at the foot of Blackstairs, the latter on a rock on a wooded bank in the same county (Carlow). *Linaria repens* and *Eriophorum latifolium* grow E of Bagenalstown—both very local species.

The explorer of the Barrow, where some good plants may still be expected, will find accommodation at Mountmellick, Portarlington, Athy, Carlow, Bagenalstown, Borris, Graiguenamanagh, New Ross, and Waterford.

H. C. HART: "The Botany of the Barrow." Journ. Bot., 1885, 9–18.

THE RIVER SLANEY.

260. The Slaney rises on the W side of Lugnaquilla (3,039 ft.), the highest of the Wicklow Mountains (**261**), and flows first W through the fine wide Glen of Imaal. Then turning S by Baltinglass and Tullow, it cuts in a fine valley through the Leinster Chain at Newtownbarry, and crosses Co. Wexford to enter the sea below Wexford town (**277**). Its course lies mainly through fertile and pretty country; its botany—not yet thoroughly known—is not remarkable. *Allium vineale* at Aghade, *Lathræa, Cephalanthera ensifolia,* and *Equisetum hyemale* at Ballintemple, *Lamium Galeobdolon* at Kilgarry, *Festuca sylvatica* at Newtownbarry, *Viola hirta* at Clohamon, are among the few rarer plants recorded from the upper reaches. The valleys of the feeders Urrin and Boro, which descend from the high ridge of Mount Leinster and Blackstairs (**271**), yield *Pinguicula lusitanica, Wahlenbergia, Orobanche major, Lastrea æmula, L. montana, Osmunda;* and beyond Camolin *Corydalis claviculata* grows. About **Enniscorthy** *Nasturtium sylvestre, Barbarea verna, Lepidium Draba, Lamium Galeobdolon, Potamogeton densus* are recorded. Then we reach the long tidal portion of the river, from Enniscorthy down, with extensive marshes which will repay further work. At Macmine *Callitriche truncata* (only Irish station), *Crepis paludosa, Allium vineale, Leucojum aestivum* (**161**) (believed native here), *Potamogeton densus,* and *Cladium* grow at sea-level. Below Wexford town the river expands into a broad muddy bay, almost closed by a sand-spit (**277**).

H. C. HART: "A Botanical Ramble along the Slaney and up the East Coast of Wexford." Journ. Bot., 1881, 338–344.

THE COUNTY OF WICKLOW.

261. One of the most beautiful areas in Ireland, and interesting to the botanist, who has a choice of an extensive mountain-mass and a varied shore-line (**272**) yielding a number of rare plants, some of which are confined in Ireland to the Wicklow-Wexford coast.

THE WICKLOW MOUNTAINS.

Almost the whole of the county of Wicklow is occupied by mountains, the main part of the Leinster Chain. These are the result of a very ancient folding (**8**), pre-Devonian in age, which forced the Ordovician rocks into a great elongated dome with a SW and NE axis, and allowed a mass of molten granite to rise underneath them. Sollas has suggested that the granites formed an immense laccolite, with an Ordovician cover and a Cambrian floor. The stripping off of the Ordovician covering by long-continued denudation has left a series of broad granite domes now peat-covered, with the altered Ordovician rocks lapping round their flanks. On the E side the schist area is wide, and cut up by Glacial and other denudative action into picturesque glens and deep valleys, embosoming some lovely lakes. Contrary to what one might expect, the schists appear to be more resistant to the agents of denudation than the granite, and consequently to have had narrower valleys worn into them. In Glacial times the ice, pushing down the wider granite valleys, became congested when the schist area was reached, with the result that its erosive power was increased, and it cut deep trenches into the latter rocks. This is held to account for the deep V-shaped valleys in the schists, and the broader nature of the same valleys in their upper part, where the granite prevails.[3] The schist district, which extends from Bray to Arklow, drained by the Bray River and the Ovoca River, is full of lovely scenery. On the W side of the range, the valleys (in which the Liffey (**237**) and Slaney (**260**) have their origin) are wider, and without the lakes and rich woods which give beauty to the E side, and which vie in charm with the best which the English Lake district can offer. The granite domes, sometimes capped by Ordovician outliers, culminate in Lugnaquilla (3039 ft.), which rises towards the S end of the range. The hills run on to the

[3] A. FARRINGTON: ''The Topographical Features of the Granite-Schist Junction in the Leinster Chain.'' Proc. Roy. Irish Acad., **39**, B, 181–191, Pl. XIV. 1927.

S edge of Co. Wicklow, but the granite continues far down across the low grounds of Carlow, occasionally rising high, as in Mount Leinster (2610 ft.), and Brandon in Kilkenny (1694 ft.). The N end of the chain projects almost to the suburbs of Dublin. This is the largest continuous area of high land in Ireland, the 1000-ft. contour enclosing 205 sq. miles of peat and rock.

262. The typical succession of vegetation is farmland, merging at about 900 ft. into hill-pasture, of which the lower part is much colonized by *Ulex europœus* and the upper part mostly almost monopolized by *U. Gallii.* This gives way, usually about 1250 ft., to *Calluna* moor, which continues to the highest points, *Scirpus cæspitosus* or *Eriophorum angustifolium* becoming dominant on flat summits, much moss (*Racomitrium lanuginosum, Sphagnum* spp., etc.) being present at the higher levels.[4] The *Scirpus* and *Eriophorum* associations, when pure, have each a limited and characteristic flora, as shown by the following :—

Scirpus Association

(Combination of lists from five stations).[5]

Scirpus cæspitosus, 5
Calluna vulgaris, 5
Eriophorum
      angustifolium, 5
Erica Tetralix, 5
Narthecium ossifragum, 5
Sphagnum *spp.*, 4
Drosera rotundifolia, 2

Andromeda Polifolia, 2
Erica cinerea, 1
Empetrum nigrum, 1
Eriophorum vaginatum, 2
Racomitrium lanuginosum, 1
Cladonia rangiferina, 2

Eriophorum Association

(Combination of lists from three stations).

Calluna vulgaris, 3
Eriophorum
      angustifolium, 3
   vaginatum, 3
Empetrum nigrum, 3
Vaccinium Myrtillus, 3

Scirpus cæspitosus, 3
Cladonia rangiferina, 2
Vaccinium Vitis-Idæa, 1
Erica Tetralix, 1
Sphagnum *spp.*, 1

---

[4] See footnote to **235**.
[5] The number following each species shows in how many of the five stations it was present.

The summit of **Lugnaquilla** (3039 ft.), the loftiest point in the area and in eastern Ireland, is remarkable among Irish mountain-tops in presenting a considerable flat expanse, which is continued in several directions in the form of high shoulders. A close felt of vegetation prevails, mostly dominated by *Racomitrium lanuginosum,* with *Vaccinium Myrtillus, Galium saxatile, Juncus squarrosus, Carex rigida, Polytrichum,* and colonies of *Lycopodium alpinum.* H. C. Hart's list of species occurring at 3000 ft. on Lugnaquilla is—*Cerastium vulgatum, Potentilla erecta, Galium saxatile, Calluna, Vaccinium Myrtillus, Rumex Acetosa, R. Acetosella, Anthoxanthum, Agrostis tenuis, Deschampsia flexuosa, D. cæspitosa, Festuca ovina.*[6]

263. The alpine flora of Wicklow is poor, as is usual in Ireland, especially E Ireland. Great tracts of featureless peat-bog, with only a few plants of interest, cover the higher grounds. The flora of that portion of the mountain bogs which lies in County Dublin, 2000–2450 ft., has been studied by Colgan,[7] who remarks that it differs to only a trifling extent from the flora of an Irish lowland bog. His list of species from this upper zone is as follows:—

Potentilla erecta	Juncus effusus
Galium saxatile	Luzula sylvatica
Vaccinium Myrtillus	Scirpus cæspitosus
Vitis-Idæa	Eriophorum vaginatum
Calluna vulgaris	angustifolium
Erica Tetralix	Carex echinata
Melampyrum pratense	Aira præcox
Rumex Acetosella	Deschampsia flexuosa
Empetrum nigrum	Agrostis tenuis
Listera cordata	Festuca ovina
Narthecium ossifragum	Nardus stricta
Juncus squarrosus	Lycopodium Selago

Only occasionally, as particularly on Lugnaquilla, are cliffs of any extent to be met with. There alone grow *Saxifraga hypnoides* and *Sedum roseum,* while *Saxifraga spathularis* (**114**) is found on Lugnaquilla and on Conavalla. The occurrence of the last-named is of very special interest

---

[6] H. C. HART: "On the Range of Flowering Plants and Ferns on the Mountains of Ireland." Proc. Roy. Irish Acad., 3rd s., 1, 512–570. 1891.

[7] N. COLGAN: "Flora of County Dublin," xlviii.

as it is one of the most characteristic of the Lusitanian species of the west and south; its next station is on the Comeraghs in Waterford (see Irish Nat., 1924, 60). On Thonelagee, a rocky scarp impends over Lough Ouler (1829 ft.) and is the only Wicklow station for *Alchemilla alpina* and *Saussurea*—the remaining two Irish stations for the former being in Kerry (**311, 324**). Other plants of "Highland" type less restricted in their local range are *Saxifraga stellaris, Vaccinium Vitis-Idæa, Salix herbacea, Listera cordata, Lycopodium alpinum, Selaginella* (lowland, rare). As on all the Irish mountain-ranges on which it occurs, *Cryptogramme* is extraordinarily rare; five tufts of it, widely scattered, are all that have been found.

In most of the lakes are *Lobelia Dortmanna* and *Isoetes lacustris*; in Upper L. Bray the latter occurs as the var. *maxima* Blytt (*I. Morei* Moore, **206**) with fronds up to 26 inches in length. *Ranunculus scoticus* grows with *R. Flammula* by Lough Dan. *Epilobium angustifolium*, mostly a rare mountain plant where native in Ireland, grows only (in Wicklow) on high cliffs over L. Nahanagan and on Lugnaquilla. *Andromeda Polifolia* is also found high up, mostly favouring deep peat-bogs at about 1500 ft. *Pinguicula lusitanica, Malaxis, Hymenophyllum peltatum, Cystopteris fragilis, Phegopteris polypodioides* and *Lycopodium clavatum* are also usually mountain plants here.

264. Many of the Wicklow glens, cut mostly in the schists which surround the central granite mass, are of exceptional interest. The most widely known is **Glendalough**, famous for the remains of an extensive ecclesiastical settlement, founded by St. Kevin in **617** A.D. A perfect round tower still rises in the centre of the deep mountain valley, and numerous early churches and other buildings provide a feast for the antiquary. The adjoining Glenmalur, though devoid of archæological interest, is even finer from other standpoints. The beautiful valley of the Bray River is referred to later on (**268**). The Scalp, and the Glen of the Downs, both near Bray, are fine examples of Glacial overflow channels, cut N to S across ridges while the W to E drainage was blocked by ice in the Irish Sea.

Ranunculus Lenormandi, Meconopsis, Corydalis claviculata, Viola lutea, Vicia sylvatica, Prunus Padus (? native), *Crepis paludosa, Wahlenbergia, Orobanche major, Mimulus Langsdorffii, Scutellaria minor, Cephal-*

anthera ensifolia, Festuca sylvatica, Lastrea montana, L. œmula, Osmunda, Equisetum variegatum are among the plants (mostly upland) which grow about the skirts of the mountains in varying abundance. *Lastrea spinulosa* is unusually plentiful in woods between Shillelagh and Tinahely. *Drosera anglica, Utricularia intermedia, Eriophorum latifolium, Phegopteris Dryopteris* (an unverified record for Glendalough, **219**) are very rare. *Hieracia* are not abundantly represented: they include *H. lucidulum, H. proximum, H. farrense, H. rubicundiforme, H. strictum,* and *H. dumosum* (**128**). *Sorbus porrigens* is rare.

265. *Saxifraga umbrosa* (the cultivated hybrid), introduced in Altadore Glen near Newtownmountkennedy, has run wild, covering a large area, and having all the appearance of a native. *Carex aquatilis* (**181**) has appeared in recent years on two artificial sheets of water—Glenasmole reservoir (**235**) and the lake at Humewood. The nearest known stations for the plant are in Meath and Westmeath, about 50 miles away, but it is probable that it has migrated from some less distant and as yet undiscovered locality, presumably by the agency of birds. *Nitella gracilis* has its only Irish stations in Lough Dan and Lough Tay, the latter lying in the precipitous hollow of Luggela. An interesting feature of the spring vegetation is the occurrence in abundance of blue-flowered forms of *Anemone nemorosa*. These are specially well developed in the valley of the Ow River above Aughrim, extending W to beyond Ballinabarry Gap; they occur also by the Liffey, as above Ballysmuttan Bridge. In tint they vary from light blue to a colour as deep as and very similar to that of *Anemone Hepatica*. With them other colour-forms occur, white inside with a crimson reverse (see Irish Nat., 1917, 120). (Blue forms have been recorded also from Cappoquin and Lismore in S Ireland, and A. G. More noted plants with sepals of a "rich dark purple" at Loughgall in Armagh.) A large-flowered *Pinguicula*, stated to resemble *grandiflora*, reported as growing freely on the Wicklow hills south of Lugnaquilla (Irish Nat., 1898, 199), invites investigation.

266. Among the more interesting lowland plants of the skirts of the mountains (see also **264**) are *Geranium pyrenaicum, Erodium moschatum, Galium uliginosum* (a Central Plain species), *Senecio erucifolius* (in Ireland confined to a limited area with Dublin as centre), *Crepis taraxacifolia, Lactuca muralis* (a very local alien), *Black-*

stonia, Calamintha ascendens, Mimulus moschatus (naturalized in Powerscourt deer-park), *M. Langsdorffii* (quite
naturalized by streams), *Orobanche Hederæ, O. major,
Mentha Pulegium, Lamium Galeobdolon, Scleranthus
annuus, Juncus macer* (several stations in the Aughrim
district[8]), *Carex strigosa, C. Bœnninghauseniana, Asplenium
lanceolatum* (wall at Glendalough, most northerly station
in Ireland), *A. Adiantum-nigrum* var. *acutum* (**195**)
(between Glendalough and Rathdrum, and at Dunran
Wood), *Lycopodium inundatum* (shore of Lower Lake,
Glendalough, very rare in Ireland). Old records for
Geranium sanguineum (Killincarrig and Bray, 1833),
Trichomanes radicans (Powerscourt waterfall, 1806, and
Hermitage Glen, 1825), and *Lastrea Thelypteris* (Glencree,
1825) should be borne in mind by the explorer.

The botany of the River Liffey, which has the upper
part of its course within the County of Wicklow, has been
referred to already (**237**). The Slaney also (**260**) rises in
Wicklow, on the W side of Lugnaquilla.

Along the valleys of the main eastern river-system (the
Ovoca and its tributaries), from Arklow up to Glendalough
and Aughrim and on towards Shillelagh, extensive woods
are found still mainly aboriginal, with *Quercus* as the
leading tree. A wood near Glendalough (on the way to
Glenmalur) is pure *Quercus*, with *Corylus* and *Ilex* as a
second stratum and a limited ground flora of *Oxalis,
Vaccinium Myrtillus, Digitalis, Melampyrum pratense,
Teucrium Scorodonia, Luzula sylvatica, Athyrium, Lastrea
Filix-mas, L. montana, L. aristata,* and *Blechnum.* An oak
wood in Glenmalur has a ground vegetation of practically
only *Teucrium* and *Oxalis.*

267. Bray Head, projecting boldly into the Irish Sea
15 miles S of Dublin, is the most easterly spur of the
mountains, and is much older than the main mass. It is
composed of Cambrian slates and quartzites, and is the
seaward termination of a very ancient ridge which here
breaks the continuity of the Silurian and granitic rocks,
and forms the Great Sugarloaf and Little Sugarloaf to the
W—two pointed hills of quartzite which give character to
the country S of Dublin, contrasting sharply with the broad
domes of the newer granite hills behind. Bray Head
harbours some uncommon plants on its precipitous face—

[8] J. P. Brunker in Irish Nat. Journ., 4, 221. 1933.

Corydalis claviculata, Spergularia rupicola, Ornithopus perpusillus, Trifolium scabrum (both the last extremely rare in Ireland), *T. striatum, T. filiforme, Vicia lathyroides, Rubia peregrina, Euphorbia portlandica.* The thin skin of peaty soil with occasional pockets of loam and outcrops of rock support a calcifuge flora very typical of such ground throughout Ireland :—

Viola Riviniana
Oxalis Acetosella
Ulex europæus
  Gallii.
Cratægus Oxyacantha
Sorbus Aucuparia
Sedum anglicum
Cotyledon Umbilicus-
  Veneris
Ilex Aquifolium
Galium saxatile

Vaccinium Myrtillus
Calluna vulgaris
Erica Tetralix
  cinerea
Teucrium Scorodonia
Betula pubescens
Scilla non-scripta
Festuca ovina
Pteris aquilina
Lastrea aristata
      etc.

268. The richly wooded valley of the **Bray River** is interesting. Though carved out of non-calcareous rocks (Ordovician near the sea, granite higher up) it yields a number of calcicole plants, the result of shelly drift and limestone detritus pushed up the valley by ice from the N : these give way above Enniskerry to a strictly calcifuge flora in the upper part of the valley, which terminates abruptly in Lough Bray (Upper and Lower, 1,453 and 1,225 ft.), lying in Glacial coombs—the former the home of the long-leaved *Isoetes lacustris* var. *maxima* (**206, 263**). A deep branch valley to the S contains Powerscourt deer-park, where is situated the high waterfall where *Trichomanes* was found for the first time in Ireland (**192**). It has been long extinct there.

Other plants of the valley (mostly about Bray and Enniskerry) are *Linum bienne, Geranium pyrenaicum, G. pusillum, Erodium moschatum, Rosa stylosa, Senecio erucifolius, Mimulus moschatus* (naturalized at Powerscourt deer-park), *M. Langsdorffii, Verbena, Mentha Pulegium, Calamintha ascendens, Lamium Galeobdolon, Ophrys apifera, Carex strigosa, C. Bænninghauseniana* (Kilruddery). Some of these have been already referred to (**266**) when mentioning the plants of the skirts of the Wicklow Mountains.

269. Considering the remarkable beauty of its scenery, which is comparable to the best of the English Lake District or of Wales, Wicklow is (perhaps fortunately) not over-supplied with hotels. There is no part of the area, however, which cannot be visited by a one-day motoring-*cum*-walking trip; and good accommodation will be found at Bray, Enniskerry, Woodenbridge, Ovoca, Glendalough, Wicklow, Ashford, Rathdrum, etc. **Glendalough**, famous alike for its scenery and its early Christian antiquities (round tower, stone-roofed oratory, many primitive churches, etc.), is a good centre for exploring Glenmalur, Glendasan, Glenmacnass, L. Dan, L. Ouler, and all central Wicklow. All these lakes and streams drain SE down the lovely Vale of Ovoca to meet the sea at Arklow (**276**). Lugnaquilla and most of the higher summits of Wicklow are within striking distance of Glendalough. Other central places where one may stay are Ovoca, Roundwood, Lough Dan, and Glenmalur. The coast (**272**) may be explored from Bray, Greystones, Wicklow, or Arklow; the W side (where in the N the Liffey and in the S the Slaney have their source) and southern areas are less frequented, but accommodation will be found at Dunlavin, Tullow, Shillelagh, Aughrim, and Tinahely.

Literature.—The Wicklow mountain district, as such, has very little botanical literature — a curious fact, considering its interest and beauty, as well as its proximity to Dublin, ever the centre of Irish field botany. Most of the records of its rarer plants, whether old or recent, will be found in the two editions of ''Cybele Hibernica''; in MOORE and MORE: ''Catalogue of the Flowering Plants and Ferns of Dublin and Wicklow'' (reprinted with corrections from British Association Guide, 1878), in Sci. Proc. R. Dublin Soc., N.S., **1**, 190–227, 1878; and in ''Irish Topographical Botany'' and its Supplements. A number of upper limits of species are recorded in H. C. HART's paper ''On the Range of Flowering Plants and Ferns on the Mountains of Ireland'' (Proc. Roy. Irish Acad. (3), **1**, 512–570, 1891), and the northern end of the district is described in PETHYBRIDGE and PRAEGER: ''The Vegetation of the District lying south of Dublin'' (Proc. Roy. Irish Acad., **25**, B, 124–180, 5 pl., coloured map). See also J. P. BRUNKER in Irish Nat., **33**, 114–6.

THE COUNTY OF WEXFORD.

270. Wexford includes the marked SE angle of Ireland.
It is a fertile area, mostly of Cambrian rocks, often hilly,
but devoid of mountains save on its W margin, where
Mount Leinster and Blackstairs (**271**) form a high granite
ridge.   The River Slaney (**260**) crosses the county to reach
the sea in the broad estuary of Wexford Harbour (**277**),
and in the SW the R. Barrow (**259**), opening into the large
submerged valley of Waterford Harbour, offers some good
ground.   But botanical interest centres on the coast, where
many rare maritime plants are congregated (**276–279**).
The large inland area E of Enniscorthy and Wexford is
generally well farmed, and its botany so far as known
offers little of interest.

Wexford is the Irish headquarters of *Erythræa pulchella,*
a local and mostly maritime plant, and also of *Ranunculus
parviflorus,* an ''English'' type plant which may be a
natural immigrant to this part of Ireland; and it is the
only county from which *Nuphar* is absent—a culmination
of the thinning out of both it and *Nymphæa* in SE Ireland.
Parnassia is another unexpected absentee.

271. Mount Leinster and Blackstairs.—These fine hills
(2,610 and 2,409 ft. respectively) form a N–S ridge, an
outlier of the granite chain of Wicklow, and they separate
Carlow and the Barrow basin from Wexford and the valley
of the Slaney.   Their heathery slopes are steep, but there is
very little rock, and the flora is poor.   Ordovician strata
on the E side of the summit furnish a home for *Saxifraga
stellaris.*   On both mountains *Carex rigida* grows, with
Lycopodium clavatum and *Hymenophyllum peltatum*
lower down; Blackstairs alone harbours *Vaccinium Vitis-
Idæa.*   *Hypericum elodes, Pinguicula lusitanica, Lastrea
montana,* etc., are found about the base of the ridge.

Pinguicula grandiflora (**145**), planted at Killanne at the
foot of Blackstairs (half a dozen roots) in 1879, was
reported to have increased to 27 plants in 1896 (Irish Nat.,
1896, 212).   C. B. Moffat states that in June, 1913, the
colony had increased to 95 plants, but that on revisiting
the spot—the marshy margin of a stream flowing across a
rough field—in June, 1920, no plants could be found.   The
cause of its disappearance is obscure.

H. C. HART: ''Notes on the Plants of some of the Mountain
Ranges of Ireland.''   Proc. Roy. Irish Acad., Ser. 2, **4** (Science),
236–8.   1884.

THE SOUTH-EASTERN COAST.

272. The SE coast of Ireland is characterized by great stretches of sand and gravel, sometimes backed by low cliffs of Boulder-clay, but seldom interrupted by rock. From Bray Head in Wicklow down as far as Waterford— a distance of more than 100 miles—stretch after stretch of shingle and sand extend, often with marsh or salt-marsh behind, as at the Murrough of Wicklow, and sometimes nearly closing the mouths of inlets still or once marine, as at Lady's Island Lake, Tacumshin Lake, and Bannow Bay, all in Wexford. The peculiarities of the flora of this region have in part an edaphic cause—the unrivalled habitat afforded by these lonely stretches for strand plants, which number among them some species especially sensitive to human interference. But climatic influences and geographical position also without doubt play a part, for this region most approximates, both as regards climate and position, to the coast of the S part of Britain and to Wales, which lies only 50 to 70 miles to the E. These Wicklow-Wexford shores form the focus in Ireland of the strand flora. Several plants find here their only Irish station ; others, while spreading slightly beyond the boundaries mentioned, have here their headquarters ; and others again, while ranging N to Louth or Down or S to Cork or Kerry, still find here their centre of distribution.

273. Beginning with the species of most restricted range, we have *Euphorbia Peplis* (**151**), once (1839) found at Garraris Cove near Tramore, Co. Waterford (in Britain a very rare SW species) ; *Trifolium subterraneum*, in Ireland growing only at Wicklow (in Britain widespread in S England) ; *T. glomeratum* (**102**), which grows with it at Wicklow, is elsewhere found only at Brittas Bay a few miles S of Wicklow, and at Rosslare (in Britain it also is southern English). The puzzling *Equisetum Moorei* (**203**) is known nowhere but on the coasts of Wexford and Wicklow ; *Diotis maritima* (Waterford and Wexford, **123**) and *Asparagus maritimus* (Waterford, Wexford, and Wicklow, **162**) are both in Britain southern and south-western, the former now nearly extinct. Then there are *Carex divisa* (Wexford and Dublin), *Glyceria Borreri* (Waterford, Wexford, Dublin, **189**), *Ornithopus perpusillus* (W Cork, Wexford, Wicklow, Dublin, Down), *Trifolium scabrum* (E Cork, Wexford, Wicklow, Dublin), *Trigonella* *ornithopodioides* (Cork to Down), *Inula crithmoides* (Kerry to Dublin), *Juncus acutus* (W Cork to Wicklow, **166**). A few non-maritime plants conform to the same type of distribution *e.g., Festuca uniglumis* (Wexford to Louth), *Asplenium lanceolatum* (**194**) (Kerry to Wicklow).

274. One or two of the rarest plants found on this coast have a second Irish station well outside the district, giving a quite discontinuous range. Thus *Matthiola sinuata* (**84**) is confined to Wexford and Clare (with a doubtful old record from Kerry), and *Scirpus nanus* (**174**) grows only at Arklow in Co. Wicklow and near Ballybunnion in Kerry. *Chara connivens* and *Tolypella nidifica* (**207**), the latter at present known only from Wexford, will probably be found elsewhere when the Charophyta are better worked.

Among the maritime plants, the following are scattered along these shores and most of them are frequent, some abundant :—

Thalictrum dunense
Glaucium flavum
Cochlearia anglica
Raphanus maritimus
Viola Curtisii
Spergularia rupicola
Linum bienne
Geranium columbinum
Erodium moschatum
　maritimum
Crithmum maritimum

Rubia peregrina
Dipsacus sylvestris
Carlina vulgaris
Blackstonia perfoliata
Calystegia Soldanella
Limonium binervosum
Euphorbia Paralias
　portlandica
Epipactis palustris
Scirpus filiformis
Hordeum nodosum.

Much of the coast is remote from railways and hotels, and before the days of the motor-car was difficult to work. As most of the exploration was done in pre-motor days, it follows that further examination under the more favourable conditions of to-day is desirable.

275. The Murrough, the most approachable part, presents a twelve-mile gravel-beach backed by marshes and lagoons. The railway from Greystones to Wicklow runs along its crest. *Mertensia maritima* has (or had) here its most southerly Irish station, and the Wicklow end is the habitat of *Trifolium subterraneum* and *T. glomeratum* (referred to above), which grow on gravelly raised beach with *T. striatum, T. filiforme, Trigonella ornithopodioides,* and *Scilla verna,* the last an E coast plant which finds its S limit at Rockfield, a few miles S of Wicklow, where

Allium vineale also grows. *Papaver Argemone* and *Geranium columbinum* occur. Half-way along the Murrough *Crambe* flourished formerly, but has not been seen for sixty years; *Calamintha Acinos* also appears now gone.

The marshes which lie behind the beach are very wet, with much standing water. There are large groves of *Cladium* (one of its rare E coast stations), *Carex riparia, C. Hudsonii, Eriophorum angustifolium*; of *Juncus bulbosus* + *Ranunculus Flammula* and of *J. bulbosus* + *Carex Goodenovii*. *Epipactis palustris, Orchis majalis,* and *Carex Pseudo-Cyperus* are present. *Potamogeton coloratus* is in drains, where *Chara aculeolata* also occurs: the presence of· these two hydrophytes of calcicole tendencies and of *Juncus subnodulosus* suggests that the marshes partake of the nature of fen. *Rhamnus catharticus, R. Frangula, Lathyrus palustris* and some other species recorded from the Murrough have not been seen for many years. Where salt-marsh supervenes, *Cochlearia anglica, Trifolium fragiferum, Atriplex portulacoides,* etc. are found.

The **Vartry River**, which debouches at Wicklow, rises on the bleak moorland of Calary Bog S of the Great Sugarloaf. The flattish valley which succeeds encloses the reservoirs which form the water-supply of Dublin City. The fine gorge of the Devil's Glen follows, harbouring *Meconopsis, Prunus Padus, Sorbus porrigens,* etc. At Rathnew is the sheltered garden of Mount Usher, where plants like *Todea superba* and those mentioned in **303** flourish in the open air. Wicklow Head with its three lighthouse towers projects boldly two miles S of the town of the same name : *Vicia lathyroides* grows with *Trifolium striatum* on its rocky summit.

276. Arklow was for a long time the only known habitat in Ireland of *Scirpus nanus*[9] (**174**), since found in Kerry.[10] *Vicia lathyroides* is on the sand-dunes here, also *Equisetum Moorei* (**203**) which has many stations along the coast to the N and S; and *Eleocharis uniglumis* occurs. The Ovoca River, draining the most picturesque part of Wicklow, debouches here into the Irish Sea. *Montbretia*

[9] A. G. MORE: "Discovery of *Scirpus parvulus* R. & S. in Ireland." Journ. Bot., 1868, 254. See also A. W. STELFOX in Irish Nat. Journ., 1926, 75.

[10] R. W. SCULLY: "Plants found in Kerry, 1889." Journ. Bot., 1890, 110–116.

Pottsii has run wild along the last few miles of its course. *Asplenium lanceolatum* (**194**) has recently been found on Arklow Head, and *Asparagus* grows S of Mizen Head. At Kilmichael Point, a few miles further S, across the Wexford border, one of the rarest of Irish plants, *Matthiola sinuata* (**84**), grows vigorously on low Boulder-clay cliffs thinly covered with sand (Irish Nat. Journ., 1926, 96). On sandhills at Kilgorman a remarkable scandent form of *Calystegia Soldanella* is abundant, with elongated stems making up to six convolutions round the stems of *Ammophila*, etc., and thus climbing to a height of 1 to 1½ ft. (Journ. Bot., 1932, 50). *Chenopodium rubrum* is here too.

Courtown Harbour, a little further S, easily reached from **Gorey**, is a pretty spot among the dunes in the middle of the *Equisetum Moorei* district, with masses of naturalized *Hippophae* binding the sands, and *Thalictrum dunense, Calystegia Soldanella, Ophrys apifera, Carex strigosa*, etc.; *Ruppia maritima* is in pools.

277. Wexford and **Wexford Harbour** offer a fine variety of tidal river, salt-marsh and sand, with *Trifolium fragiferum, Galium uliginosum, Calamintha ascendens, Stachys officinalis, Chenopodium rubrum, Potamogeton panormitanus, Eleocharis uniglumis, Scirpus rufus, Glyceria distans,* the rare *G. Borreri* (**189**), *Elymus, Equisetum Moorei* (**203**), *Chara connivens, C. canescens, Tolypella nidifica* (**206**) (only Irish station at present known for the last—see Irish Nat., 1917, 134-5).

The extensive marshes by the R. Slaney above Wexford are interesting ground and will repay further study (**260**).

The lonely shore extending N from Wexford Harbour to Cahore Point yields in several or many places *Matthiola sinuata* (**84**), *Thalictrum dunense, Spergularia rubra, Euphorbia Paralias* (**152**), *E. portlandica* (**153**), *Juncus acutus* (**166**), *Festuca uniglumis, Elymus, Equisetum Moorei* (**203**).

278. Rosslare (hotel), the packet harbour for Fishguard, on the railway S of Wexford, is an attractive place; the low projection on which it stands forms the well-marked SE corner of Ireland. Close to Rosslare the American *Sisyrinchium californicum* (**160**) has its only European station in flat damp fields, where it grows in great profusion. It was claimed as native by the finder, E. S. Marshall, who studied its habitat here throughout a number of years (see Journ. Bot., 1898, 47), but is usually looked on as having been introduced in some unexplained way. *Trifolium glomeratum* (**102**) is here too in its second Irish station, accompanied by *T. striatum, Trigonella, Sagina ciliata, Linum bienne, Salvia horminoides, Erythræa pulchella, Chenopodium rubrum, Ophrys apifera, Lemna polyrrhiza, Elymus. Atriplex maritima* and *Erodium moschatum* also occur. *Asplenium lanceolatum* (**194**) grows close to Carnsore Point near by, also *Trifolium scabrum*, and at Churchtown *Lemna gibba. Œnothera odorata* is established.

At **Carnsore** the coast, which for some 130 miles has run almost due N and S, turns abruptly W. A patch of granite at Carnsore Point acting as a "sort of resisting corner-stone," has protected the softer rocks behind it from the inroads of the sea.

Rosslare makes the best base for exploration of the interesting though desolate shore which stretches W, past **Lady's Island Lake** and Tacumshin Lake—shallow inlets of the sea which have become closed or almost closed by great bars of sand and gravel. The beach which dams the former is the headquarters in Ireland of *Diotis maritima* (**123**), which may be seen growing over a mile or so of sand—a very remarkable sight (Plate 7). It forms silvery bushes spaced rather close together on almost bare wind-swept sand, its concomitants being scattered plants of *Ammophila, Calystegia Soldanella*, and *Euphorbia Paralias*.[11] At Lady's Island Lake *Rumex maritimus* flourishes, and *Polygonum nodosum* occurs in some quantity—the only place in Ireland where it has the appearance of a native. *Zostera nana* is also recorded.

279. At Crossfarnoge (whose appropriate alternative name is Forlorn Point), a ridge of rock running N and S breaks the monotony of the endless beach and produces a brief change of scenery and flora. Here modest accommodation may be had (at Kilmore Quay). The ridge continues seaward and forms the Saltee Islands (**280**), 3 and 4 miles out. From Crossfarnoge another great stretch of beach with strangled sea-inlets leads to the long rocky projection of Hook Head and to Waterford Harbour; *Solanum*

---

[11] C. P. HURST: "On the Range of Diotis candidissima in England and Wales, and in Ireland." Manchester Memoirs, 46, 1–8, 2 plates. 1902.

PLATE 7.

DIOTIS MARITIMA Cass.

At Lady's Island Lake, Co. Wexford. *Euphorbia Paralias* and *Ammophila* are also seen.

W. Andrews & Son, Photo.

278.

PLATE 8.

H. G. Tempest, *Photo.*

COUMSHINGAUN, CO. WATERFORD.

Looking across the lake to the 1000-ft. cliff at the head of the cirque.

nigrum seems well established on this part of the coast. The sands at Ballyteige and Bannow Island furnish habitats for *Asparagus* (**162**) and *Chenopodium rubrum*; *Trifolium scabrum* and *Mentha Pulegium* grow near Fethard on the way to Hook Head, where *Inula crithmoides* and *Limonium binervosum* are abundant on limestone rocks.

H. C. HART: "A Botanical Ramble along the Slaney and up the east coast of Wexford." Journ. Bot., 1881, 338–344; "Report on the Flora of the Wexford and Waterford Coasts." Sci. Proc., Roy. Dublin Soc., N.S., 4 117–146 (1883), 1885. G. E. H. BARRETT-HAMILTON and L. S. GLASCOTT: "Plants found near New Ross, Ireland." Journ. Bot., 1889, 4–8; "Plants found near Kilmanock, Co. Wexford." *Ibid.* 1890, 87–89. G. E. H. BARRETT-HAMILTON and C. B. MOFFAT: "Notes on Wexford Plants." Journ. Bot., 1892, 198–200; "The Characteristic Plants of Co. Wexford." Irish Nat., 1892, 181–3. E. S. MARSHALL: "Irish Plants collected in June, 1896." Journ. Bot., 1896, 496–500: "Some Plants observed in Co. Wexford, 1897." Journ. Bot., 1898, 46–51.

280. The Saltees.—There are two small islands, the Great and Little Saltee, both with a rocky shore of gneiss and a drift-covered surface rising 100 to 200 ft. The Great Saltee (216 acres), formerly cultivated, has been derelict for about 20 years. Both are famous sanctuaries for sea-birds of many kinds. The interest possessed by the vegetation centres in the larger island. Here the flora is still in course of reorganization after a period of tillage, and is being altered also by the encroachment of the enormous bird-colonies—Puffins, Razorbills, Guillemots, Herring Gulls, Lesser Black-backed Gulls, Manx Shearwaters—and also by the great increase in the rabbit population. The tramping and guano kill out many species, but in the bird-colonies a few flourish exceedingly, such as *Spergularia rupicola*, *Silene maritima*, *Rumex Acetosa*, *R. Acetosella*, which form solid masses of almost pure cultures. *Pteris* with a dense undergrowth of *Scilla non-scripta* (with scapes bearing up to 50 flowers) occupies many acres where the bird population is thinner. On the area of former tillage *Potentilla Anserina*, *Carex arenaria* (there is no sand on the island), *Erodium maritimum* form each dense swards of considerable extent.

The more interesting plants are mostly of maritime habitat —*Spergularia rupicola*, *Erodium maritimum*, *Crithmum*, *Inula crithmoides*, *Scirpus filiformis*, *Asplenium marinum*; and a few others—*Hypericum elodes*, *Cotyledon Umbilicus-*

Veneris, Sedum anglicum—are, like the majority of the above, of Watson's "Atlantic" type—a not unexpected feature, in view of the proximity of the Saltees to western Britain.

Ranunculus parviflorus is the only non-native plant of interest.

H. C. HART: "Report on the Flora of the Wexford and Waterford Coasts." Sci. Proc. Roy. Dublin Soc., N.S., **4**, 117–146. 1883.
R. LL. PRAEGER: Phanerogamia, in "Notes on the Flora of the Saltees." Irish Nat., 1913, 181–191.

THE RIVER SUIR.

281. The Suir, the third of the trio of south-eastern rivers, rises on the E slope of Devil's Bit (**251**) in Tipperary within a couple of miles of the source of the Nore (**252**), and flows down a mountain glen and out into the low-lying limestone plain, where its course lies S across Tipperary (**249**) and past the E end of the Galtees to the foot of the Knockmealdowns (**284**). There it turns abruptly N for a few miles, following the edge of the limestone, and then E down a limestone trough, becoming tidal at Carrick-on-Suir; it passes Waterford, cutting through a barrier of Ordovician and Devonian rocks to meet the combined Barrow and Nore a few miles further on, and 120 miles from its source. A good deal of its middle course, as about **Thurles**, is very flat, and liable to inundation. Here are found *Trifolium fragiferum* (in one of its few inland localities in Ireland), *Galium uliginosum, Leucojum æstivum* (**161**), *Eriophorum latifolium* (at Holycross). Further down, as at Golden, and from Cahir to Clonmel, where the Galtees, Knockmealdowns, and Comeraghs rise to W, S, and E, the scenery is very beautiful. *Orobanche Hederæ* is at Cahir. As in the case of the Barrow and Nore, detailed botanical work along the Suir is still required; at present few plants of interest are on record from its valley. *Euphorbia hiberna* (**154**) attains its NE limit in Ireland about Clonmel and the valley of the Nier, and *Rosa stylosa* has been found at Newcastle.[12] *Nasturtium sylvestre, Linum bienne, Geranium columbinum, Calamintha ascendens, Ceratophyllum demersum, Allium vineale, Potamogeton densus, Lemna polyrrhiza, Lastrea*

[12] Crépin in Bull. Soc. Bot. Belgique, **31**, pt. 2, 149.

montana, Lycopodium clavatum have been found near **Clonmel**; the same *Geranium* and *Lemna* and also *Juncus diffusus* and *Butomus* at **Carrick-on-Suir**, which harbours a local weed in *Mercurialis annua*: *Cephalanthera ensifolia* grows at Curraghmore.

The lower tidal portion of the river yields *Cochlearia anglica* (**86**), *Trifolium fragiferum, Potamogeton densus, Scirpus filiformis, Lepturus,* and so on; and *Juncus acutus* (**166**) on the N bank below Waterford.

Holycross, one of the most beautiful of Irish abbeys, is situated on the river-bank below Thurles, and at Cahir and elsewhere imposing old castles still stand guard. The **Rock of Cashel**, on which rises the most famous group of early Christian edifices in Ireland, is only a few miles distant from the river near Golden; a sharp fold of the limestone, forming a rocky knoll, has been seized on by the old architects, who have crowded it with beautiful buildings in the Hiberno-romanesque style.

The explorer of the course of the Suir will find accommodation at Thurles, Cahir, Clonmel, Carrick-on-Suir, and Waterford.

H. C. HART: "On the Botany of the River Suir." Sci. Proc. Roy. Dublin Soc., N.S., 4, 326–334, 1885.

THE COUNTY OF WATERFORD.

282. A fine varied county, formed of several types of rock, but, like Wexford, with very little limestone. In the E, the city of Waterford stands, with many miles of tidal waters. The coast (**286**) is often precipitous, with several muddy inlets blocked by sand-bars. In the W the tidal Blackwater (**287**) runs seaward through a fine gorge to Youghal (**289**). The mountain-group of the Comeraghs (**285**), with great cirques and tarns, lies wholly in the county, and the less interesting Knockmealdown group (**284**) forms the boundary to the NW. The occurrence locally of *Saxifraga spathularis* (Comeraghs) and *Euphorbia hiberna* (Dungarvan, etc.) is a foretaste of the Lusitanian element which becomes so marked a feature of the flora further west.

WATERFORD.

283. The City of Waterford is situated strategically by the junction of the Suir (**281**), Nore (**252**), and Barrow

(**257**)—the trio of rivers which drains almost the whole SE portion of Ireland—and close to the point where the counties of Waterford, Kilkenny, and Wexford meet. Being served with railways in six directions it makes an excellent centre, but apart from the flora of the rivers just mentioned and estuary its immediate environs offer little to the botanist. The land is mostly tilled, and there are frequent rocky places with a dry calcifuge flora with much *Erica cinerea*. *Leucojum æstivum* (**161**) grows in Kilbarry Bog. The most attractive spot is **Tramore** (hotels), on the coast 8 miles S. Here a great stretch of sand-dunes encloses a shallow bay; on the other side cliffs rise which continue for many miles. The sands are the home of *Asparagus maritimus* (**162**) and *Cuscuta Epithymum*; also of *Brachypodium pinnatum*, in Ireland often a railway-bank plant, but in this station above suspicion (Irish Nat., 1898, 253). These grow among a vast profusion of *Rubus cæsius*. An old note of *Asperula cynanchica* awaits verification. *Ligustrum*, indigenous here, grows abundantly, quite prostrate, in large patches. On the salt-marshes are *Erythræa pulchella, Trifolium fragiferum, Juncus acutus* (**166**). A puzzling record—discredited though backed by specimens—for *Erica vagans* is attached to Islandicane, on the coast a few miles W (**211**). The plant is not there now. The ghost of *Euphorbia Peplis* (**151, 215**) haunts Garraris Cove close by: it has not been seen here (its only Irish station) since 1839. Other inhabitants of this rocky shore are *Linum bienne, Lavatera* (native), *Rubia peregrina, Crithmum, Dipsacus sylvestris, Inula crithmoides, Limonium binervosum, Spiranthes spiralis.*

The fine sea-inlet of Waterford Harbour, into which the combined Barrow, Nore, and Suir discharge, has few plants recorded from it: but *Trifolium glomeratum* (**102**) has one of its rare Irish stations near Arthurstown.

H. C. Hart: "Report on the Flora of the Wexford and Waterford Coasts." Sci. Proc. Roy. Dublin Soc., N.S., **4**, 127–128. 1885.

KNOCKMEALDOWN.

284. Knockmealdown forms a bold E–W range (highest point, 2609 ft.) between the valleys of the Suir and Blackwater—an anticline of Old Red Sandstone, breaking through the Carboniferous limestone of the low grounds and presenting from the N a highly picturesque row of steeply-rising conical peaks, on the S dropping in a long heathery slope to the Blackwater. A good road running N from Lismore on the latter river to Cahir climbs across the centre of the ridge and provides easy access. These hills belie their promise. There are no cliffs or lakes, only heathery slopes, and interesting plants are few, repeating with many omissions the flora of the Comeraghs (**285**) which adjoin on the E. *Saxifraga spathularis* (**114**), *S. sponhemica, Salix herbacea, Listera cordata, Carex rigida, Hymenophyllum peltatum* occur, all running up to over 2500 ft. Probably a few other species of interest will yet be found.

H. C. HART: "Notes on the Plants of some of the Mountain Ranges of Ireland." Proc. Roy. Irish Acad., Ser. 2, 4 (Science), 231–6, 1884.

THE COMERAGHS.

285. The Comeraghs, lying in Co. Waterford, are the most easterly of the several mountain-groups, formed of inlying masses of Old Red Sandstone and Silurian rocks, which, with the great rounded mass of Slievenaman (2364 ft.) immediately to the N, break up the continuity of the Limestone Plain along its S margin. Massive horizontal beds of slate and conglomerate form a table-land of about 2000 ft. elevation, in the edges of which the ice of the Glacial Period has excavated a series of great coombs embosoming tarns, with cliffs up to 1000 ft. high and moraine-choked valleys. The best botanical ground centres around these tarns, all of which lie close to the highest ridge (2597 ft.) and within a couple of miles of **Coomshingaun** (plate 8), the finest and best known of the cirques. This last is easily reached from the Dungarvan road from Carrick-on-Suir, which town forms the best base.

The flora is poor save on the cliffs; there, though not rich in alpines, it is luxuriant and interesting. Of the Lusitanian species so characteristic of Kerry and Cork, *Saxifraga spathularis* (**114**) alone comes so far E: it is not uncommon in a rather small strongly serrate-leaved form. *Saxifraga stellaris, S. hirta* and *S. sponhemica, Salix herbacea, Vaccinium Vitis-Idæa* (here reaching its S limit in Ireland), *Hymenophyllum peltatum* are characteristic cliff species: at Coomshingaun they are joined by *Meconopsis, Sedum roseum, Crepis paludosa, Hieracium, anglicum, Cystopteris fragilis.* On the high moors *Carex rigida* is common, with *Listera cordata* rather lower down.

Isoetes lacustris is in Coomduala L.; *Lastrea æmula* and *L. montana* occur round the flanks of the hills.

H. C. HART: "Notes on the Plants of some of the Mountain Ranges of Ireland." Proc. Roy. Irish Acad., Ser. 2, 4 (Science), 231–6. 1884.

THE WATERFORD COAST.

286. The coast of County Waterford, stretching from Waterford city and the mouth of the Barrow to Youghal and the mouth of the Blackwater, is mostly bold, fine ranges of cliffs having been carved by the Atlantic out of the rhyolites and other volcanic rocks which prevail in the E part, and out of the Old Red Sandstone which occupies the W. In the middle, at Dungarvan, a limestone trough (occupied by the Blackwater (**287**) except in its lower part) comes in from the W, resulting in low land and a deep bay, across the inner part of which the ocean has thrown a narrow sandy spit. The coast is beautiful and lonely, as its precipitous character does not encourage the presence of villages. On the long stretch from Tramore (**283**) to Dungarvan, *Rubia peregrina* and *Raphanus maritimus* are abundant; with them on the cliffs are *Linum bienne, Lavatera arborea* (indigenous on sea-stacks), *Vicia sylvatica, Dipsacus, Inula crithmoides, Ligustrum* (frequent and native), *Limonium binervosum, Orobanche Hederæ, Osmunda,* with a great show of *Silene maritima, Ulex Gallii, Anthyllis, Armeria,* and *Erica cinerea.* A similar flora colonizes the rocks from Dungarvan to Youghal (**289**).

H. C. HART: "Report on the Flora of the Wexford and Waterford Coasts." Sci. Proc. Roy. Dublin Soc., N.S., 4, 117–146. 1885.

Dungarvan, with its sheltered bay bisected by a long sand-pit, offers considerable variety of ground. Its flora includes *Euphorbia hiberna* (**154**) in its most easterly Irish station. Here also (at Ballynacounty) *Glyceria Borreri* (**189**), very local both in Ireland and in Great Britain, has been found recently. *Cochlearia anglica* grows in salt-marshes, *Geranium pusillum* and *Carex Pairœi* on limestone near Cappagh, a few miles westward. An old record (1845) for *Diotis* (**123**) from the sands at Dungarvan should be kept in mind by the visiting botanist.

THE BLACKWATER.

287. The Blackwater rises on Knockanefune (1441 ft.), the highest point of the wide expanse of bog-covered uninteresting hills, formed of Upper Carboniferous shales, etc., which occupy the Castleisland area of Kerry and the adjoining parts of Cork and Limerick. Flowing first S, it soon assumes the eastern course characteristic of the rivers of southern Ireland (**18**). Finally turning S again at Cappoquin, it reaches the sea at Youghal after a course of 85 miles. In its upper course, S of the turn from S to E, about **Millstreet**, there are high hills (Caherbarnagh, 2239 ft., The Paps, 2273, 2284 ft.) the N end of the mountain mass of central Kerry. (The Paps, as seen from the E, form the loveliest pair of cones in Ireland.) *Hieracium hypochœroides* grows here with other Hawkweeds (on cliffs over Gurtavehy L.), and *Equisetum variegatum* (typical!) occupies a cliff-ledge at the unusual elevation of 1500 ft. Save for the presence of a few of the more widely-ranging Lusitanian species—*Pinguicula grandiflora* (**145**) at Gurtavehy L. and *Euphorbia hiberna* (**154**)—and of plants such as *Osmunda*, the river-valley has little interest until Mallow is reached. Here, and again lower down at Cregg Castle, *Allium Scorodoprasum* grows in thickets, looking native—a plant in Ireland almost confined to Kerry and Cork, where it is not infrequent, but probably in all cases escaped from cultivation. *Mentha Pulegium* also occurs.

288. From **Mallow** onwards the river is very pretty, mostly with steep well-wooded banks. About **Fermoy** *Nasturtium sylvestre*, *Crepis paludosa*, *Orobanche Hederæ*, *Bartsia viscosa* (**142**), *Calamintha ascendens*, *Spiranthes spiralis*, *Allium vineale*, *Butomus*, *Lastrea æmula* grow. *Corydalis claviculata* is recorded from Ballyduff. At **Lismore**, a delightfully situated little town, *Nasturtium sylvestre*, *Corydalis claviculata*, *Pimpinella major*, *Bartsia viscosa*, *Butomus*, *Festuca sylvatica* have been noted. A few miles lower down, at Cappoquin, the river leaves its W–E limestone trough and, turning suddenly S, becomes tidal, and cuts through high ridges of Old Red Sandstone to the ocean at Youghal (**289**). *Nasturtium barbarœdes* (*amphibium* × *sylvestre*), *Verbena*, *Stachys officinalis*, *Ophrys apifera*, *Potamogeton nitens* are found at Cappoquin; *Euphorbia hiberna* (**154**) follows the W–E trough on to Dungarvan (**286**), its most easterly station in Ireland. The Blackwater gorge from Cappoquin to Youghal (**289**) deserves further working: it probably contains some good plants: and as for attractiveness, "there is not a more lovely bit of scenery than this in the British Isles" writes H. C. Hart—and he was no mean judge. At the old town of Youghal the river rushes through a narrow gate of Old Red Sandstone into the Atlantic.

The valley from Mallow downwards is by no means thoroughly explored botanically. In this connection an old vague record by G. H. Kinahan—"Blackwater valley"— for *Trichomanes radicans*, should be kept in mind.

The Blackwater can be worked from Millstreet, Kanturk, Mallow, Fermoy, Lismore, Cappoquin and Youghal; all of which possess railway connections and hotels.

H. C. HART: "Botanical Notes along the rivers Nore and Blackwater, etc." Journ. Bot., 1885, 228–233.

289. Youghal forms an excellent centre for botanical work. To the N lies the Blackwater gorge, with its salt-marshes, steep banks, and woods (**288**). The salt marshes yield *Limonium humile*, *Atriplex portulacoides*, *Festuca procumbens* (little more than casual in Ireland), etc. Southward sandy beaches extend, with *Glaucium*, *Trifolium scabrum*, *T. filiforme*, *Erodium moschatum*, *Erythræa pulchella*, *Calystegia Soldanella*, etc. To the W there is an extensive bog, where grow *Potamogeton Friesii*, *Carex Pseudo-Cyperus*, *Osmunda*, and so on. Other plants of the vicinity are *Linum bienne*, *Geranium columbinum*, *Calamintha ascendens*, *Ophrys apifera*, *Carex gracilis*. About the town are *Kentranthus*, *Geranium pusillum*, *G. rotundifolium*, *Salvia horminoides*, *Rumex pulcher*, *Mercurialis annua*. Eastward across the ferry the Waterford coast (**286**) at once rises into cliffs, which continue with few interruptions to Dungarvan, and thence to Tramore, and yield among other plants *Papaver Argemone*, *Vicia sylvatica*, *Rubia peregrina*, *Inula crithmoides*.

At Ardmore, on the coast 6 miles E of Youghal, is a notable group of Early Christian buildings—a fine round tower, a very early oratory, a Hiberno-romanesque cathedral, etc., which will recompense a visit to this pretty spot.

THE COUNTY OF CORK.

290. Cork, the largest county in Ireland (2890 sq. miles), is mostly hilly, with great ridges of Old Red Sandstone alternating with valleys of Carboniferous slate and limestone. As one goes W across southern Ireland, two of the Cantabrian plants, *Saxifraga spathularis* and *Euphorbia hiberna* make their appearance (in Waterford) before the Cork boundary is reached, and the species composing this interesting group come in successively as one proceeds, till on the Kerry boundary almost all are present. The ground gets more hilly also towards the W, and is eventually a mass of picturesque and rugged mountains. The River Lee (**291, 299**) crosses the county from W to E, turning S near Cork city to form the complicated land-locked Cork Harbour. The Blackwater (**287**) has a parallel course further N. The coast-line is very extensive and mostly bold, with many sheltered inlets. In the W, Bantry Bay and Dunmanus Bay form much deeper indentations. Save for tarns in the W, lakes are practically absent from the area. Apart from the vicinity of Cork city (**291**) with a large flora peculiarly rich in adventive species, interest centres in the picturesque western area (**296**).

CORK CITY AND NEIGHBOURHOOD.

291. The River Lee (**290, 299**), flowing E from Kerry down a limestone valley bounded by ridges of Old Red Sandstone, enters near its mouth a complicated archipelagic area, the sea-water flooding over foundered E–W limestone troughs which it enters through ancient narrow N–S gorges cut through the Old Red. The city of Cork lies at the NW extremity of this area, where the river becomes tidal; Cobh (pronounced *Cove*) or Queenstown stands on an island in the middle of the archipelago, facing a large area of water, with the open Atlantic peeping through beyond bold headlands of sandstone some miles to the S.

Cork was founded on an island in the river, but has long since invaded the land on either side, climbing up the hills— in some places steep—which rise on all sides. An ancient centre of trade, and possessed of one of the mildest climates in Ireland, the feature of the vegetation of the neighbourhood of the city which immediately strikes the botanical visitor is the variety and luxuriance of plants, mainly southern, not native to the country, which have run wild, and now occupy walls, rocks, waste ground, etc., often in abundance. These include

Ranunculus parviflorus
Papaver Argemone
Barbarea verna
Diplotaxis tenuifolia
   muralis
Lepidum latifolium
Geranium pusillum
Hypericum calycinum
   hircinum
Sedum album
   rupestre
Fœniculum vulgare
Valerianella rimosa
Kentranthus ruber
Senecio squalidus
   squalidus × vulgaris
   (125)
Crepis biennis
Lactuca muralis

Tragopogon porrifolium
Symphytum tuberosum
Verbascum virgatum
   Blattaria
Linaria Elatine
   minor
Antirhinum majus
   Orontium
Erinus alpinus
Verbena officinalis
Lamium molucellifolium
Orobanche minor
Rumex pulcher
Mercurialis annua
Stratiotes Aloides
Allium triquetrum
   Scorodoprasum
Spartina Townsendii[13]
Azolla filiculoides

Sedum dasyphyllum, usually looked on as alien where it occurs on old walls, occupies in abundance the bare limestone which covers Carrickshane Hill near Midleton amid a purely native vegetation : R. A. Phillips is I believe right in his suggestion that it is indigenous there. (See Journ. Bot., 1934, 70–1.)

For the rest, the neighbourhood of Cork, presenting mainly well-tilled or grazed ridges of Old Red Sandstone, with its debris covering much of the limestone in the valleys, does not offer a flora of special interest, but the city forms a very convenient centre for excursions; the ramifications of the harbour, the old towns of Youghal and Kinsale, as well as Bandon, Macroom, Mallow, etc. are easily reached by train or otherwise, and the indented open coast-line both E and W possesses much of interest.

[13] Introduced in 1925 at Dunkettle and near Glanmire—see H. A. CUMMINS in Econ. Proc. R. Dublin Soc., 2, 419–421, Pl. xxviii. 1930.

292. Of the plants especially characteristic of SW Ireland, several extend E as far as or beyond Cork. *Euphorbia hiberna* (**154**) is abundant in many places; *Bartsia viscosa* and *Pinguicula grandiflora* (**145**) are more local. *Saxifraga spathularis* (**114**) comes as far east as Inniscarra, to re-appear on the mountains of Waterford and S Tipperary. *Trichomanes radicans* (**192**) is recorded from stations to the W and N of Cork city, also *Hymeno-phyllum tunbridgense*. *Asplenium lanceolatum* (**194**) occurs, as at Coachford and Kinsale—in Ireland a very local fern of the S and SE.

293. The maritime flora of the Cork neighbourhood (harbour and open coast) is varied, including as it does

Cochlearia anglica (**86**)
Crambe maritima
Viola Curtisii
Spergularia rupicola
Lavatera arborea (**97**)
Erodium maritimum
Trigonella
    ornithopodioides
Trifolium scabrum
Crithmum maritimum
Œnanthe pimpinelloides
    (**120**)
Rubia peregrina (**121**)
Limonium humile (**132**)
    binervosum
Calystegia Soldanella
Atriplex portulacoides
    littoralis
Euphorbia Paralias (**152**)
    portlandica (**153**)
Scirpus filiformis
Carex punctata (**184**)
Glyceria distans

Of the above, *Crambe* is doubtfully native here; *Trigonella*, very rare, is found at Currabinny; *Œnanthe pimpinelloides* has its only known Irish station at Trabolgan.

294. Other plants worthy of note which occur in the vicinity of Cork are

Linum bienne
Geranium columbinum
    rotundifolium
Erodium moschatum
Rubus hesperius
    Colemanni
    iricus
Rosa micrantha
    stylosa (*var.* systyla)
Pimpinella major
Galium erectum
    uliginosum
Dipsacus sylvestris
Carlina vulgaris
Carduus nutans
Crepis paludosa
Picris echioides
Crepis taraxacifolia

Wahlenbergia hederacea
Blackstonia perfoliata
Centunculus minimus
Orobanche rubra (**143**)
    Hederæ
    major
Pinguicula lusitanica (**146**)
Calamintha ascendens.
Scutellaria minor
Chenopodium rubrum
Rumex maritimus
Ceratophyllum demersum
Allium vineale
Spiranthes spiralis

Ophrys apifera
Potamogeton coloratus
    obtusifolius
Carex Pseudo-Cyperus
Festuca sylvatica
Brachypodium pinnatum
Hordeum nodosum
Asplenium Adiantum-
        nigrum *var.* acutum
        (**195**)
Lastrea æmula
Equisetum trachyodon
Chara aculeolata
Nitella translucens

Of these, *Rosa micrantha*, very local in Ireland, is frequent; *Orobanche rubra* is found only at Crosshaven; of *Rumex maritimus* the old record for Kilcoleman was verified in 1904; *Equisetum trachyodon* (**204**) grows at St. Anne's.

T. Allin: "The Flowering Plants and Ferns of the County Cork." 1883. R. A. Phillips: "The Characteristics of the Flora of County Cork," *in* Smith's "Antient and Present State of the County of Cork" (reprinted 1893–4 by the Cork Hist. and Arch. Soc.), 2, chap. vii.

THE BANDON-KINSALE AREA.

295. The Bandon River, a considerable stream coming from W Cork, turns S near Bandon in the manner characteristic of the South Irish streams to enter the Atlantic through a long winding estuary at the ancient port of Kinsale. There is pretty and interesting country about here. Bandon has long been known as the home of *Linaria sepium* of Allman (*L. repens* × *vulgaris*, **136**); it grows on both sides of the river. *L. repens* has here one of its few Irish stations, and *Euphorbia amydaloides*, not native, occurs. Many of the plants listed above for the area about Cork are found in this district also.

WEST CORK.

296. The W part of this area is included below under the heading THE KERRY-CORK HIGHLANDS (**300**), but a brief account of the flora of W Cork in general is given here.

West Cork is to a great extent an echo of S Kerry, but with lower hills, a greater amount of farmland, and a flora in which many of the Kerry rarities persist, but in diminishing quantity as one goes E. *Saxifraga Geum* (**113**), *Carum verticillatum* (**119**), *Arbutus Unedo* (**129**), *Microcala filiformis* (**134**), *Spiranthes gemmipara* (**155**), *Sisyrinchium angustifolium* (**159**), *Juncus macer* (**165**) do not in Co. Cork venture far beyond the Kerry border. *S. spathularis, Bartsia viscosa, Euphorbia hiberna, Trichomanes,* on the other hand, penetrate across Cork and on into Co. Waterford. *Pinguicula grandiflora* does not range so far as these last; while abundant in W Cork, it dies out about the line joining Cork and Mallow.

297. The Cork-Kerry boundary runs along the high rugged ridge of the Caha Mountains (Hungry Hill, 2251 ft., Knockowen, 2169), whence the Cork slope drops into lovely Bantry Bay, the most southerly of the great south-western sea-inlets. At its head, facing full S and embowered in woods, is **Glengarriff** (good hotels), entirely sheltered by the mountains from the prevailing winds, and producing a wealth of moisture-loving and heat-loving vegetation excelled only by Killarney. Glengarriff forms a splendid centre for the botanist. Northward, the winding main road climbs to more than 1000 ft., tunnels under the final rock-ridge, and drops down into Kenmare over a bridge across the head of the inlet. Westward, the coast road runs between the sea and the hills to Adrigole, a beautiful land-locked bay, and on to Berehaven, the original European station for *Spiranthes gemmipara* (**155**). Southward of Glengarriff, a road runs round the indented head of the bay to the rail-head at Bantry, some 13 miles distant. A highly picturesque road runs N through the mountains to Kilgarvan, and another through the Pass of Keamaneigh by Gouganebarra and along Lough Allua to Macroom. The hills to the W are very wild, full of coombs and tarns, and will repay further exploration. *Saxifraga stellaris* is reported to assume a very large hairy form here, and is often viviparous. From Glengarriff S to Cape Clear (the most southern point of Ireland), E to **Clonakilty** and N to the Derrynasaggart Mountains, a wide area of hilly country with a much indented coastline extends, full of interest but as yet seldom visited by the botanist. Two plants, *Ranunculus tripartitus* (**82**), very rare in England, and *R. lutarius,* have in West Cork their only Irish stations.

The first occurs at Baltimore, the second at Adrigole,[14] where also *Arbutus, Trichomanes,* and *Asplenium lanceolatum* have been found. *Eryngium campestre,* not usually admitted as indigenous in Ireland, grows freely on Sherkin Island, and is looked on by R. A. Phillips as possibly native there (Irish Nat., 1901, 173). *Asplenium lanceolatum* (**194**), very local in Ireland as in Britain, has in this region its head-quarters, occurring in a number of places. Elsewhere it is known from isolated stations in Kerry, Wexford, Carlow, and Wicklow. Another very rare local plant is *Helianthemum guttatum* (**89**), which grows abundantly on the Calf Islands near Cape Clear, and also, with its var. *Breweri,* at Three Castle Head (its only other Irish station being Inishbofin in Galway). The next station for the type is the Channel Islands, though the variety grows in Anglesea. *Fuchsia gracilis* is quite naturalized in wild places in the Adrigole and Berehaven districts, as *F. Riccartonii* is in many areas.

298. Among plants which are more or less frequent in West Cork the following may also be mentioned :—

Elatine hexandra
Lavatera arborea (**97**)
Althæa officinalis
Linum bienne
Geranium columbinum
Erodium maritimum
Drosera anglica
    longifolia
Lobelia Dortmanna
Wahlenbergia hederacea
Centunculus minimus
Linaria Elatine
Antirrhinum Orontium

Pinguicula lusitanica (**146**)
Verbena officinalis
Scutellaria minor
Calamintha ascendens
Ophrys apifera
Spiranthes spiralis
Sparganium minimum
Scirpus filiformis
Carex lasiocarpa
    punctata
Glyceria distans
Lastrea æmula
Osmunda regalis

The following are local or rare :—*Cochlearia anglica* (**86**), *C. alpina* (Hungry Hill), *C. grœnlandica* (Sherkin I.), *Subularia, Viola lutea* (**92**), *V. lactea* (Inchigeela, hybridizing too with *V. canina*), *Sagina subulata, Spergularia rubra, Geranium rotundifolium, Ornithopus* (at Crookhaven, extremely rare in Ireland), *Trifolium fragiferum, Rubus*

[14] See H. and J. GROVES, Journ. Bot., 1896, 277; R. A. PHILLIPS, Irish Nat., **1896, 167.**

iricus, Rosa micrantha (in Ireland entirely southern),
Pimpinella major, Rubia peregrina, Inula crithmoides,
Crepis paludosa (curiously rare in SW Ireland), *Hieracium*
hypochœroides (Millstreet Mtns., in the form *saxorum*),
Erythrœa pulchella, Orobanche major (very local in
Ireland), *O. rubra* (Crosshaven), *Salvia horminoides,*
Mentha Requienii (naturalized at Castletownsend), *Listera*
cordata, Cephalanthera ensifolia, Juncus acutus (**166**),
Malaxis, Eriocaulon (**173**), *Eriophorum latifolium, Carex*
limosa, Brachypodium pinnatum (native on cliffs near
Courtmacsherry), *Cystopteris fragilis, Phegopteris poly-*
podioides, Hymenophyllum peltatum, H. tunbridgense,
Asplenium Adiantum-nigrum var. *acutum* (**195**), *Lastrea*
montana, Lycopodium inundatum (very rare in Ireland),
Isoetes lacustris, Nitella flexilis. Stachys officinalis, though
frequent in Kerry, has its only Cork station near Berehaven.

The visitor will be struck by the unusual abundance of
some species, such as *Crithmum* and *Anthemis nobilis,* and
by the unexpected rarity of some well-nigh universal plants,
such as Equisetums and Polystichums.

One of the most remarkable places in the district is
Lough Hyne (or Ine), a deep land-locked bay joined to the
Atlantic by a very narrow shallow channel, so that tides
are reduced to a few feet. A kind of bathymetrical
telescoping prevails, deep-water animals living in a few
fathoms or even a few feet of water among littoral forms.
Zostera nana of extraordinary size occurs, and also its
hybrid with *Z. marina* (see Nature, 1934, 912).

The calcicole flora is poor, since, though the geological
structure is that familiar in Cork and Kerry, the synclinal
valleys are occupied not by limestone, but by the older
Carboniferous Slate. Calcareous sands supply a habitat
for some species.

299. The inland parts of the area are drained by the
upper waters of the Blackwater (**287**) and the Lee (**290,
291**). The latter rises among highly picturesque surroundings above the little lake of Gouganebarra, which is over-
hung by the cliffs of Bealick (1764 ft.), and flows E through
Lough Allua and past Inchigeela to **Macroom**. Here, in a
flat-bottomed valley, it spreads into a network of clear
anastomosing streams and wooded islets untouched by
grazing animals—a place of singular charm, known as The
Geeragh. Twenty-five miles further on it reaches tidal
waters at Cork (**291**).

The highly dissected coastline is indicative of a long period during which the land stood higher than at present and rain and rivers carved the surface of the country, followed by depression, which allowed the sea to flow up the valleys thus formed. The result is a charming mixture of land and sea, with little towns nestling in sheltered inlets; such are Kinsale, Clonakilty, Rosscarbery, Glandore, Skibbereen, Baltimore, Schull, Crookhaven. Baltimore is noted for being the most northern place plundered by the Algerian corsairs during the period of their activities: they sacked the town in 1631.

T. ALLIN: "The Flowering Plants and Ferns of the County Cork." 1883. R. A. PHILLIPS: "The Characteristics of the Flora of County Cork," *in* Smith's "Antient and Present State of the County of Cork" (reprinted 1893–4 by the Cork Hist. and Arch. Soc.), 2, chap. vii. R. A. PHILLIPS: "Rare Plants in West Cork." Irish Nat., 1894, 205–206.

THE COUNTY OF KERRY.

Kerry (excepting the N) and West Cork form a homogeneous area of a very high order of beauty, and a flora of unsurpassed interest; they are conveniently treated together (**300**). North Kerry falls far short in botanical interest (**330**).

THE KERRY-CORK HIGHLANDS.

300. The south-west of Ireland, with its mountain-ranges projecting into the Atlantic, presents an area unsurpassed in any portion of Europe as regards both the loveliness of its scenery and the interest of its botany. Especially impressive is its combination of mountain and ocean. "The County of Kerry," says Camden, "near the mouth of the Shannon, shoots forth like a little tongue into the sea roaring on both sides of it. This County stands high, and has many wild and woody hills in it; between which lye vallies, whereof some produce corn, others wood." Many longer descriptions might be less comprehensive.

The strong compression to which this region was subjected during the Hercynian (early Permian) epoch threw the Carboniferous and underlying Devonian strata into a series of great folds running WSW and ENE (**11**)

Subsequent prolonged denudation has laid bare over most of the area the conglomerates and slates of the latter period, which form high mountain-ridges; the limestone, often with Carboniferous Slate below, persists still in many of the valley bottoms, and its solution has played a part in allowing the ocean to penetrate far up the valleys in a series of deep parallel inlets, 25 to 30 miles in length. At the inland edge of the region, in its central part, where the highest land in Ireland (Macgillicuddy's Reeks, 3414 ft., **319**) breaks down abruptly into a low narrow plain of limestone, are situated the Lakes of Killarney, world-famous for their beauty and of special botanical interest (**320**).

301. The area in general consists of four rugged and lofty mountain promontories, separated by the three sea-inlets of Bantry Bay, Kenmare River, and Dingle Bay. Islands and pinnacled rocks, the summits of submerged mountains, continue the ridges seaward; they are interesting especially on account of their great colonies of breeding birds, and as furnishing some of the wildest scenery to be found in the British Isles. Almost the whole area is mountainous, moorland and hill-pasture occupying three-quarters of the surface. The limit of the most interesting region, whether from the point of view of the botanist or of the tourist, may be defined by a line drawn from Clonakilty in W Cork to Millstreet and thence to Tralee. This is the area dealt with in the succeeding account. Eastward, across the county of Cork, the ground steadily lessens in elevation and the amount of farmed land increases, but many of the peculiar plants remain abundant (**292**). Northward, beyond a zig-zag band of low limestone, lies a mass of Upper Carboniferous rocks stretching to the Shannon mouth — an area devoid of striking scenery and of little attraction to the botanist (**330**).

The region is one of rocky hills, thinly peat-covered, with lowlands also mostly peaty. The valleys are often deep, and, where shelter exists, aboriginal forest is still occasionally found in greater quantity than elsewhere in W Ireland. References to the composition of these woods will be found in subsequent sections (**312, 313, 320**). Those which are so richly developed about Killarney are of special interest on account of the plentiful admixture in them of *Arbutus Unedo*. Farmland occupies most of the lower valleys and a good deal of the coastal region where exposure is not too great. Small lakes abound. Rivers are short and rapid. The coast is intricate and of great extent, often low, sometimes rising into grand precipices. The western promontories and islands plunge into deep water, and are more mercilessly exposed to the open Atlantic than any other part of the Irish coast.

302. The climate of this essentially oceanic district is remarkably mild and humid (**22, 303**). The close aggregation of the hills and consequent narrowness of the valleys provides greater shelter than is to be found for instance in the companion region of Galway-Mayo. To these features are due, at least in part, the presence of many rare plants, and particularly the extraordinary variety, profusion and luxuriance of moisture-loving Phanerogams and also groups such as the Ferns, Mosses, and Liverworts. The native species of *Drosera, Pinguicula, Hymenophyllum*, and such plants as *Osmunda, Lastrea œmula*, etc., are present in remarkable abundance.

The same conditions are probably responsible for a converse effect—the absence or rarity of a number of plants which both in Ireland and in Britain have a wide general range. Some examples follow; italics show species completely absent from Kerry; the remainder occur, but are extremely rare. It will be noted that, as might be expected, the majority are characteristic of dry or light soils.

Ranunculus auricomus	*Carlina vulgaris*
Lingua	*Tragopogon pratense*
Papaver Argemone	*Gentiana Amarella*
Fumaria officinalis	*Lycopsis arvensis*
Hypericum perforatum	Linaria vulgaris
Ononis repens	Lamium amplexicaule
Trifolium medium	album
arvense	*Sceleranthus annuus*
Parnassia palustris	*Euphorbia exigua*
Pimpinella Saxifraga	*Agropyron caninum*

Conditions in W Cork, adjoining on the E, are not quite so extreme; there is less mountain, less peat, less precipitation; a greater number of the above species occurs there and some become commoner; the light-soil flora increases eastward till in the district around Cork city it is quite well represented (**291**).

303. Rainfall in the area varies from under 40 inches annual average on the low coast of N Kerry to about 87 inches on low ground at the Upper Lake of Killarney, and 97 inches at 1760 ft. above sea on Mangerton hard by (where 141 inches was recorded in 1903!). As to temperature, at Valencia the average mean for 30 years for January, the coldest month of the year, is 44·5° F. (the same as at Mentone), while the average for August, the hottest month, is 59·5° F., giving an annual range of only 15°. But extremes of temperature— especially the lower limit—have more effect on vegetation than averages. At Valencia during 30 years only six times has 7° of frost been reached or exceeded, the lowest temperature during this period having been 20° F. in 1893–4. Under these circumstances Kerry gardens boast fine specimens of many shrubs and herbs very seldom seen in the open air in the British Isles save in Cornwall and Scilly, such as *Asophila australis* (Australia), *Banksia serratifolia* (Australia), *Clethra arborea* (Madeira), *Corynocarpus lævigata* (New Zealand), *Dicksonia antarctica* (Australia, etc.), *Metrosideros* spp. (New Zealand), *Myoporum lætum* (Australia), *Templetonia retusa* (Australia).

The district is well equipped for the visitor, whether as regards railways, roads, or hotels. The strong NE–SW folding allows easy penetration from the landward side; it impedes communication in the NW–SE direction, and the roads which cross the ridges between Glengarriff and Kenmare, and between Kenmare and Killarney, have to climb 1000 ft. or so to surmount the great Devonian rock-ribs. By rail, one can travel from Cork to Bantry (near Glengarriff). From Mallow (on the main line 3–4 hours from Dublin) one passes up the picturesque valley of the Blackwater to Killarney and on to Tralee (and thence N to the Shannon). Three branch lines penetrate SW to Kenmare, Valencia, and Dingle, the latter two close to the extremities of the two greatest of the mountain-promontories. Buses and tourist coaches connect the rail-heads with each other and with more distant places.

304. Kerry and W Cork form the headquarters in Ireland of many of the most interesting plants found in the country, notably of the Hiberno-Lusitanian group (**35**, **37**) and its concomitant, the Hiberno-American (**35**, **37**). While some of the species belonging to these range E across Cork (**291**) or on into Waterford (**282**),

and N far along the W coast, others and a few additional
species find here their only habitats.   Such are

Ranunculus tripartitus (**81**)		*A*	Spiranthes gemmipara
lutarius (**81**)			(**155**)
Lathyrus maritimus (**104**)			Simethis planifolia
Microcala filiformis (**134**)			(**163**)
Sibthorpia europæa (**137**)			Carex hibernica

Among the rare plants which, though also found elsewhere in Ireland, have here their Irish headquarters, are

S	Saxifraga Geum (**113**)	*A*	Sisyrinchium angusti-
S	S. spathularis (**114**)		folium (**159**)
	Carum verticillatum	*A*	Juncus macer (tenuis
	(**119**)		*auct.*) (**165**)
S	Arbutus Unedo (**129**)		Scirpus nanus
	Bartsia viscosa (**142**)	*A*	Naias flexilis (**172**)
S	Pinguicula grandiflora		Trichomanes radicans
	(pl. 9, **145**)		(**192**)
	Euphorbia hiberna		Asplenium lanceo-
	(pl. 10, **154**)		latum (**194**)

The majority of the plants in these two lists are of
either southern (*S*) or American (*A*) origin.   Their distribution, both in Ireland and elsewhere, has been discussed
already (**37**).

305. Probably on account of the mildness of the climate
and the consequent failure of the regular winter covering
of snow which they love, alpine species are not so numerous
as the height of the hills and abundance of elevated precipitous ground might lead one to expect.   This indeed
applies to Ireland generally.   But as compared with most
Irish mountain regions the list of plants usually looked on
as alpine is rich, including

Thalictrum alpinum	Oxyria digyna
Cochlearia alpina	Salix herbacea
Draba incana	Carex rigida
Alchemilla alpina	Poa alpina
Saxifraga stellaris	Juniperus sibirica
Sedum roseum	Asplenium viride
Hieracia (**128**)	Polystichum Lonchitis
Lobelia Dortmanna	Lycopodium alpinum,
Polygonum viviparum	

and no less than seven hypnoid Saxifrages (**117**).

PLATE 9.

PINGUICULA GRANDIFLORA.

In the author's garden in Dublin, in May. Roots from Co. Cork.

R. Welch, Photo.

PLATE 10.

EUPHORBIA HIBERNA.

In woods at Kenmare, Co. Kerry, growing among Brambles and Ferns in May.

R. Welch, Photo.

But, as observed elsewhere,[15] the usual grouping in Great Britain of certain species as mountain plants (Watson's Highland Type) breaks down in Ireland. In Kerry, for instance, the following Highland Type plants all have their lower limit between sea-level and 70 ft. :—

Subularia aquatica	Hieracium iricum
Sedum roseum	Carex aquatilis
Galium boreale	Juniperus sibirica
Hieracium anglicum	Isoetes lacustris

On the other hand, the vegetation of even the highest of the mountains is made up mainly of common lowland plants. Of 54 species found in Kerry above 3000 ft., all but ten or twelve are quite lowland normally, or at least are found indifferently at all elevations in Ireland.

306. On account of extreme exposure, the maritime vegetation tends to be limited as regards the rocky coasts, and most of the rarer plants are found in quiet inlets. The seaside flora includes

Arabis Brownii (**85**).	Rumex maritimus
Cochlearia grœnlandica (**87**)	Euphorbia Paralias (**152**) portlandica (**153**)
Spergularia rupicola	Zostera nana
Lavatera arborea (**97**)	Scirpus nanus (**174**)
Lathyrus maritimus (**104**)	Carex punctata (**184**)
Inula crithmoides	

As a consequence of the orography, streams are in general short and rapid, mostly maintaining a steep course down to their mouths. The chief exception is the Flesk, which in its lower course from the Killarney Lakes to Castlemaine Harbour is comparatively broad and slow, meandering through low limestone country. Lough Leane, or the Lower Lake of Killarney, 5 miles long by a couple of miles wide, only 81 ft. above sea-level, is the largest sheet of water in the district; of small lakes and mountain tarns there is great abundance, up to 2300 ft., many of them romantically situated among cliffs and peaks of purple

---

[15] *E.g.* PRAEGER: "Tourist's Flora of W. Ireland," p. 19. Also **67**, *supra*.

slates and red conglomerates, and owing their origin to intense local glacial action. The rarer hydrophytes include

Ranunculus tripartitus
(**81, 297**)
    lutarius (**81**).
Subularia aquatica
Elatine hexandra
Callitriche autumnalis
Utricularia intermedia
    ochroleuca
    Bremii
Ceratophyllum demersum
Naias flexilis (**172**)

Eriocaulon septangulare
(**173**)
Scirpus nanus (**174**)
Carex aquatilis (**181**)[16]
Isoetes lacustris
    echinospora
Pilularia globulifera
Chara canescens
Nitella batrachosperma
(**207**)

Although the non-calcareous rocks of the Old Red Sandstone and Dingle Beds prevail over the greater part of the area, a sufficiency of Carboniferous limestone has been left to provide a suitable substratum for a considerable calcicole flora, and calcareous coastal sands help to extend the range of lime-loving plants into otherwise unsuitable areas. But none of the rarer west coast calcicole plants, such as *Gentiana verna, Euphrasia salisburgensis, Ajuga pyramidalis, Neotinea intacta, Epipactis atropurpurea*, has been induced to migrate southward into the habitats available. The most southern-ranging of these, and the only one which crosses the Shannon, is *Euphrasia salisburgensis* (at Askeaton in Limerick (**338**)).

307. Among the rarer plants occurring in the area which do not fall within any of the above categories or are not mentioned above are

Teesdalia nudicaulis
Helianthemum guttatum
(**89**)
Erodium moschatum
Drosera anglica
    longifolia
Ornithopus perpusillus
Rubia peregrina (**121**)
Galium boreale
Asperula cynanchica
Wahlenbergia hederacea
Pyrola minor

Monotropa Hypopitys
Erythræa pulchella
Linaria repens
Orobanche rubra (**143**)
    Hederæ
Pinguicula lusitanica
(**146**).
Malaxis paludosa
Cephalanthera ensifolia
Allium triquetrum
    Scorodoprasum
Rhynchospora fusca (**177**)

[16] Very far south for this northern plant.

Scirpus filiformis
Carex Bœnninghauseniana
  limosa
Hymenophyllum peltatum
  tunbridgense
Cystopteris fragilis
Lastrea æmula
  montana

Osmunda regalis
Equisetum trachyodon
  (**204**).
Lycopodium inundatum
Nitella translucens
  flexilis

It will be noted that this list is composed almost wholly of indigenous plants. The only ones which are certainly not native are the two species of *Allium*; the standing of *Teesdalia, Erodium moschatum,* and *Linaria repens* is doubtful.

The most striking absentee from the area is *Selaginella selaginoides,* a plant which in Ireland ranges freely from sea-level to over 2000 ft., and is widespread save in Munster, where it is almost confined to the extreme N. *Parnassia* also, widespread in Ireland, is absent from the whole of Kerry and also Cork.

The mountainous nature of the country gives scanty opportunity for husbandry, and only some 15 per cent. of Kerry is under tillage. This fact reacts on that important section of the flora which has been introduced by farming operations or owes its presence in other ways to man's operations. Dr. Scully contrasts the County of Dublin, with 30 per cent. tillage and one-fifth of its flora non-indigenous, with Kerry which has 15 per cent. of tillage and only one-seventh of its flora falling under this category.

308. It is impossible to conclude a general account of the botany of this area without reference to its Mosses and Hepatics, of which a large number of very rare species occur. Details cannot be given here, but among the mosses may be mentioned *Campylopus Shawii, C. Schimperi, C. introflexus, Dicranum flagellare, Fissidens polyphyllus, Leptodontium recurvifolium, Trichostomum hibernicum, T. fragile, Ulota Ludwigii, Philonotis Wilsoni, P. rigida, P. seriata, Daltonia splachnoides, Hookeria laetevirens, Sematophyllum demissum, S. micans, Hypnum imponens, H. dilatatum, Hylocomium umbratum,* and among the Hepatics *Petalophyllum Ralfsii, Fossombronia Dumortieræ, Haplomitrium Hookeri, Acrobolbus Wilsoni, Leptoscyphus cuneifolius, Cephalozia hibernica, Cephaloziella Turneri, Bazzania Pearsoni, Scapania nimbosa, Radula Carringtonii,*

R. Holtii, R. voluta, Madotheca Porella, Cololejeunea Rossettiana, C. minutissima, Lejeunea flava, L. Holtii, L. diversiloba, Plagiochila ambagiosa, and many others—a list to make the bryologist's mouth water.

309. Literature.—The study of the south-western flora has been placed on a most satisfactory basis by the publication in 1916 of Dr. Scully's "Flora of County Kerry,"[17] the most comprehensive and accurate floristic study as yet produced in Ireland. Its existence has allowed of a much fuller treatment of this region in the present work than would have been possible otherwise. This is very fortunate, as the south-west is from almost every point of view the most interesting region in Ireland for the botanist, and is the area where the special features of the Irish climate and vegetation attain their most pronounced expression. In Dr. Scully's Flora, literature up to 1916 will be found fully listed. Since that date not much has been published: where subsequent papers are utilized in these pages, references are given. In so large and mountainous an area a complete survey is very difficult, and Kerry still offers opportunities of further additions to its varied flora. For instance, Dr. Scully thinks that the fine Mullaghanattin range, lying 10 miles SW of the Reeks, may yield some good plants.

KERRY IN SPRING.

310. To see the peculiar flora of Kerry in its most characteristic aspect the time to visit the area is not the usual holiday time of July and August, beloved of botanists, but the month of May and early June. Then *Pinguicula grandiflora,* the glory of Kerry, is at its best, its great purple blossoms nodding in the wind over hundreds of square miles of bog and rocky mountain. *Saxifraga spathularis* and *S. Geum* are in full bloom too, decking the rocks, woods and glens with their slender red stems starred with white flowers. *Euphorbia hiberna* is at its loveliest stage, just attaining full growth, and forming luscious clusters of greenish gold along the streams and in all the

---

[17] R. W. SCULLY: "Flora of County Kerry, including the Flowering Plants, Ferns, Characeae, etc." 6 plates [of Robertsonian Saxifrages], map. Dublin, Hodges, Figgis & Co. 1916. 12s. 6d. See also R. LL. PRAEGER: "Arbuti Corona" (a review of same). Irish Nat., 1917, 21–27.

rough pasture-land. By the end of June it will be wholly green, with tubercled fruit and a straggling habit. *Osmunda* also is at its loveliest, golden brown, its strong shuttlecocks of fronds half expanded, with loose rags of brown fluff clinging to their straight smooth glaucous stems; it lines the stream-banks and fences everywhere, and lights up the brown bogs. Alone among the rare species, *Arbutus* is not at its best. Growth is only just beginning, the old leaves are yellowing and falling, and the half-grown fruit, so beautiful at a later stage, is still inconspicuous.

Everywhere *Ulex* is blazing, with *Cytisus scoparius* following close behind. In the woods the delicate green of the young *Betula* contrasts with the old gold of the *Quercus* and the grey-green of the *Salices*; *Ilex,* though full of flower, looking very dark in comparison. The shaggy young stems of *Lastrea Filix-mas* add again a touch of golden brown, while half-expanded *Athyrium* and *Lastrea œmula* are full green. The two species of *Hymenophyllum* are still in excellent condition, *H. tunbridgense* separating itself at once from its ally by its bluish-green tint, and the broader rounded tip of its tiny fronds.

In the lakes there is as yet little life. *Lobelia* alone is sending flower-stems up towards the light. *Subularia, Littorella* and *Eriocaulon* are still dormant, the curiously persistent old fruit-stems of the last lying criss-cross like straws across the fresh green mats of foliage.

By the sea, the white sheets of *Cochlearia* have already passed away, but the rocks are bright with *Armeria* and *Silene maritima,* often in remarkable profusion, while the dunes are decorated with thousands of the yellow and blue flowers of *Viola Curtisii.*

THE KENMARE AREA.

311. The little town of Kenmare, prettily situated at the head of the Kenmare River (so-called), the longest and narrowest of the great sea-inlets of Kerry, forms an excellent centre. Good roads radiate from it in five directions—south-westward, one down the S shore to Cloonee and Derreen, and another down the N shore to Parknasilla, Sneem, Derrynane, Waterville; south up the Sheen River and across a high pass, the boundary of Cork and Kerry, to Glengarriff and Bantry; north across another high pass and by the Upper and Lower lakes of Killarney to the town of that name; and NE beside the railway up the Roughty River to Kilgarvan, Morley's Bridge and on to Headfort Junction. All of these roads are interesting and picturesque, and afford the botanist abundant material. Cross-country work is also fruitful; mountains (2000–2700 ft.) rise on every side, save where the widening sea-inlet opens towards the Atlantic, and much of the ground is only partially explored. There is plenty of hotel accommodation.

The most uncommon halophile species of the region is *Carex punctata* (**184**), which is found at intervals along the shore. *Cochlearia anglica* (**86**) and *Limonium humile* (**133**) are frequent, the latter in a very large form at Kenmare, with upright leaves nearly a foot long and inflorescence to match. *Allium triquetrum* (near Kenmare) is one of the more interesting alien plants; *A. Scorodoprasum* also occurs, both evidently escapes. *Verbascum virgatum,* not native, has grown in one spot near Kenmare for over a century, and *Verbena officinalis* occurs. The miscellaneous plants of the neighbourhood include *Barbarea verna* (alien), *Geranium columbinum, Linum bienne, Rubus adenanthus, R. iricus, Rubia peregrina, Wahlenbergia hederacea, Microcala filiformis, Orobanche Hederæ, Pinguicula lusitanica, Mentha Pulegium, Calamintha ascendens, Stachys officinalis, Taxus baccata, Juniperus sibirica, Malaxis, Eriophorum latifolium, Juncus macer.* But the main interest lies in the abundance and wide distribution of many of the special plants of the south-west and some others :—

Saxifraga Geum (**113**)	Utricularia intermedia
spathularis (**114**)	(**144**)
Carum verticillatum (**119**)	Bartsia viscosa (**142**)
Lobelia Dortmanna	Euphorbia hiberna (**154**)
Pinguicula grandiflora	Hymenophyllum peltatum
(**145**)	tunbridgense
lusitanica (**146**)	Osmunda regalis

Trichomanes (**192**) appears to be extinct in the district, the last record being now over 60 years old.

Kenmare is not one of the best centres for the exploration of the higher Kerry mountains except with the assistance of a car; few of the rare plants of the hills occur close by, but *Alchemilla alpina,* elsewhere in Ireland known only

PLATE 11.

ERIOCAULON SEPTANGULARE,
At Cregduff Lough, Connemara, July. (The plant has been raised above water-level.)

PLATE 12.

SISYRINCHIUM ANGUSTIFOLIUM AT WOODFORD, CO. GALWAY.

Growing in gravelly stream-bed, accompanied by *Senecio aquaticus*, *Prunella vulgaris*, *Juncus*

R. Welch, Photo.

from Brandon, 30 miles to the NW (**325**), and from Thonelagee in Wicklow (**263**), grows with *Salix herbacea* along the rocky scarp of Boughil overlooking Lough Barfinnihy. The immediate vicinity of Kenmare, being mostly tilled land, is not so exciting as certain other spots; but one has only to cross the bridge below the town and go a few hundred yards to the right to find one's-self in a wood with rocks clothed with sheets of *Hymenophyllum tunbridgense* and *Saxifraga spathularis,* while the ground is full of *Euphorbia hiberna* and of *Lastrea œmula* with fronds two ft. long.

312. Cloonee.—One of the most interesting and typical places for the botanist is Cloonee, 9 miles from Kenmare along the S shore, where a string of four low-lying lakes runs from the sea into the hills. Here many of the characteristic Kerry plants grow in abundance—*Saxifraga spathularis, S. Geum, Carum verticillatum, Bartsia viscosa, Pinguicula grandiflora, Euphorbia hiberna, Carex punctata.* The waters yield *Elatine hexandra, Lobelia Dortmanna, Utricularia intermedia, Eriocaulon* (plate 11), *Isoetes lacustris, Nitella translucens.* On the shores or on the islets are *Arbutus* (plate 13), *Microcala, Sisyrinchium angustifolium* (plate 12), *Juncus macer, Rhynchospora fusca.* The islets, ungrazed by sheep and cattle, provide a glimpse of the natural flora, untouched by human influence. Thus a rock in Inchiquin Lough is tenanted by *Quercus sessiliflora* and *Arbutus,* with some *Sorbus Aucuparia, Populus tremula, Ilex, Salix,* and a dense undergrowth of *Vaccinium Myrtillus, Calluna, Luzula sylvatica,* and *Osmunda.* Uragh Wood, overlooking Inchiquin, the uppermost of the lakes, is a remnant of native forest. *Quercus sessiliflora* and *Betula pubescens* are dominant, with some *Ilex, Sorbus Aucuparia, Cratægus, Populus tremula, Corylus.* The ground flora includes much *Saxifraga spathularis* and *S. Geum* with hybrid intermediates, *Lastrea aristata, L. œmula,* and the two species of *Hymenophyllum; Neottia* is also recorded. The name "Uragh" testifies the former abundance of *Taxus,* Irish *Iubhair* (pronounced *Ure*), Yew; it is now rare here.

About Cloonee too are *Linum bienne* and *Calamintha ascendens,* which are usually associated with less wild country. Nearer to Kenmare, the **Sheen River** yields *Rubus adenanthus, Hieracium argenteum,* and also *Sisyrinchium* and others in the above lists.

313. Roughty River.—An excursion up the limestone valley of the Roughty River gives us *Wahlenbergia* at intervals; about Kilgarvan *Hieracium argenteum* (Slaheny River) and *Equisetum litorale* (**201**) (railway station, also at Reen Bridge W of Kenmare); further up near Morley's Bridge are *Hieracium Scullyi* (endemic, only known station) and *H. sparsifolium,* and by the Upper Roughty *H. orimeles* (**128**).

The Loo and Clydagh Rivers (**Glenflesk**).—Further on, the watershed, a deep trench in the hills, is crossed, and we get into the head-waters of the Flesk River. Here, about Poulgorm Bridge, the endemic *Hieracium subintegrum* grows in its only known station. About Loo Bridge, well seen from the railway, are fine native woods clothing the sheltered side of the valley. *Quercus* and *Betula* again predominate, with *Ilex, Corylus,* and a little *Fraxinus, Alnus,* and *Salix.* At Glenflesk Castle is *Hieracium lepistoides,* in the form *sublepistoides.* The Flesk valley forms the Kerry headquarters of *Wahlenbergia.*

Blackwater.—Another interesting place is the gorge of the Blackwater, seven miles from Kenmare along the Parknasilla road. The river, which has flowed for miles through an open valley, in the last half-mile of its course cuts through a rib of rock, falling in a cascade into the head of a deep narrow inlet, filled with sea-water and densely wooded, so that *Fucus ceranoides* in a peculiar form drapes vertical rocks under the shade of moss-covered oaks on which *Saxifraga spathularis* and other plants grow as epiphytes; above the bridge *Wahlenbergia* is abundant, and one may study an excellent series of hybrid Saxifrages (*Geum* × *spathularis*)—see plate 4. The curious form of *Osmunda* called *decomposita* was found 2 miles E of Blackwater Bridge (Irish Nat., 1902, 159).

PARKNASILLA AND WATERVILLE.

314. Parknasilla (large hotel) is near the village of Sneem on the N shore of Kenmare River, 14 m. below Kenmare, well sheltered by an archipelago of islets, and offering a considerable variety of ground. Waterville lies 13 m. W, facing the open Atlantic across Ballinskelligs Bay, with the large Lough Currane immediately behind it. Derrynane lies between the two, where the coast turns

N—a place long famous as the only Irish habitat of *Simethis planifolia* (**163**), which has had its sole British station near Bournemouth almost destroyed by building operations. The standing of the plant in its English station has been called in question, but no suspicion attaches to it at Derrynane. Here it occupies a tract of rough rocky furzy heath stretching 8 or 9 m. along the coast from Derrynane eastward, and for 1 to 2 m. inland. This ground in no portion has been nor could be cultivated, and indeed the plant seems very shy of approaching any spot subject to man's influence. In view of its Irish and its Continental range (SW France, N Spain, and along the Mediterranean to Italy), and the inclusion of SW England in the range of the geographical plant-group to which *Simethis* belongs, the Irish station strongly supports the plant's claim to be indigenous at Bournemouth. This area also yields *Spiranthes gemmipara* (**155**), whose only other European station is at Castletown Berehaven, on the shore of Bantry Bay about 12 miles to the SE. It has recently been found in several places about West Cove near Waterville, and may prove to be not uncommon in this neighbourhood. This fragrant orchid, vanilla-scented, is, with its near ally *S. stricta* (**156**) of N Ireland and Colonsay, the most attractive as well as the rarest of the group of N American plants whose presence lends so strong an interest to the flora of Ireland. There is another N American plant which, like the last, is confined to the area under discussion, but which has no claim to be considered indigenous— *Polygonum sagittatum*, which flourishes over about 12 acres of wet ground near the head of the Castlecove stream E of Derrynane, and thence down the stream to its mouth; it has maintained its territory for at least 40 (probably at least 70) years. It is looked on as certainly introduced, perhaps during the famine year of 1847, when American grain was distributed to the starving people in this district and elsewhere (Scully, *l.c.*, 245). Druce was told a story of its origin through a cargo of maize ground in a mill near by after shipwreck (Irish Nat., 1907, 150).

315. Of the plants which are associated especially with Kerry, the two Saxifrages, *S. Geum* and *S. spathularis*, *Pinguicula grandiflora* and *Euphorbia hiberna* are abundant in the Parknasilla-Waterville region. *Sisyrinchium angustifolium* (5 m. W of Sneem) and *Eriocaulon* (L. Fadda W of Blackwater) are very local; *Arbutus* lingers about L. Currane; it occurs also in the Sneem and Derrynane areas, but was probably planted there. Of maritima species the critical *Arabis Brownii* grows on sand-dunes at Derrynane, and *Carex punctata* is frequent along the coast. *Cochlearia anglica, Spergularia rupicola, Asperula cynanchica, Limonium humile, Euphorbia Paralias, E. portlandica* and *Scirpus rufus* also occur. Other plants of the area are *Althœa officinalis* (a long-established alien), *Cochlearia alpina, Saxifraga hirta* and *S. incurvifolia* (on the inland hills), *Carum verticillatum, Rubia peregrina, Lobelia Dortmanna, Centunculus, Bartsia viscosa, Orobanche Hederœ, Utricularia intermedia, Pinguicula lusitanica, Mentha Pulegium, Stachys officinalis, Juniperus sibirica, Juncus macer* (abundant along old roads), *Eriophorum latifolium, Carex lasiocarpa, Osmunda, Isoetes lacustris, I. echinospora* (both the last at sea-level). *Trichomanes radicans* had formerly a number of stations about Loch Currane, but has now been almost exterminated by collectors.

This is a picturesque and varied area, with a fringe of rough low ground rising to rugged hills up to 2200 ft. in height, an extremely broken coast-line, and in the inland part many lakes. Railheads at Kenmare and Cahirciveen, and plenty of hotel accommodation. The visitor to Derrynane ought not to miss the opportunity of seeing Staigue Fort, one of the finest prehistoric stone-built fortresses in Ireland. It lies two or three m. NE of Castlecove.

VALENCIA AND CAHIRCIVEEN.

316. Valencia Island, the largest island on the Kerry coast, forms the NW corner of the great mountain-promontory which separates Kenmare River from Dingle Bay. A narrow channel with a sharp turn in it cuts it off from the mainland and the rail-head, whence a ferry communciates with Knightstown, which is the place usually meant when Valencia is mentioned. For the botanist, Cahirciveen, on the mainland 2 m. back along the railway, forms a better centre than Valencia, as it is the focus of the local roads. Thence one may conveniently explore the Glenbeigh Mountains mentioned below (**317**). The scenery along the Atlantic coast is particularly fine, with grand precipices impending over the ocean, as at Bolus Head (940 ft.) and Doulus Head (921 ft.), and outlying islands of which the romantic **Skelligs**, mere pinnacles of rock, are especially interesting. The Great Skellig rises like a cathedral to 714 ft.; a group of ruined bee-hive huts, the dwellings of early anchorites, clings to its precipitous slopes. The Little Skellig, also lofty, is the breeding-place of thousands of Gannets (*Sula bassana*); its sparse flora includes *Lavatera arborea* (**97**), undoubtedly indigenous here, though of doubtful standing in most of its frequent Kerry stations. S of Cahirciveen, Ballinskelligs Bay, Lough Currane, and the valley of the Inny invite exploration.

While most of the characteristic rare plants of Kerry (**304–307**) are abundant in this area, one that merits special mention is *Inula crithmoides,* for which an old record exists which calls for confirmation—"It grows on the rocks near Bolus-head in Iveragh and on other parts of the sea cliffs" (Smith's "History of Kerry," 1746). *Malaxis,* rare in Kerry as elsewhere in Ireland, has been found near Cahirciveen, and *Spergularia rubra* (very local in Ireland) at Valencia. The alien *Althæa officinalis* is naturalized and frequent in the area.

GLENCAR AND CARAGH LAKE.

317. The River Caragh, draining much of the mountainous country W of the Reeks, flows N through the winding Caragh Lake and thence by a short flat course to meet the sea among the sands which block the head of Dingle Bay and enclose Castlemaine Harbour. Hotels situated in Glencar and at the N end of Caragh Lake permit of sojourn in one of the loveliest parts of Kerry. For the botanist this is an ideal centre, with the Reeks lying to the SE, the wild lake-strewn Glenbeigh range, which terminates in Drung Hill, with peaks up to 2500 ft., to the W, and with the estuarine (mud and sand) flora and low-lying peat-bogs of Castlemaine to the N. In or about Caragh Lake are

Subularia aquatica	Utricularia intermedia
Elatine hexandra	(**144**)
Linum bienne	Cephalanthera ensifolia
Rubus regillus	Sisyrinchium angusti-
Rubia peregrina	folium (**159**)
Lobelia Dortmanna	Naias flexilis (**172**)
Centunculus minimus	Eriocaulon septangulare
	(**173**)

Carex aquatilis (**181**)	Equisetum variegatum *var.*
gracilis	Wilsoni (**205**).
lasiocarpa	Isoetes lacustris
limosa	echinospora
Asplenium Adiantum-	Nitella batrachosperma
nigrum *var.* acutum	translucens
(**195**).	flexilis;

also (as colonists) *Aster* sp. and *Galium erectum*; while Glencar yields—

Rubus iricus	Cephalanthera ensifolia
Hieracium orimeles	Juncus macer (**165**)
Pyrola minor	Potamogeton prælongus
Pinguicula lusitanica	Eriophorum latifolium
Utricularia intermedia	Rhynchospora fusca
Malaxis paludosa	

All over the area are found of course *Saxifraga spathularis, S. Geum, Pinguicula grandiflora, Euphorbia hiberna,* and other wide-spread rarities of Kerry. Additional plants of the neighbourhood include *Carum verticillatum, Bartsia viscosa, Lastrea œmula.* Now very rare, and to be treated with reverence, is *Trichomanes radicans.*

The picturesque **Glenbeigh Mountains**, half-way between Caragh Lake and Valencia, provide their share of alpine plants, including *Saussurea,* and also many of the characteristic Kerry species. They furnish fine climbing and walking, and invite further exploration.

CASTLEMAINE HARBOUR AND INCH.

318. The upper end of Dingle Bay, unlike the other great sea-inlets of Kerry, is choked with sand, and the terminal 8 m. of it are almost cut off by the long sandy peninsula of Inch on the N and two smaller similar projections on the S, enclosing the land-locked Castlemaine Harbour, half of which runs dry at low water. The largest Kerry river, the Laune, and other streams, discharge into it, so that estuarine conditions, rare in Kerry, prevail. On the S shore broad stretches of low-lying bog, another local rarity, extend. The flora for these reasons differs from that which generally characterizes Kerry, but at the same time it contains fewer noteworthy plants. The best species is *Lathyrus maritimus* (**104**), which finds on the sands of Rosbeigh opposite Inch its only Irish habitat. It was first noted from here in 1756 (Smith's "History of Kerry") and was collected at intervals until 1845. At some time after that it apparently vanished, and was sought in vain until 1918, when it was found in fair quantity, perhaps owing to seeds long buried in the sand and at length brought again under the influence of suitable conditions for growth (see Scully in Irish Nat., 1918, 113–5), perhaps only because its quite restricted station was overlooked. It grows luxuriantly here, with prostrate stems a yard long, over a limited area of the non-stable zone of the dunes, forming deep green patches among a continuous growth of *Ammophila*, the only other plants present in the sand being *Eryngium* and a few clumps of *Lotus corniculatus*. The presence of young plants on ground not yet fully colonized by *Ammophila* provides satisfactory evidence that the plant is holding its own. In Great Britain it has a wide but very discontinuous range along the E coast, and elsewhere has an arctic and northern distribution embracing Europe, Asia, and America.

Castlemaine forms the Kerry headquarters of *Sisyrinchium angustifolium*, which is extremely abundant over a wide area. Other plants of the neighbourhood include *Cochlearia anglica, Erodium moschatum, Carum verticillatum, Mentha Pulegium, Bartsia viscosa, Pinguicula grandiflora* and *P. lusitanica, Euphorbia portlandica* and *E. Paralias, Potamogeton nitens, Ruppia maritima, Eleocharis acicularis, Rhynchospora fusca.*

MACGILLICUDDY'S REEKS.

319. "The Reeks" (*i.e.* ricks) form a compact group of tall conical peaks, cut off from Killarney on the E by the deep Gap of Dunloe (a very fine Glacial overflow channel) and Purple Mountain, and on the S dropping steeply into the head of the valley in which the Upper Lake of Killarney lies. The highest point (Carrantual) is 3414 ft. and a number of summits exceed 3000 ft. The slopes are steep, often grandly precipitous, and embosom numerous picturesque tarns (up to 2238 ft.). Green and purple grits prevail, and the vegetation is often grassy. Considering the elevation and extent of cliffs suitable for mountain plants, the alpine flora is poor. No "Highland"

plant has here its sole Irish station, and *Draba incana* is the
only one which does not occur elsewhere in Kerry.   Among
the alpines which are found at high levels (above 3000 ft.)
are

Cochlearia alpina	Carex rigida
Sedum roseum	Deschampsia alpina
Saxifraga stellaris	Asplenium viride
Oxyria digyna	Lycopodium alpinum
Salix herbacea	

Lower down, some additional "Highland" plants come
in—

Draba incana	Polystichum Lonchitis
Hieracia	Isoetes lacustris
Juniperus sibirica	

and, as already mentioned (**305**), some of the species of
"Highland" type extend down to sea-level.
The summit flora of Carrantual (3414 ft., the highest
point in Ireland) consists of

Potentilla erecta	Juncus squarrosus
Saxifraga stellaris	bulbosus
spathularis	Luzula sylvatica
Galium saxatile	Agrostis canina
Vaccinium Myrtillus	tenuis
Calluna vulgaris	Deschampsia cæspitosa
Armeria maritima	flexuosa
Thymus Serpyllum	Sieglingia decumbens
Rumex Acetosa	Festuca ovina
Acetosella	Lycopodium Selago

Among the more interesting of the other plants (in
addition to the alpines already named) which occur above
3000 ft. are :—

Saxifraga rosacea	Empetrum nigrum
hirta	Cystopteris fragilis
incurvifolia	Hymenophyllum peltatum
Melampyrum pratense *var.*	
montanum	

In spite of an unexpected want of variety, the alpine
flora is abundant and luxuriant in suitable places, the most
attractive spots being the cliff-ranges south of L. Eagher

at the head of Cumloughra and the series of coombs N of L. Googh. The higher parts form magnificent ground for the mountaineer, and the walk along the main ridge, from above the Gap of Dunloe to Cummeenacappul, forms, in the estimation of H. C. Hart, a most competent judge, "the grandest bit of mountaineering to be met with in Ireland."

The flora increases rapidly in variety as one descends the hills. While some of the following ascend into the alpine zone, all are better developed at lower levels. The upper limit which each attains on the Reeks is given :—

Subularia aquatica, 1250
Elatine hexandra, 1125
Saxifraga Geum, 2100
Drosera anglica, 850
Carum verticillatum, 700
Lobelia Dortmanna, 1550
Arbutus Unedo, 525
Bartsia viscosa, 700
Pinguicula grandiflora, 2800
   lusitanica, 1450
Euphorbia hiberna, 1350
Scirpus filiformis, 850
Malaxis paludosa, 800
Carex aquatilis, 510
Trichomanes radicans, 500
Hymenophyllum tunbridgense, 1000
Cystopteris fragilis, 3150
Lastrea æmula, 2100
  montana, 1600
Osmunda regalis, 850

All of these, it may be noted (with the possible exception of *Malaxis*), occur in Kerry down to sea-level.

The Reeks are not very readily accessible, as few hotels are situated close to their base, and all are on low ground. Glencar forms the most convenient jumping-off place, the hotel being only 5 miles in a direct line from the summit of Carrantual. Killarney and Caragh Lake (good hotels) are about twice that distance from the summits. The Killarney-Waterville road viâ Ballaghisheen Pass passes at Lough Acoose within 3 miles of Carrantual summit at an elevation of 545 ft.

H. C. HART: "Report upon the Botany of the Macgillicuddy's Reeks, County Kerry." Proc. Roy. Irish Acad., 2nd ser., **3** (Science), 573–593. 1882. R. W. SCULLY: l.c.

Mangerton.—The great broad mass of Mangerton (2756 ft.), which rises S of Killarney, beyond Muckross Lake, would be dull ground were it not for the grand coombs on the N side known as the Horse's Glen and the Devil's Punch-bowl. On the cliffs surrounding there are concentrated the interesting plants of Mangerton—*Saussurea,*

Hieracium orimeles, Deschampsia alpina, Polystichum Lonchitis, etc., and of course the usual Robertsonian Saxifrages, Pinguiculas, Hymenophyllums, etc., that one associates with Kerry. *Lycopodium clavatum,* strangely rare in SW Ireland, has one of its few local stations about the mountain.

KILLARNEY.

320. The town of Killarney is built in a low-lying valley of Carboniferous limestone, a mile N of the Lower Lake or Lough Leane (much the largest of the Killarney lakes) on the side opposite to the grand wooded hills of Old Red Sandstone that slope abruptly to its further shore, with Macgillicuddy's Reeks rising behind Purple Mountain and Tomies Mountain, and the broad mass of Mangerton more to the E, across the deep cleft in which is situated the Upper Lake. Limestone forms the promontories and islands of much of the Lower Lake and of Muckross Lake (which is little more than a continuation of it), while the Upper Lake, with its islands and surrounding woods, lies well within the Old Red Sandstone region: it is this variety of rock and consequent change of topography which produce the charm of scenery and (coupled with climatic conditions) the richness of flora which render Killarney a paradise for the botanist.

While a good deal of planting up, especially of *Pinus sylvestris,* has been done in places, this district, and especially the region around the Upper Lake, still yields the largest area of native woodland now to be found in Ireland; the shelter and exceptional moisture produce there a remarkable undergrowth, almost tropical in its luxuriance. *Quercus sessiliflora* is the dominant tree, with much *Arbutus* and *Ilex,* and some *Betula pubescens, Fraxinus, Populus tremula, Taxus, Sorbus rupicola,* and *S. Aucuparia;* the dense undergrowth includes much *Calluna, Vaccinium Myrtillus, Melampyrum pratense, Euphorbia hiberna, Luzula sylvatica, Pteris, Blechnum,* with the two species of *Hymenophyllum,* masses of *Sphagnum* and many mosses and hepatics covering the rocks and tree-trunks. On the tops of boulders *Hymenophyllum peltatum* and *Sedum anglicum* form a mat together—a strange partnership!

321. I cannot refrain from quoting here the vivid

description of these woods given by Rübel (in the Results of the First International Phytogeographical Excursion[18]) :—‘‘The dominant tree is Quercus sessiliflora which reminds us that we are in northwestern Europe. This is confirmed by a few of its associates such as Betula pubescens, Salix cinerea, Prunus spinosa, Corylus, etc. But these are quite accessory, the striking feature in this oak wood being the glittering, shining effect produced by the reflection of light which takes place in the second layer of the wood. The smaller trees and the shrubs in fact have shiny leaves and you fancy yourself in the southern country of evergreen *Laurel woods.* The holly, Ilex aquifolium, is the general dominant and specially contributes to this effect. Frequently interspersed and locally even dominant we find one of the most interesting woody plants of Ireland, the strawberry-tree, Arbutus Unedo, directing our thoughts to the West Mediterranean. This species embellishes the Killarney woods, not only as a shrub of 3–5m. in height, as we are used to seeing it in Corsica, but as a tree of considerable height . . . Another laurel-leaved plant, Rhododendron ponticum, catches your eye by its luxuriance. It is not native, but it spreads subspontaneously and evidently feels quite at home here. The many epiphytes give a nearly tropical aspect to the woods; the oaks bear epiphytically Polypodium vulgare, Geranium Robertianum, Ilex Aquifolium and Saxifraga umbrosa [spathularis]. Even Rhododendron ponticum ascends the lofty habitat of a branch of Prunus spinosa. Very striking also are the masses of thin-leaved Hymenophyllum tunbridgense and peltatum; and in this fern sward we find a plant of Sedum album.[19] When we look at the luxuriant heathy undergrowth, other countries come to mind. Calluna bushes I measured had a height of about 2m., the tallest I found was 220cm. high, giving the illusion of a ‘Monte verde’ of the Canaries, or of the mountain heaths of Corsica, where Erica arborea and other species cover great tracts of country.’’ He goes on to draw a close parallel between the Killarney woods and those of the mountains of Teneriffe.

322. The tree-limit, which here (unlike most places in Ireland) is quite natural and undisturbed, lies at 600–800

[18] E. Rübel: ‘‘The Killarney Woods.’’ New Phytol., 1912, 54–56.
[19] Corrected on p. 404 to *anglicum.*

ft., the species which range highest being *Quercus, Betula, Ilex,* and *Sorbus Aucuparia.* The effect on the "Baum-grenze" of aspect, wind and so on, can be excellently studied here.

Corylus, Cratægus, Euonymus, Rhamnus catharticus and *R. Frangula, Juniperus communis,* favour the lime-stone woods, while *Juniperus sibirica, Myrica,* etc., prefer the Old Red Sandstone. The smaller islands in the Lower Lake, formed of limestone often fantastically carved by the action of water, have a native vegetation in which *Arbutus* forms a striking and pleasing feature, growing with *Taxus* and *Sorbus* spp., and many calcicole species occur. A conspicuous plant is a bright yellow form of *Orobanche Hederæ.* The glory of Killarney is the *Arbutus* (**129**), the finest of the Hiberno-Mediterranean species (plate 13). It finds here its Irish focus; in much diminished quantity it extends as far SW as L. Currane, and S as far as Adrigole on Bantry Bay. See also **420.**

The most characteristic area is the wooded slopes and broken lower grounds of the Old Red Sandstone hills, especially around the Upper Lake (plate 14). Here many of the more famous plants—*Saxifraga Geum, S. spathularis, Arbutus, Pinguicula grandiflora, Euphorbia hiberna,* and the two species of *Hymenophyllum,* attain a delightful abundance and luxuriance, the last-named, for instance, clothing tree-trunks and wrapping rocks in continuous and luscious sheets of dark green which may cover several square yards. The rich and local limestone flora on the other hand, which is best developed along the N shore and islands of the Lower Lake, includes a group of plants of comparatively xerophytic habitat :—

Teesdalia nudicaulis
Rhamnus Frangula
  catharticus
Vicia sylvatica
Sorbus Mougeotii *var.*
 anglica

Sorbus rupicola
 porrigens.
Galium sylvestre
 boreale

all of which have here their only Kerry station. The first named *Sorbus* is not at present known elsewhere in Ireland and is extremely rare in Britain.

The same conditions of efficient drainage are probably responsible for the occurrence of *Silene maritima*

ARBUTUS UNEDO IN WOODS AT KILLARNEY,
Forming trees 25–30 feet in height.

R. Welch, Photo.

PLATE 14.

THE UPPER LAKE OF KILLARNEY,
Looking NE. Muckross Lake in the distance.

R. Welch, *Photo.*

Cerastium tetrandrum, C. semidecandrum, and *Armeria* on the shores of the Lower Lake—a group very rare inland at low levels, and rivalling the similar group found on the L. Neagh shores (**464**). But suitable habitat alone will not account for the presence of *Asplenium marinum* on rocks near the Upper Lake, its only truly inland station in Ireland. Woods on the limestone yield the humus-loving *Monotropa, Cephalanthera ensifolia, Lathræa.*

323. The local hydrophytes and marsh plants include *Caltha radicans, Lobelia, Polygonum laxiflorum, Utricularia intermedia* and *U. ochroleuca, Ceratophyllum, Potamogeton prælongus, P. densus, P. nitens, P. angustifolius, P. Seemenii* (= *gramineus* × *polygonifolius,* River Laune), *Naias flexilis, Eriophorum latifolium, Lastrea Thelypteris, Isoetes lacustris, I. echinospora, Pilularia, Nitella batrachosperma, N. translucens, N. flexilis.* Of these much the most interesting is *Naias flexilis.*[20] It grows near the W end of the Upper Lake, and in a number of spots in the Muckross and Ross Island part of the Lower Lake. In Kerry it occurs also in Caragh Lake: in Ireland also in W Galway and W Donegal. Its distribution elsewhere has been dealt with already (**172**). The American element is represented here by *Sisyrinchium angustifolium* (**159**) and *Juncus macer* (**165**).

In the "Long Range" (the stream which joins the Upper to the Lower Lake), a curious Pondweed which catches the attention by its abundance and has been the subject of discussion is considered to be a barren form of the common *Potamogeton natans.*

Other plants of the Killarney neighbourhood which deserve mention are:—

Rubus adenanthus	Rubia peregrina
regillus	Crepis paludosa
iricus	Hieracium orimeles
Linum bienne	iricum
Rosa micrantha (**108**)	Wahlenbergia hederacea
Carum verticillatum (**119**)	Centunculus minimus
Pimpinella major	Microcala filiformis (**134**)
Galium erectum	Bartsia viscosa (**142**)
uliginosum	Orobanche Hederæ
sylvestre	Pinguicula lusitanica (**146**)

[20] See Journ. Bot., 1886, 83.

Mentha Pulegium
Calamintha ascendens
Stachys officinalis
Ophrys apifera
Potamogeton coloratus
Rhynchospora fusca (**177**)
Carex Bœnninghauseniana
    aquatilis (**181**)
    hibernica (**182**)
    Hudsonii
    gracilis
    limosa
    lasiocarpa
Festuca sylvatica

Asplenium Adiantum-
    nigrum *var.* acutum
    (**195**)
Cystopteris fragilis
Lastrea æmula
    montana
Phegopteris polypodioides
Osmunda regalis
Equisetum trachyodon
    (**204**)
      variegatum *var.*
    Wilsoni (**205**)
Lycopodium inundatum
    clavatum

Of these, *Carex hibernica* (**182**) was described from its Kerry station (above Galway's Bridge on the old Killarney-Kenmare road) by Arthur Bennett in 1897 (Journ. Bot., 250)—a curious plant, unknown elsewhere, believed to be *C. aquatilis* × *Goodenowii*. The var. *Wilsoni* (**205**) of *Equisetum variegatum* is considered endemic in Ireland; it grows on the shores and islands about Ross Bay on the Lower Lake, also in Muckross demesne and by Caragh Lake.

Trichomanes radicans (**192**) formerly grew in some abundance on the Killarney hills, as about Torc; but here more than anywhere else in Kerry it has suffered from the greed of collectors both commercial and botanical, and it is now extremely rare. *Utricularia Bremii* has its only Irish station, so far as at present known, in the Gap of Dunloe. A few plants of the Flesk valley above the lakes of Killarney are mentioned in **313**.

Among the rather large alien flora *Aster* sp., *Lactuca muralis*, *Mimulus Langsdorfii* are quite naturalized. *Dianthus plumarius* has grown on a limestone promontory on the Lower Lake for over 60 years. The presence of many gardens in this favourite spot has resulted in a considerable enrichment of the native flora from outside sources.

The most fascinating part of the Lower Lake, from both the botanical and the scenic point of view, is the limestone area on and around Ross Island. Further E Upper Carboniferous rocks come in, and the shore is low, with extensive gravelly beaches and a poor calcifuge flora.

The southern shore is occupied by the Old Red, in which the deep valley of the Upper Lake is excavated. Ross Island itself, largely in its primitive condition in spite of planting up, is a very interesting study, with limestone ridges colonized by *Arbutus, Taxus,* the rare *Sorbus anglica, Rhamnus* spp., *Rubia,* etc.; swampy jungles with *Lastrea Thelypteris, Polygonum laxiflorum,* and great thickets of *Osmunda* six feet high; dry places densely occupied by *Silene maritima and Armeria maritima*; and rubbly stony shores with *Caltha radicans, Equisetum Wilsoni, E. trachyodon,* and so on. It will be noted that almost all the Lusitanian species, so abundant on the Old Red Sandstone area, are absent here on the limestone.

N B

THE DINGLE PENINSULA AND BRANDON.

324. The most northerly of the imposing mountain-promontories of Kerry differs from the others in having a comparatively narrow neck, and its back-bone of mountains ending abruptly in a great transverse mass, rising in Brandon to 3127 ft. and dropping down in precipices into the ocean on the N. The town of Dingle (good hotel) lies on the shore S of this mass, but the better place to stay for the exploration of Brandon, the most interesting of the Kerry mountains, is at Cloghane (small inn) at the E base of the hill, and at the head of a sandy creek which opens to the northern sea. A light railway starts along the N shore from Tralee on the main line, and crossing the peninsula at 691 ft. descends to the S shore at Dingle, giving off a branch northward to Castlegregory, a very useful centre (small inn). The narrow neck of the peninsula contains the fine range of Slieve Mish (Baurtregaum, 2796 ft.) and great accumulations of sand fringe both its N and its S coast (of which the latter has been dealt with already (**318**)).

The prevailing rocks are slates and sandstones, as elsewhere in Kerry, but in the W half of the peninsula they consist of the problematical "Dingle Beds," usually placed at the base of the Lower Devonian.

325. On the E side of Brandon a tremendous cliff-range drops into a great coomb in which a series of lakes lie one above the other like steps, the highest (2350 ft.) yielding the most elevated stations in Ireland for *Myriophyllum*

alterniflorum and *Isoetes lacustris*. This coomb and its
cliffs form for the botanist the richest alpine ground in
the county, and for mountain scenery are unsurpassed anywhere in Ireland. Of the alpine plants which occur
Alchemilla alpina (found also on Boughil, see **311**) and
Poa alpina (**187**) have each only one other Irish station
outside Kerry (in Wicklow and Sligo respectively);
Thalictrum alpinum[21] (not seen recently), and *Polygonum
viviparum* are rare west coast mountain species. All but
four of the Kerry "Highland" plants are gathered
together on Brandon, the species present including, in
addition to the before-mentioned, *Cochlearia alpina*,
Subularia, *Saussurea*, *Oxyria*, *Salix herbacea*, *Juniperus
sibirica*, *Deschampsia alpina*, *Polystichum Lonchitis*, etc.
An old record for *Saxifraga aizoides* from Connor Hill
near by is much in need of verification. On the lower
grounds, *Trichomanes radicans* still survives the greed of
collectors, and here, or among the alpines, the characteristic
Kerry plants come in in unusual abundance—*Saxifraga
spathularis*, *S. Geum*, *Pinguicula grandiflora*, *Bartsia
viscosa*, and many other things such as *Sagina subulata*,
Saxifraga Drucei, *S. Sternbergii*, *S. rosacea*, *S. hirta*, *S.
affinis* (not seen since 1805), *S. incurvifolia* (a galaxy of
rare Hypnoid Saxifrages unequalled elsewhere in Ireland
or in Britain—see **117**), *Sedum roseum*, *Lycopodium
alpinum*. One or two of the most notable Kerry plants,
it is interesting to observe, are in this region unexpectedly
rare or absent—*Euphorbia hiberna*, for instance, and
Eriocaulon. The special plant of the Dingle Peninsula is
the little *Sibthorpia* (**137**), which is abundant along the N
slopes from Brandon to Tralee, and from sea-level to
1700 ft. In Ireland it is not found elsewhere.

326. The town of **Dingle** offers the best accommodation
for the traveller to be found on this wild peninsula, but
the mountain-slopes on this S side are featureless and dull,
and not to be compared in interest and picturesqueness to
those of the N side. An old record for *Inula crithmoides*,
from the coast near Dingle, requires confirmation; it is a
very local plant in Ireland, and exceedingly rare in Kerry.
Other plants reported from the neighbourhood of Dingle
are *Arabis Brownii*, *Spergularia rupicola*, *Erodium
moschatum*, *Rubia peregrina*, *Asperula cynanchica*, *Bartsia*

---

[21] See A. Ley in Journ. Bot., 1887, 374.

viscosa, Utricularia intermedia, Atriplex portulacoides, Juncus subnodulosus, J. macer (Connor Hill), *Rhynchospora fusca, Carex limosa, Trichomanes radicans.*

At **Cloghane,** by the sea at the E base of Brandon, *Rubus iricus* occurs on roadsides, *Erodium moschatum* and *Euphorbia portlandica* favour sandy ground, and *Zostera nana* grows in abundance on mud-flats. *Cochlearia grœnlandica* and its hybrid with *officinalis* are on the estuary and at Caher Point, and *Eriophorum latifolium* in bogs.

The W part of the Dingle Peninsula, a wild wind-swept area surrounded on three sides by the Atlantic, is remarkably rich in pre-Christian and early Christian remains, which the visitor should not fail to see—the extraordinary settlement W of Ventry, with its promontory forts, raths, crosses, inscribed stones, souterrains and hundreds of stone bee-hive huts;[22] the famous Oratory of Gallerus; the interesting 12th-century church of Kilmalkedar; and many other monuments. The extreme W, facing the Blasket Islands, possesses magnificent coast scenery; and here, curiously enough, *Orobanche rubra* has its only station in SW Ireland. Other plants of this storm-swept area are *Althœa officinalis* (an early introduction), *Trifolium fragiferum, Pimpinella major, Blackstonia* (at Ventry), *Epipactis palustris, Carex diandra.* Deelick Point (on Brandon Head) and the "Three Sisters" near Smerwick yield the peculiar dwarf form of *Ophioglossum vulgatum* usually now set down as var. *polyphyllum* Braun (var. *ambiguum* C. & G.).

The craggy Blasket Islands deserve a separate section (**329**); the NE part of the Dingle Peninsula, with its extensive sands and comparative shelter, has now to be dealt with.

H. C. HART: "Notes on the Plants of some of the Mountain Ranges of Ireland." Proc. Roy. Irish Acad., 2nd ser., **4** (Science), 211–231. 1884.

CASTLEGREGORY AND CAMP.

327. At Castlegregory a flat peninsula of sand and limestone extends far out to Rough Point, with the Magheree islets beyond, while close to Castlegregory lies

[22] See R. A. S. MACALISTER in Trans. Roy. Irish Acad., **31**, 209–344. 1899.

the shallow Lough Gill. This is a rich area botanically, and very different from the rest of the Dingle Peninsula. Among its plants are

Arabis Brownii
Erodium moschatum
Rubus iricus
Carum verticillatum
Pimpinella major
Trifolium fragiferum
Asperula cynanchica
Wahlenbergia hederacea
Centunculus minimus
Cuscuta Trifolii
Bartsia viscosa

Juncus subnodulosus
Potamogeton nitens
Ruppia maritima
Eleocharis uniglumis
Scirpus rufus
Rhynchospora fusca
Carex diandra
    limosa
Asplenium lanceolatum
Chara canescens

Of these, the last two are the most interesting. The *Asplenium* has its only Kerry station here, at Camp; and the *Chara* is abundant in the lake at Castlegregory. It appears to be the sandy soil, rather than the fact that the underlying rock is limestone, that induces the presence of a number of plants which, though widespread in Ireland, are rare under the peculiar edaphic conditions prevailing in Kerry.

TRALEE AND SLIEVE MISH.

328. The last remark applies also to the neighbourhood of Tralee and the limestone area to the NW of that town, but here the limestone lies on the surface and forms a rocky shore. All the same, the rarest local plants are not calcicole—*Rumex maritimus* and *Chenopodium rubrum* near Fenit, *Zostera nana* on the shore at Spa. *Crambe*, found some years ago at the latter place, appears to have been almost certainly an outcast. The immediate neighbourhood of Tralee is flat and estuarine. *Cochlearia anglica, Althœa, Galium uliginosum, Verbena, Calamintha ascendens, Epipactis palustris, Allium Scorodoprasum, Potamogeton densus* occur. At Barrow Harbour, to the NW, *Arabis Brownii* is on the sand-dunes, and *Althœa, Trifolium fragiferum, Rubia peregrina, Asperula cynanchica, Limonium binervosum, Juncus subnodulosus*, etc., re-appear. The Tralee district invites further exploration, for in addition to the sands, limestones and sheltered waters of or around Tralee Bay, there is the fine ridge of Slieve Mish (2796 ft.) with a rather limited alpine flora, and most of the characteristic Kerry plants—*Saxifraga spathularis, S. Geum,* and their hybrids (**115**), also *S. hirta* (**117**), and *S. sponhemica; Pinguicula grandiflora* (**145**), *Asplenium Adiantum-nigrum* var. *acutum* (**195**), *Cystopteris fragilis, Osmunda,* and *Sibthorpia* (**137**) in its most easterly Irish station. The grass known as *Deschampsia alpina* is here also, but the identity of the Irish form with the plant of Roemer and Schultes is not quite clear. Though the hills are high, the surface is mostly rather smooth, and cliffs are rare. Among other plants of the Tralee area are *Limonium humile, Bartsia viscosa, Stachys officinalis, Euphorbia Paralias, Carex diandra.*

Among plants which are frequent in the Dingle Peninsula (some of which have been already named in connection with particular stations) *Lobelia Dortmanna, Mentha Pulegium, Pinguicula lusitanica, Eleocharis uniglumis, Scirpus rufus, Lastrea montana, L. œmula* may be mentioned also.

H. C. HART: "Notes on the Plants of some of the Mountain Ranges of Ireland." Proc. Roy. Irish Acad., 2nd ser., 4 (Science), 211–231. 1884. SCULLY: Fl. Kerry.

THE BLASKETS.

329. The Blasket Islands, which are the peaks of submerged hills forming the seaward continuation of the high mountain-range of the Dingle Peninsula, form the most westerly part of Ireland and of Europe. They are a group of six small islands with many outlying rocks, and are remarkable for their bold craggy outlines and wild scenery. With open ocean on three sides, and deep water all round, they are exceedingly exposed. Interest centres on the Great Blasket, much the largest of the group, a knife-edge four miles long and half a mile wide, rising at two points to nearly 1000 ft. It alone is inhabited, a group of cottages and small patches of cultivation (potatoes and oats) occupying the steep butt-end which faces NE towards the mainland. Elsewhere cliffs and one-to-one slopes prevail, and botanizing on the Great Blasket is very like botanizing on the steep roof of a church (plate 15). The vegetation is closely shorn by storms and by grazing. The S slope is occupied mainly by short grass full of *Plantago maritima*

and *P. Coronopus,* with patches of *Pteris* in the hollows.
The N slope in its upper part is tenanted by dwarf *Calluna,*
which occupies much of the high dry back-bone of the island.
In the extreme W, where the ridge falls towards the ocean,
there is a little plateau covered with *Juncus squarrosus*
heath. Beyond that, the *Plantago* sward is replaced by a
nearly pure *Armeria* association, which gives way, about
100 ft. above sea-level, to spray-swept rocks with *Armeria,
Spergularia rupicola,* and *Crithmum.*

The flora in general is very stunted and limited, and
all but a few hardy species must be looked for in nooks
where exposure reaches a minimum, or on steep ground
where the ubiquitous sheep cannot penetrate. A few
trickling springs are the only habitat for water-loving
species. Altogether 208 species have been recorded from
the Great Blasket. The remaining islands of the group,
several of which support great bird colonies, add only
8 species more. This is an extraordinarily small total even
for the exposed islands of the west coast, only equalled by
the 147 species of Tory in Donegal (**449**). Rare plants are
almost absent: *Saxifraga spathularis* is the sole representative of the southern group that characterizes W Ireland:
among the more interesting species are *Cochlearia grænlandica* (**87**), *Cerastium arvense* var. *Andrewsii, Listera
cordata, Hymenophyllum peltatum, Ophioglossum vulgatum*
var. *polyphyllum*; the last forms a large colony on the
ridge over the village at 600 ft., the plants very small and
crowded together—about 100 to the sq. ft. (see **200**). In
spite of the extreme exposure, 11 species of ferns are
present, including *Lastrea æmula* and the shade-loving
Athyrium Filix-fæmina. Asplenium marinum grows on the
old signal tower 750 ft. above the sea; *Carex arenaria* and
Ammophila arenaria on the edges of cliffs at an almost
equal elevation, where there is no trace of sand (see also
448). The presence of these plants and the fact that
Senecio aquaticus, the prevailing Ragweed on most of the
western islands, is here replaced by *S. Jacobæa,* are an
indication of the rapid drainage of the steep slopes. The
proportion of introduced species to the total flora is 22½ per
cent.

On several islands of the group, including some never
inhabited, *Lavatera* (**97**) grows, undoubtedly native. This
fine plant is such a favourite with cottagers that its
standing along the coast is usually much obscured.

PLATE 15.

THE GREAT BLASKET,
Looking NE, towards Croaghmore, 961 ft.

A. W. Stelfox, Photo.

PLATE 16.

SCIRPUS TRIQUETER BY THE SHANNON AT LIMERICK.

The substratum is black tidal mud. At high water the plant is submerged. It is accompanied by S. Tabernaemontani, Alisma Plantago-aquatica, Apium nodiflorum, Polygonum Persicaria, Potomogeton densus, Chara sp., and green Algae. July.

R. Welch, Photo.

The flora of the Blaskets is interesting on account of its very poverty, and the scenery throughout the group is magnificent, rendering a visit a most exhilarating experience. Landing on the Great Blasket (from Dunquin in a curragh) is easy in ordinary summer weather, but all the other islands are difficult of approach, and need a very calm day. Rough accommodation is obtainable.

R. M. BARRINGTON: "Report on the Flora of the Blasket Islands, Co. Kerry." Proc. Roy. Irish Acad., Ser. 2 (Science), 3, 368–391. 1881. R. LL. PRAEGER: "Notes on the Flora of the Blaskets." Irish Nat., 1912, 157–163.

NORTH KERRY.

330. From Tralee to the Shannon (and on northward far into Clare), save for a ridge of Old Red Sandstone projecting at Kerry Head, Upper Carboniferous rocks, mainly shales and sandstones, prevail, rising inland into featureless hills reaching 1000 ft. or more. The hills are deeply peat-covered, and the monotony of their vegetation is equalled only by that of the flattish lowlands, where rushy pastures extend for mile after mile. In comparison with the rest of Kerry this is a very dull area both scenically and botanically, interesting chiefly as showing the rapid decrease of the characteristic Kerry flora under changed conditions. *Saxifraga spathularis, S. Geum, Sisyrinchium angustifolium, Juncus macer, Trichomanes radicans,* and all the alpines are absent: *Carum verticillatum* and *Bartsia viscosa* soon die out as one goes N: *Rhynchospora fusca* follows; *Pinguicula grandiflora* (**145**) struggles on to the Shannon sparingly; *Euphorbia hiberna* (**154**) alone maintains a local abundance up to the N limit of the county. Against this, the region harbours a number of species which, though widespread in Ireland, are very rare in Kerry. These are largely inhabitants of sandhills and marshes. The rarest plant is *Scirpus nanus* (**174**), which grows in great abundance on tidal mud for several miles along the River Cashen. From above Ferry Bridge to below Dysert church it forms a thin sward an inch high from half-tide level downwards, and its roots and creeping stems consolidate the sandy mud into a somewhat compact layer an inch thick. Often it is covered by a thin deposit of fresh fine mud, and is then difficult to see. When hidden by high tide its presence may often be detected by examination of the storm debris on the banks, frequently composed of a felt of this plant mixed with *Callitriche* and *Ulva*. *S. nanus* has only one other station in Ireland (Arklow, on the E coast) and in Great Britain occurs in one spot in N. Wales and in a very few localities along the S English coast. The Cashen enters the sea at the S end of a range of dunes which extend N to **Ballybunnion,** a watering-place of high local repute.

Among the plants of the area are *Arabis Brownii, Spergularia rupicola, Erodium moschatum, Althœa officinalis, Trifolium striatum, T. fragiferum, Rubus iricus* (Tarbert), *Pimpinella major, Rubia peregrina, Asperula cynanchica, Dipsacus, Crepis paludosa, Limonium humile, Centunculus, Blackstonia, Pinguicula lusitanica, Utricularia intermedia, Mentha Pulegium, Chenopodium rubrum, Euphorbia portlandica, E. Paralias, Epipactis palustris, Ophrys apifera, Scirpus rufus, S. filiformis, Eriophorum latifolium, Carex punctata* (its N limit in Ireland), *C. limosa, Potamogeton coloratus, Festuca sylvatica, Lastrea œmula, Osmunda. Valerianella carinata* (not seen recently) is or was at Ballyheige.

The most interesting of several old records which have never been verified are those for *Matthiola sinuata* (**84**)— near Beal Castle (1756) and at Banna (1878). The terrain is suitable, and some of these strand plants are notoriously irregular in their appearance. Possibly the *Matthiola* may still be restored to the Kerry flora, which is highly desirable in view of its rarity in Ireland (Clare and Wexford only). The case of *Lathyrus maritimus* (**318**) encourages this hope.

Tarbert, Listowel, Ballybunnion, Tralee, etc., offer accommodation to the visitor.

THE SHANNON ESTUARY.

331. The Shannon, the largest river in Ireland and larger than any in Great Britain, 214 miles in length, rises among the Cavan hills, and drops 186 ft. in 10 miles to L. Allen (**428**). Thence it flows southward across the Limestone Plain for 130 miles, falling only 51 ft., and forming at intervals great lake-like expanses, with winding shores and numerous islands. At the lower end of L. Derg the river cuts through high hills of Old Red Sandstone and Silurian rocks, and drops 97 ft. in 18 miles to reach sea-level at Limerick, whence it forms a great estuary, meeting the Atlantic 55 miles further on, between Loop Head and Kerry Head. The estuary is river-like and muddy in its upper part, with high dykes protecting low marshy meadow-land. The middle portion is lake-like and much wider, with stony, gravelly or muddy shores; and the lower reaches are quite marine, with clean strands, rocks, cliffs, and sand-dunes, and a full maritime flora.

In the upper estuarine reaches, at and below Limerick, the stream, a few hundred yards wide, with a rise of tide of over 20 ft., is fringed with *Phragmites, Scirpus Tabernæmontani, S. maritimus,* and the rare *S. triqueter* (**176**), here in its only Irish station, forming with *Typha angustifolia* and other tall plants a reed-zone among which grow *Cochlearia anglica, Nasturtium sylvestre, Potamogeton densus.* In the marshy meadows which fringe the river are *Leucojum æstivum* (**161**), here abundant and native, *Allium vineale, Carex riparia, Hordeum nodosum.*

A few miles below Limerick the river begins to widen, and there are broad expanses of water, with islands and often gravelly beaches. The R. Fergus comes in from the N through an islet-studded estuary which is long and wide (**342**). A very rare grass grows here, *Glyceria Foucaudii* (**188**), occupying in abundance the foreshore of some of the islands. *Spartina Townsendii* has in recent years been planted in a number of places—see Proc. R. Irish Acad., **41** B, 119–120, 1932. *Atropa Belladonna* is naturalized and abundant on Coney Island. Other characteristic plants of the middle section are *Cochlearia anglica, Apium graveolens, Œnanthe Lachenalii, Dipsacus, Artemisia maritima, Limonium humile, Beta, Allium vineale. Erodium moschatum* is at Glin.

Beyond the narrow turn at Tarbert the conditions become quite marine. There are rocks and cliffs with *Spergularia rupicola* and *Crithmum,* and sands with *Glaucium, Raphanus maritimus, Euphorbia portlandica,* and so on. At its entrance into the Atlantic the Shannon is nine miles wide. In the tidal portion there are over 200 miles of shore-line, much of which has not been thoroughly explored.

The hinterland on both the S (Limerick) and N (Clare) side of the Shannon estuary offers interesting ground. This is chiefly in the upper part, where Carboniferous limestone prevails.

WEST LIMERICK.

332. West of Foynes (**337**) Upper Carboniferous rocks, mainly shales, occupy the ground, rising into barren hills and moors with a calcifuge flora, rushy or heathery. This area is a continuation of the N Kerry region dealt with already (**330**), and it continues across the river far up into Clare. The Limerick portion (W of Newcastle, Rathkeale, and Foynes) presents extensive rushy grass-heaths, rising into peat-covered hills of over 1000 ft. high. The flora is poor, as is usual on the Coal-measures, *Drosera anglica* and *Carex limosa* being among the few rarer plants recorded from the higher grounds. In the W, along the Galey R., *Euphorbia hiberna* (**154**) grows in abundance, a northern outpost of its great Cork-Kerry area. On the E slope, *Epilobium angustifolium,* very rare as a native in S Ireland, grows at Glenastar waterfall. The extensive woods of Mount Trenchard and the district W of Foynes are more productive, and harbour *Neottia, Carex strigosa, Milium, Festuca sylvatica.* Along the White R. near by grows a rare hybrid rose, *R. canina* × *spinosissima* (*hibernica* var. *glabra*) : and *R. stylosa* is not infrequent in this area. *Erodium moschatum* is at Newcastle, and *Orchis majalis* on the hills.

S. A. STEWART: ''Report on the Botany of South Clare and the Shannon.'' Proc. Roy. Irish Acad., Ser. 3, **1**, 343–369. 1890. M. C. KNOWLES and C. G. O'BRIEN: ''The Flora of the Barony of Shanid.'' Irish Nat., 1907, 185–201.

THE COUNTY OF LIMERICK.

333. The County of Limerick, lying between Cork and Kerry on the one side and Clare on the other, might be expected to yield a number of the plants characteristic of those favoured regions—particularly the Lusitanian and American species; but, no doubt largely owing to its low elevation and inland position, it is quite poor in this respect; *Euphorbia hiberna* is the only Lusitanian species present. On the other hand, half of it is composed of Carboniferous limestone, which connects with the great area of that rock which forms the Central Plain; but Limerick is much cut off from the plain by masses of older non-calcareous rocks, and the Central Plain flora also is sparsely represented. Remains of old volcanoes SE of

Limerick city form hills of hard igneous rocks of about the same age as the limestone which surrounds them, but the change has no marked effect on the flora.

Interest centres in the N, where the estuary of the Shannon extends (**331**), with good botanical ground adjoining, mostly situated on the low-lying limestone (**338**). The seaward portion of the estuary, carved out of Upper Carboniferous non-calcareous deposits, is more elevated, but more inhospitable, and poor in its flora (**332**). In the extreme SE, the W part of the lofty Galtee range (**250**) lies within the county, and in the N the Slieve Felim hills (**251**) rise along the border. In spite of the deficiencies mentioned, the flora of the county is rich and interesting.

LIMERICK CITY.

334. The visitor tarrying for a little botanizing in Limerick and the surrounding area may spend an interesting day close to the city. The district is low-lying, formed of Carboniferous limestone with patches of volcanic ash; and the solid geology is obscured by wide tracts of alluvium. The more interesting plants are either hydrophile or halophile, haunting wet meadows and ditches or the muddy banks of the Shannon. To the latter category belongs the most interesting of the local species—*Scirpus triqueter* (plate 16), which grows in great profusion on the mud of the foreshore from Limerick down for some miles (**176**). With it are associated *Cochlearia anglica* (**86**), *Nasturtium sylvestre, Eleocharis uniglumis, Potamogeton densus. Typha angustifolia* grows 11 ft. high on the tidal foreshore. The adjoining marshes yield *Lathyrus palustris* (**103**), *Leucojum æstivum* (**161**, plate 17) (native here), *Hordeum nodosum,* with great abundance of *Allium vineale* and *Carex riparia.* Near the river above tidal influence are the four British species of *Lemna, Polygonum laxiflorum, Potamogeton densus, Sium latifolium, Sagittaria, Butomus, Carex aquatilis* (**181**). By roadsides and in meadows *Erodium moschatum, Pimpinella major, Crepis biennis, Blackstonia, Ophrys apifera* are characteristic species. A plant much in need of re-finding is *Scrophularia alata*, of which a specimen collected by Isaac Carroll, an accurate botanist, and labelled "near Limerick, Aug. 1848" is in the British Museum (Journ. Bot., 1877, 238, and 1909, 385; Irish Nat., 1909, 222).

Certain old quarries and waste ground about Limerick have yielded a remarkable crop of alien plants, owing to the dumping of rubbish of various sorts. While most of these are but fleeting casuals, *Lepidium latifolium, L. Draba, Aster* sp., *Tragopogon porrifolium* and *Mercurialis annua* have established themselves.

Cratloe Wood, on the Clare side of the Shannon, a famous hunting-ground for insects, is a bit of original forest. Its plants include *Neottia* and probably other things.

R. A. PHILLIPS: ''Some Notes on the Flora of Limerick.'' Journ. Limerick Field Club, 3, No. 9, 32–35. Plate. 1905.

335. Lough Gur.—A ten-mile drive S from Limerick brings one to the interesting lake called Lough Gur, the flora of which includes *Nasturtium sylvestre, Chenopodium rubrum, Rumex maritimus, Ceratophyllum demersum,* and other species rare in Ireland. Bogs near by yield *Sparganium minimum, Potamogeton coloratus, P. obtusifolius, Lemna polyrrhiza.* About L. Gur numerous skeletons of the ''Irish Elk'' (*Cervus giganteus*) and objects of early human manufacture have been dug up, and in its vicinity there is a wealth of Bronze Age and also Early Christian monuments. About the same distance WSW of Limerick is the fine old demesne of Curragh Chase, the flora of which includes *Monotropa, Neottia, Carex Pseudo-Cyperus.*

Adare (*Ath Dara,* the ford of the oaks), situated on the River Maigue, still tidal here though 8 miles from the Shannon, is famous for the beautiful ruins of three monasteries and a castle, and for its fine demesne. It is a small village with a pleasant hotel. *Euphorbia hiberna* (**154**) is recorded from here, but not seen recently. *Hieracium lepistoides* (var. *sublepistoides*) is abundant at Adare Manor; *Rhamnus catharticus, Crepis biennis, Ophrys apifera* also occur.

336. Castleconnell, six miles up the river from Limerick, where the Shannon falls picturesquely over a series of limestone steps, is the lower limit for some of the characteristic Shannon plants, such as *Teucrium Scordium* (**148**) and *Galium boreale;* its flora includes *Rhamnus catharticus, Rubus hesperius, Apium Moorei* (**118**), *Butomus, Potamogeton nitens.* There is an extensive peat bog close at hand—one of the few in Co. Limerick—with *Drosera anglica, Andromeda,* etc. Good hotel.

PLATE 17.

LEUCOJUM ÆSTIVUM.

At Ballinacurra Creek below Limerick, May, forming large clumps in
wet clayey meadow-land.

G. Fogerty, Photo.

PLATE 18.

GLYCERIA FOUCAUDII NEAR FOYNES, CO. LIMERICK.

On Trummera Bir. Forms large upright tufts 2 ft. high at and below high-water mark,

Glenstal is a fine demesne on the hills 10 miles E of Limerick. *Trichomanes* (**192**) has been found in more than one spot, and *Hymenophyllum tunbridgense.* The glen yields several local species such as *Prunus Padus, Crepis paludosa, Festuca sylvatica*; and some plants are naturalized there, including *Geranium sylvaticum, G. phæum, Saxifraga granulata.*

R. A. PHILLIPS: "Some Notes on the Flora of Limerick." Journ. Lim. Field Club, 3, 32–35. 1905. R. LL. PRAEGER: "Notes on the Limerick Flora." Irish Nat., 1900, 260–5.

337. Foynes, a small village, forms a good centre for work in W Limerick and along the Shannon. It offers a fine contrast of vegetations. East of the town are low-lying limestones, frequently bare of covering, and intersected with extensive muddy channels. West, the ground rises at once into hills formed of shales and sandstones, with fine woods and calcifuge flora. Some of the plants to be found about Foynes are mentioned in **331, 332, 338.** Among the more interesting maritime species which grow here are *Glyceria Foucaudii* (**188**, plate 18), *Artemisia maritima* (plentiful), and *Cochlearia anglica*; *Papaver Argemone, Viola hirta, Geranium columbinum, Linum bienne, Rubia peregrina, Crepis biennis, Tragopogon porrifolium, Juncus subnodulosus, Potamogeton densus, Hordeum nodosum* also occur. A plot of *Spartina Townsendii* was laid down at Foynes in 1931.

M. C. KNOWLES and C. G. O'BRIEN: "The Flora of the Barony of Shanid." Irish Nat., 1907, 185–201.

THE ASKEATON LIMESTONES.

338. The low-lying bare limestones of Central Clare (**344**) are continued S across the Shannon into Limerick, and form a conspicuous feature about Askeaton and Mullough. The Upper Carboniferous Limestone, which forms the 'crags' of Clare, is here replaced by the Lower Limestone. This area exhibits a flora of the characteristic facies of Burren, but very much reduced. *Euphrasia salisburgensis* (**141**) is still abundant, and with it are *Rubia peregrina* (**121**), *Asperula cynanchica, Ceterach*, etc. *Viola hirta, Geranium columbinum, Cornus sanguinea, Carlina, Spiranthes spiralis, Anacamptis pyramidalis,* are calcicole plants which are also present in greater or less quantity.

The noticeable feature of the Limerick limestones is the great reduction of the characteristic crag flora as compared with even the most S portion of the limestones of Clare, immediately to the N. This may be the result of the barrier offered to plant-migration by the wide estuary of the Shannon, or the explanation may be lithological rather, depending on the relative suitability for this group of plants of the Upper and the Lower Limestone.

There is a good deal of scrub, consisting of *Cratægus, Ilex, Prunus spinosa, P. avium, Pyrus Malus, Euonymus, Viburnum Opulus, Cornus,* and *Corylus.*

Askeaton.—About Askeaton, a straggling village built on a tidal stream running between rocky limestone banks decorated with interesting mediæval ruins, will be found *Linum bienne, Geranium columbinum, Dipsacus, Silybum Marianum, Fœniculum, Rhinanthus major* (only Irish station), *Calamintha ascendens, Salvia horminoides, Verbena, Allium vineale, Festuca Myuros*; and on walls *Euphrasia salisburgensis* (**141**) and *Orobanche Hederæ.* The neighbourhood is distinctly interesting to the botanist.

M. C. Knowles and C. G. O'Brien: "The Flora of the Barony of Shanid." Irish Nat., 1907, 185–201.

THE COUNTY OF CLARE.

339. The large county of Clare offers a bold contrast of three geological formations, each with a characteristic vegetation. The whole SW, W of a line drawn from Fisherstreet opposite the Aran Islands to the estuary of the Fergus, is occupied by the Upper Carboniferous shales sandstones, etc., already referred to (**330**) as stretching N from Kerry. The centre, from the Shannon to Galway Bay, is Carboniferous limestone, low and lake-strewn except in the N, where it rises into the bare hills of Burren (**346**). E Clare is occupied largely by sandstones and slates of Devonian and Gotlandian age, forming hilly country rising in Slieve Bernagh to 1746 ft., and dropping down into Lough Derg (**353**). The whole S edge of the county fronts the Shannon (**331**) and includes the extensive island-studded estuary of the Fergus (**342**); the W side faces the Atlantic, where marine denudation has produced the finest cliffs in the country (**345**). In the NW the Aran Islands (**352**), one of the most fascinating places in Ireland whether for the botanist, the archæologist, or the artist, belong politically to Galway, but geologically and botanically they are a seaward extension of the limestones of Burren.

LOOP HEAD AND MONMOR BOG.

340. Towards the SW, Co. Clare tails away in a long wedge between the Atlantic and the Shannon, terminating in the cliffs of Loop Head. This is a wind-swept, almost treeless district of low undulating Coal-measures, with a rather limited flora. Monmor may be taken as the centre of an extensive bog district, now much reduced by cutting, lying behind Kilrush, Kilkee, and Doonbeg. Here *Elatine hexandra*, *Eriocaulon* (**173**), and *Isoetes echinospora* have their only Clare station. Numerous bog plants occur, such as *Drosera* spp. (*D. longifolia* unusually abundant), *Utricularia* spp., *Oxycoccus*, *Scutellaria minor*, *Rhynchospora fusca* (**177**), *Carex diandra*, *C. lasiocarpa*, *Osmunda*. The form of *Juncus effusus* with spreading stems is common. *Lobelia* and *Centunculus* are also recorded. The dunes on the coast yield *Viola lutea*, *V. Curtisii*, *Asperula cynanchica*. About Kilrush, *Lepidium campestre* and *Rumex pulcher* appear naturalized.

S of Kilkee, bog-land and dunes give way to poor grasslands and continuous cliffs. The pastures are full of *Lythrum* and *Senecio aquaticus*. The cliffs are savage, black, and almost devoid of plant-life—the effect of the rapid weathering of the shales. Extreme exposure pushes the maritime plants inland for some distance; thus *Aster Tripolium* and *Glaux* grow by the roadside a quarter of a mile from the sea, at 300 ft. elevation. *Asplenium marinum* may be seen on rocks in a little glen the same distance inland. On spray-swept stony slopes *Suæda maritima* grows 100 ft. above sea-level, accompanied by *Spergularia salina*, *S. marginata*, *S. rupicola*, *Limonium binervosum*, *Cochleria danica*. The cliffs, on account of the horizontal bedding of the shales and their rapid denudation, are vertical or even overhanging, and in appearance very bare and forbidding.

Loop Head presents an unbroken precipice of about 200 ft. Above, over a considerable area, an almost pure *Armeria* sward occupies the ground. Elsewhere *Plantago* sward is well developed, giving way on the low hilltops (about 300 ft.) to dwarf *Calluna*. The rarer plants of the neighbourhood include *Sagina subulata* and *Juncus diffusus*.

Kilkee, a little watering-place situated on a delightful land-locked bay on the open ocean, forms an excellent centre for the botanist visiting this region.

S. A. STEWART: ''Report on the Botany of South Clare and the Shannon.'' Proc. Roy. Irish Acad., 3rd ser., **1**, 343–369. 1890.

R. LL. PRAEGER: ''Botanical Notes, chiefly from Lough Mask and Kilkee.'' Irish Nat. 1909, 32–40.

ENNIS.

341. Ennis is the chief town of Clare, situated near the lower end of the flat limestone depression which occupies the centre of this county and which is continued S as the wide estuary of the Fergus. The surrounding country is interesting botanically, and only half explored. About the R. Fergus and lakes near the town grow *Rubus iricus, Sium latifolium, Tragopogon porrifolium* (Clarecastle), *Mimulus Langsdorffii, Ceratophyllum demersum, Ophrys apifera, Hydrocharis, Typha angustifolia, Butomus, Sagittaria, Sparganium minimum, Carex axillaris, C. Pseudo-Cyperus.* The water flora will repay further work.

The limestone crag country which is so characteristic of N Clare is continued down to Ennis on a more limited scale, yielding *Gentiana verna* (**135**), *Euphrasia salisburgensis* (**141**), *Saxifraga hypnoides, Cornus sanguinea,* and other members of the Burren flora. The vicinity of Ballybay Lake adds *Trifolium fragiferum, Monotropa, Typha angustifolia.* The interesting demesne of Dromoland (**342**), five miles to the SE, adds several good plants to the above groups. W and SW of Ennis, where Coal-measure hills alter the facies of the vegetation, *Lobelia, Trichomanes* (**192**), *Phegopteris polypodioides,* and the two species of *Hymenophyllum* and three of *Drosera* have been found. For the flora of the great arm of the Shannon estuary that lies below Ennis, where the River Fergus debouches, see **342**. From Ennis the interesting lake and crag district about Corofin and Crusheen (**344**) can easily be explored, two lines of railway assisting.

THE FERGUS ESTUARY.

342. The River Fergus, draining central Clare, discharges into the Shannon below Ennis through a wide estuary, guarded at the mouth by a double rampart of limestone islands. The islands are formed of low reefs of rock, many covered with drift, others with the limestone quite bare. The outer ones have their shores open to the

tides and waves of the Shannon; the inner are surrounded
at low water by miles of mud. The majority are inhabited,
or grazed, or both. The W shore of the estuary represents
the edge of the limestone, and is backed by rising Coal-
measure country. The E shore consists of low limestones,
with woods and lakes. The calcicole flora is represented
on the islands by *Viola hirta, Cornus, Rubia, Carlina,
Blackstonia, Ophrys apifera*, etc. The rocks on the eastern
mainland shore are richer, and add *Saxifraga hypnoides,
Galium sylvestre, G. uliginosum, Crepis biennis, Euphrasia
salisburgensis* (**141**), *Epipactis atropurpurea* (**157**), *Sesleria*
(**186**). The woods, lakes, and marshes of **Dromoland**
demesne yield a rich flora, including *Epipactis atropurpurea, Carex paradoxa* (**179**), elsewhere in Ireland in
Westmeath only, *Carex Hudsonii, C. lasiocarpa, C. Pseudo-
Cyperus, Lastrea Thelypteris*, etc. The beaches of the
islands and adjoining mainland possess some interesting
plants. The Mediterranean *Glyceria Foucaudii* (**188**), elsewhere in Ireland confined to County Down, grows
abundantly in various places; *Artemisia maritima, Limonium humile, Cochlearia anglica* are characteristic species :
Atropa Belladonna is thoroughly established on the shores
of the Coney Island. Extensive plantings of *Spartina
Townsendii* have been made during recent years on the
mud-flats (see Proc. R. Irish Acad., **41**, B, 120, 1932).
Agrimonia odorata and *Hordeum nodosum* grow about
Killadysert, *Potamogeton decipiens* at Killone Lough. The
Coal-measure hills which rise over the W side of the
estuary are clayey and boggy as usual, with a calcifuge
flora. Their vegetation has been briefly referred to above
(**341**).

The district is rather inaccessible, and the islands
difficult of approach on account of the intricacy of the
channels through the slob-land, and the absence of harbours.
The explorer can stay comfortably at **Newmarket**, and
Killadysert also offers accommodation.

M. C. KNOWLES and R. D. O'BRIEN: "A Botanical Tour in the
Islands of the Fergus Estuary and adjacent Mainland." Irish Nat.
1909, 57–68.

SLIEVE BERNAGH AND KILLALOE.

343. South of the Slieve Aughty anticline, a group of
rather bold hills known as Slieve Bernagh rises picturesquely
on the W bank of the Shannon, over the little town of

Killaloe, at the S end of L. Derg. These uplands attain
an elevation of 1746 ft.; they are continued E across the
Shannon in the lower Arra Mountains and SW to Cratloe.
Like Slieve Aughty, these hills are formed of Old Red
Sandstone and Silurian rocks, which break through the
prevailing limestone; and they are clothed with heather.
So far no plants of particular interest are known from
the Slieve Bernagh group, *Crepis paludosa* being one of
the few records.

Killaloe is situated at the lower end of Lough Derg,
just where the Shannon breaks away from its long placid
course through the Central Plain. (That placid course is
now continued artificially on account of the dam some
miles further down connected with the Shannon Electricity
Works.) It is a fine centre for the exploration of the lower
part of that lake (**353**). The few plants on record from
the immediate neighbourhood of Killaloe include *Myriophyllum verticillatum, Crepis paludosa, Butomus, Potamogeton densus, Ophrys apifera, Festuca Myuros,* with *Poa
nemoralis* at Ballyvalley. A mile below Killaloe, on Friar's
I., *Equisetum litorale* (**201**) grew formerly; but the island
is now submerged.

THE CENTRAL CLARE LIMESTONES.

344. To the E and S, along a curve stretching from
Corranroe to Killinaboy, the Burren hills descend rather
abruptly into a wide lake-strewn limestone plain which
runs through Ennis S to the Shannon, with Coal-measure
uplands on the W and Old Red Sandstone hills to the E
beyond Crusheen and the railway. This is a flattish lonely
country, consisting of low ridges of bare grey rock or farm
land, with water-logged hollows occupied by mazy lakes
or turloughs.[23] The drainage is southward into the estuary
of the Fergus (**342**); and that erratic stream and its
tributaries flit like wraiths through the district, appearing
and disappearing, and springing underground from lake
to lake. This fact is accountable for the extreme clearness
and purity of the water of these apparently stagnant lakes
and marshes. Peat occurs about only a few of the lakelets;
usually there is a flat marginal fringe of white marl
colonized by grasses and sedges, which, in dry seasons,
forms a dazzling broad belt of sticky mud around the lakes.

[23] See **360**.

Interest centres in the wide expanses of bare limestone, which harbour a full "Burren" flora, including quantities of *Arabis hirsuta, Arenaria verna, Geranium sanguineum, Rhamnus catharticus, Rubus cæsius, Saxifraga hypnoides, Rubia peregrina, Galium boreale, G. sylvestre, Asperula cynanchica, Gentiana verna, Blackstonia, Euphrasia salisburgensis, Plantago maritima, Taxus, Juniperus communis, Sesleria, Ceterach.* A few other characteristic members of the same flora—*e.g. Dryas, Epipactis atropurpurea*—only occasionally descend from their headquarters on the Burren hills into this plain. In the NE, about Rockforest, *Potentilla fruticosa* and *Rhamnus Frangula* are abundant over a large area, and *Spiræa Filipendula,* which has its Irish headquarters about Gort, is widely but sparingly distributed. *Adiantum* occurs in some abundance in several localities, as well as *Neotinea.* The lakes and marshes yield an interesting flora, *Caltha radicans, Nasturtium sylvestre, Butomus, Potamogeton coloratus, Cladium, Carex Hudsonii, C. lasiocarpa, Chara aculeolata* being characteristic species. *Sium latifolium, Lemna polyrrhiza, Lastrea Thelypteris, Equisetum trachyodon* (**204**), *E. variegatum* also occur. *Limosella* has one of its few Irish stations at Inchiquin Lake; *Rubus iricus* also grows there, and *Typha angustifolia.* A patch of *Teucrium Scordium* (**148**) at Castle Lough represents one of the two colonies of this plant outside the watershed of the Shannon. *Viola stagnina* (**91**), which grows by Skaghard L. (*Loch Skeach Árd,* the lake of high whitethorns) and at Tirneevin, becomes abundant in the Gort area. *Ceratophyllum demersum* and *Potamogeton nitens* are among the plants of the district. *Euphrasia salisburgensis* (**141**), usually calcicole, flourishes on the tops of large boulders of impure limestone at Inchicronan L., among calcifuge plants such as *Digitalis* and *Athyrium.* *Orchis majalis* is frequent in the form *occidentalis,* extending into Burren and the shale country at Lisdoonvarna.

Corofin is the natural centre for this district, but accommodation there is very limited. The visitor will find in Ennis excellent headquarters, with railway communication to the NW, NE, and SE, and of course buses. Gort, in the north, is also a convenient centre.

R. LL. PRAEGER: "Notes on the Botany of Central Clare." Irish Nat. 1905, 188–193.

LISDOONVARNA.

345. Lisdoonvarna lies comparatively high (300 ft.) on the Coal-measures outside the SE margin of the Burren limestones. Its bracing air and medicinal waters have given it a high local repute as a health resort, and the straggling town consists mainly of hotels. The shales produce as usual a heavy, wet soil; and treeless tracts of marshy, moory pasture and bog extend. The streams cut down through the shales often till they reach the underlying limestone. A calcifuge upland flora prevails. Among the plants occurring are *Meconopsis, Drosera anglica, Rubia peregrina, Pinguicula lusitanica, Malaxis, Lastrea œmula, L. montana, Botrychium,* and grand masses of *Osmunda. Pinguicula grandiflora* (**145**) grows on a wet cliff close to the Spa (see Irish Nat., 1907, 242), but the habitat, though natural, is by no means free from suspicion. It has increased considerably there in the last twenty years. *Adiantum* (**193**) occurs on a small outcrop of limestone.

Glencolumbkille

Abbey Hill　　　　Bell Harbour　　　　Ballyvau

FIG. 16.—THE BURREN HILLS FROM
Drawn by S. Rosamond P

The complete difference of flora between the limestone and the shales can be studied nowhere better than about Lisdoonvarna. On the shales near Ennistymon *Neotinea intacta* (**158**) grows, 6 to 8 miles from the nearest limestone (see also **370**). G. C. Druce has recorded *Callitriche polymorpha* from Lisdoonvarna (Irish Nat. Journ., **3**, 218).

Cliffs of Moher.—From near Lisdoonvarna a high ridge of Coal-measure shales and flag-stones runs SW, and, where it strikes the coast, forms a magnificent range of cliffs, which drop 650 ft. sheer into the Atlantic. This vast wall is tenanted by *Sedum roseum* and the usual maritime species. Along the summit *Viola lutea* flourishes, and *Sagina subulata*; *Erodium moschatum* is near by; and between Roadford and the cliffs *Phegopteris Dryopteris,* so rare in Ireland, has been gathered. About Fisherstreet, at the N end of the cliffs, *Limosella* occurs. The Cliffs of Moher are within easy distance of Lisdoonvarna on the NE, and of Lehinch and Ennistymon on the SE. At **Lehinch**, *Viola lutea* (**92**) is abundant on sand-dunes; this place is situated within the Coal-measure area, and lacks the interest that pertains to the limestone country. About Lehinch and Miltown Malbay, a little to the S, *Leucojum æstivum* (**161**) grows, and looks native.

THE BURREN HILLS.

346. Looking S from the Connemara shore across Galway Bay, the hills of Burren are seen rising gaunt and bare (fig. 16). Although their sky-line is undulating and

Cappanawalla   Carn Seefin

Black Head

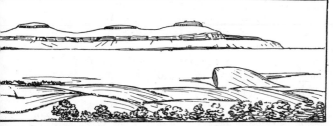

᷍MARA Shore near Galway.
a Sketch by A. B. Wynne.

unbroken, the face of the hills is ruled with parallel horizontal bands, as if painted with streaks of grey of varying shades. In the general outline we see the moulding action of the Glacial Period, which rounded the hills and passed away without leaving any friendly covering of drift. The weather has had free play on the limestone rocks, so that the horizontal bedding stands boldly out on a naked and waterless surface. Burren is, indeed, the dry skeleton of a country, and deserves the description said to have been given of it by one of Cromwell's officers, who, returning from a reconnoitring expedition, reported that it was a savage place, yielding neither water enough to drown a man, nor wood enough to burn a man, nor soil enough to bury a man (plate 19). Drainage is almost entirely subterranean, and a stream is a rare sight; the extent of the underground drainage system is shown by the cave of Slieve Elva, occupied by a river, which has been traversed for two miles without its end being reached (**350**).

This is a county of contradictions. Though little but bare rock is visible, there is an abundant vegetation, due to the rich soil which fills every crevice of the limestone. Though it is waterless—for the rain sinks immediately down the innumerable fissures—yet the vegetation is fresh and luxuriant, fed by the wet warm Atlantic winds. Though frost is here unknown, plants characteristic of the high Alps grow in great profusion, extending right down to sea-level, as in Britain they do only in the extreme north of Scotland, and mixed with them are others, unknown in Britain, whose home is on the sunny shores of the Mediterranean. Though the limestone mostly lies bare of covering, *Calluna* is everywhere, and sometimes, as in Glen Cahir, becomes dominant.

347. These hills rise from the water to over 1000 ft. (Slieve Elva, 1109 ft.), and cover an area of about 14 by 10 miles. Over this district bare limestone extends almost without interruption, harbouring a very remarkable flora. Plants usually rare attain here an immense profusion; and the sheets of flowers and foliage that turn the grey rock into a botanical paradise in spring and summer consist largely of *Arenaria verna, Geranium sanguineum, Rubus saxatilis, Dryas* (**106**, plates 2, 20), *Saxifraga hypnoides, Rubia peregrina* (**121**), *Galium sylvestre, Asperula cynanchica, Arctostaphylos Uva-ursi, Gentiana verna* (**135**, plate 23), *Euphrasia salisburgensis* (**141**), *Orchis mascula* (very abundant and luxuriant, from nearly white to deep purple), *Sesleria* (**186**), *Phyllitis Scolopendrium, Ceterach,* etc. All of these extend down to sea-level; and many of them ascend to the tops of the hills. Other interesting ingredients of the flora are *Thalictrum minus, Cochlearia grœnlandica* (Black Head and Poulsallagh), *Helianthemum canum* (**90**), *Rhamnus catharticus, Geranium columbinum, Spirœa Filipendula, Potentilla fruticosa* (**107**, plate 21), *Saxifraga Sternbergii* (plate 22), *S. hirta, Galium boreale, Hieracium hypochœroides* (Slieve Carran), *H. crebridens, Pyrola media, Blackstonia, Orobanche rubra* (**143**), *Ajuga*

PLATE 19.

A HILLSIDE IN BURREN.

Bare weatherworn limestone. The joints and chinks are filled with luxuriant mesophile vegetation.

R. Welch, Photo.

PLATE 20.

DRYAS OCTOPETALA NEAR BLACK HEAD, CO. CLARE.

The plant is completely dominant and with the aid of mosses has buried boulders three feet in

pyramidalis (**149**), *A. pyramidalis* × *reptans, Taxus, Spiranthes spiralis, Ophrys apifera, Epipactis atropurpurea* (**157**), *E. palustris, Neotinea intacta* (**158**, plate 2), *Adiantum* (**193**, plate 23), *Cystopteris fragilis, Equisetum variegatum, E. litorale* (**201**).

The commingling in these lists of northern and alpine types such as *Dryas, Arctostaphylos, Euphrasia salisburgensis, Ajuga pyramidalis,* with southern types, *e.g. Neotinea, Adiantum,* is a very remarkable feature. "I know of no similar examples in the German or north Alpine flora," writes O. Drude, "of so perverse a distribution and mixture of relict stations.[24]

An interesting feature of the Burren flora lies in the fact that it is entirely an assemblage of *indigenous* plants. No alien or species of doubtful origin takes part in its peculiar associations. And certain native plants which are abundant within the surrounding areas, in the farmland, on railway banks, and so on, cease when "the rocks" begin. Notable examples are *Ulex europæus* and *Primula veris.*

348. In certain parts of the lower grounds a good deal of scrub occurs, formed mainly of *Prunus spinosa, Cratægus, Ilex, Euonymus, Fraxinus, Corylus,* with some *Cornus* and both species of *Rhamnus.* On the heights, heath often prevails of *Calluna* and *Erica cinerea* mixed with *Arctostaphylos, Dryas, Gentiana verna,* and other calcicole plants. An example of the peculiar conditions and peculiar flora prevailing in the region is given :—

Near Black Head, 300 ft. over sea.

Two inches of peaty soil covering a large flat slab of limestone rock *in situ.*

Dryas octopetala, dominant.	Thymus Serpyllum, r.
Sesleria cœrulea, v.c.	Plantago lanceolata, r.
Neckera *sp.*, v.c.	Kœleria gracilis, r.
Carex diversicolor, c.	Hypericum pulchrum, v.r.
Empetrum nigrum, c.	Campanula rotundifolia, v.r.
Potentilla erecta, f.	
Viola Riviniana, r.	Gentiana verna, v.r.
Linum catharticum, r.	Euphrasia salisburgensis, v.r.
Lotus corniculatus, r.	
Calluna vulgaris, r.	Festuca ovina, v.r.
Rhinanthus minor, r.	Musci.

[24] New Phytologist, 1912, 246.

349. Lakes and marshes are almost absent from the district, as are streams, most of the drainage being underground; but *Potamogeton perpygmæus* (*P. lanceolatus* quoad pl. hibern. = *coloratus* × *pusillus*), has one of its three known stations in the Cahir River, S of Black Head (see **349** *infra*), and *Viola stagnina* (**91**) occurs about Turlough, SE of Ballyvaughan. *Sedum acre* affects a curious habitat, growing in small shallow saucer-shaped solution-hollows on the flat rocks amid a profusion of algæ, where it is flooded by every shower and parched at other times. *Teucrium Scordium* (**148**) has an outlying station beside Castle Lough, on the SE edge of the district. *Limosella*, in Ireland confined to Clare, is at Poulsallagh, and *Eriophorum latifolium* at Feenagh. The Cahir River, mentioned above, which runs past Formoyle, is remarkable in that it traverses this waterless region from source to sea (4 miles) without sinking into the limestone. In its middle portion it meanders through flat meadowland overlying silt, a slow deep stream. This is the head-quarters of *Potamogeton perpygmæus* (**167**), whose numerous creeping stems form a mat which entangles sediment, so that it produces cushions of vegetation and mud up to 2 ft. across, in water from $\frac{1}{2}$ to 3 ft. deep. It is here accompanied by a little *P. coloratus* and *P. natans*. Above and below, where the stream is shallow, rippling over pebbles, it persists in a few inches of water, from near Formoyle almost to the sea. The plant (at least in its earlier stage) has exactly the red colour of *P. coloratus*, which accompanies it also in the upper rippling reaches, a very unusual habitat for the former. In the same flat area, *Equisetum variegatum* and *E. litorale* (**201**) are abundant along the banks of the stream (the former up to 3 ft. in height), and both range widely up and down the river.

On the N and W the Burren hills look down on the sea. To the E and S they drop abruptly into a plain of bare limestone described in **344** and **361**. On this E slope there is some shelter, and dense *Corylus* scrub occurs over considerable areas. Below the cliffs of Slievecarran well-developed *Betula-Fraxinus* wood is found. Here the trees make a canopy 15–20 ft. above ground, sufficiently dense to preclude the presence of all but shade plants. *Oxalis* is the leading species, with *Circæa lutetiana*, *Viola Riviniana*, *Nepeta hederacea*, *Sanicula*, ferns, and very luxuriant mosses on the stones and tree-trunks. On the cliff-ledges above, an open *Betula-Fraxinus* wood is present, with a little *Cratægus, Sorbus Aucuparia, Ilex, Euonymus, Rosa canina, Viburnum Opulus, Ulmus montana.*

West of Ballyvaughan, with a northern exposure, the frequent scrub consists of *Corylus, Ilex, Cratægus, Prunus spinosa, Sorbus Aucuparia, Fraxinus* (no *Quercus*).

350. South-westward, a sharp contrast of vegetation results from the superposition on the limestone of Coal-measure shales, and undulating country of heavy clay, with a rank grassy and rushy vegetation and many marsh-loving plants, replaces the grey limestone terraces (see **345**). *Gentiana verna* extends sparingly to banks in the shale country, but most of the Burren plants stop abruptly on the edge of the limestone. The flora of the coast line includes a few interesting species, such as *Cochlearia grænlandica, Crithmum, Artemisia maritima, Limonium binervosum, L. transwallianum* (**351**), *Calystegia Soldanella.*

The Burren can be penetrated from Lisdoonvarna (**345**) on the S, or (best) from Ballyvaughan (*infra*) on the N.

The attractiveness of the neighbourhood has led to the recording of many rare critical plants, such as *Hieracium britannicum, H. cymbifolium, Limonium transwallianum* (elsewhere known only from limestone cliffs in Pembroke-shire, Journ. Bot., 1930, 316, 347), and *Geranium Robertianum* subsp. *celticum* Ostenfeld (in Report Bot. Exch. Club, 1919, and Irish Nat., 1921, 23), which last is widespread in Burren up to 1000 ft.

The visitor interested in caves should view the great cavern of Poulnagollum, near the base of Slieve Elva. The main tunnel has been explored for 2 miles without its termination being reached, and altogether 5 miles of passages have been traversed (see E. A. BAKER in Report Wells Nat. Hist. and Archæol. Soc., 1925, 40–42, map).

F. J. FOOT: "On the Distribution of Plants in Burren, County of Clare." Trans. Roy. Irish Acad., 24 (Science), 143–160. Map. 1864. T. H. CORRY: "Notes of a Botanical Ramble in the County of Clare, Ireland." Proc. Belfast Nat. Hist. & Phil. Soc., 1879–80, 167–207.

351. Ballyvaughan is a small village on Galway Bay, with sufficient hotel accommodation, excellently situated for the exploration of Burren. A grassy valley runs S into the heart of the hills, with the grey terraced lime-

stones rising on either hand, and a level road leads **W**
along the coast to Black Head and on southward.   Among
the plants which grow near Ballyvaughan are *Arabis
Brownii, Viola stagnina, V. lactea, Geranium rotundifolium, Potentilla fruticosa* (in quantity, see plate 21),
*Saxifraga Sternbergii, Galium boreale, Hieracium cre-
bridens, Linaria Elatine, Verbena, Orobanche rubra,
Neotinea* (plate 2), *Adiantum* (plate 23), and all the more
widely-spread plants of Burren, such as *Dryas* (plates 2,
20), *Gentiana verna* (plate 24), and *Euphrasia salisburgensis* (plate 28).   The maritime vegetation appears poor,
in spite of a promising variety of rock, sand, and mud,
but includes *Matthiola sinuata* (extremely rare in Ireland)
and *Artemisia maritima*.   E of Ballyvaughan, near the
ruins of Corcomroe Abbey, *Juniperus communis* may be
seen growing quite prostrate on the limestone in bushes up
to 30 ft. in diameter.

Steamer from Galway thrice a week during the summer
months, car from Ardrahan, or bus from Galway.

THE ARAN ISLANDS.

352. On Aran, the limestone pavement flora attains
its most remarkable development.   The three islands are
simply a reef of limestone stretching out from the coast
of Clare in a WNW direction, and broken across by the
Atlantic into three parts.   The largest, Inishmore or Aranmore, is 8½ miles long; the other two, Inishmaan and
Inisheer, much less; all are 2 to 3 miles wide.   Four
miles of rough water separate Inisheer from Co. Clare,
and to the N the nearest of the many islets off the Connemara coast lies across a rather broader channel.   The
Aran Islands are formed, save a little blown sand, of
fissured limestone—absolutely bare until the people,
carrying sand and seaweed in baskets on their backs, began
to cover the rock in sheltered hollows, and to grow small
patches of potatoes and oats.   Innumerable loose stone
walls, 5 or 6 ft. high, enclosing tiny fields of rock, help
to break the wind, and numerous blocks of Connemara
granite, some of them of great size, relics of the Ice Age,
lying on the limestone, add a fantastic touch to the scene.
The SW side of the islands presents a precipitous front
to the ocean; thence the ground slopes, so that the opposite

PLATE 21.

POTENTILLA FRUTICOSA AT BALLYVAUGHAN, CO. CLARE.

An almost continuous low growth, about two feet in height, extending over several acres of flat heathy limestone ground occasionally flooded. Flowering just beginning. May.

R. Welch, Photo.

351.

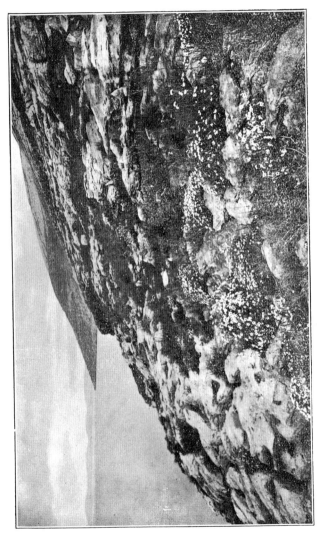

PLATE 22.

SAXIFRAGA STERNBERGII AT BLACK HEAD, CO. CLARE.

It is accompanied here by *Dryas octopetala*, *Geranium sanguineum*, *Arctostaphylos Uva-ursi*, *Gentiana verna*, *Neotinea intacta*, *Sesleria cærulea*, &c.   About 50 feet above sea level.   The hills rise to

shore is low. The highest point (on Inishmore) is 406 ft. Exposure is extremely great; mist and rain are very frequent; frost and snow almost unknown.

A large part of the flora, including many of the more interesting plants, is confined to the innumerable joints in the bare grey rock; these have become widened by centuries of rain, and filled at the bottom—often two or three ft. down—with a thin vegetable soil. The rarest Irish plants occurring on Aran include *Astragalus danicus* (**101**), in its only Irish stations, *Saxifraga Sternbergii, Ajuga pyramidalis, Calamagrostis epigejos.* The visitor will be struck by the abundance of *Helianthemum canum, Gentiana verna, Euphrasia salisburgensis, Adiantum, Ceterach.* Other abundant species are *Arenaria verna, Viola hirta, Geranium sanguineum, Cornus sanguinea, Galium boreale, G. sylvestre, Asperula cynanchica, Rubia peregrina, Pimpinella major, Carlina, Spiranthes spiralis, Sesleria.* Less widely distributed species include *Matthiola sinuata* (Straw Island, not seen since 1835), *Arabis Brownii, Crambe, Rhamnus catharticus, Sagina subulata, Saxifraga hirta, Erodium moschatum, Sedum roseum, Carduus nutans, Blackstonia, Cuscuta Trifolii, Orobanche Hederæ, O. rubra, Calamintha ascendens, Chenopodium rubrum, Ophrys apifera, Allium Babingtonii, Juncus macer, J. subnodulosus, Juniperus communis, Equisetum variegatum. Sedum anglicum,* often monopolizing small ant-heaps, earns on this account its local name of *Poureen-shingan* (ant-fold or ant-house).

Moisture-loving plants such as the ferns attain remarkable dimensions, despite the exposure. *Adiantum* is often two ft. in length, *Ceterach* a foot high and very boldly crenate (fig. 15); these grow deep down in the joints of the limestone. The wind, combined with intensive grazing, forces the arbuscular vegetation to remain in the rock-crevices among the low herbs if it is to escape destruction; and one finds *Euonymus, Rhamnus catharticus, Prunus spinosa, Cratægus, Viburnum Opulus, Quercus, Corylus* often not rising above the level floor of rock.

It will be noted that the characteristic plants listed above present in a striking degree the interesting mixture of northern and southern types that is so marked a feature of the flora of W Ireland. Curiously enough, *Neotinea intacta* is not recorded from this focus of the geographical plant-group of which it is so usual a member.

The Aran Islands are connected with Galway by steamer three times a week; but a day excursion generally leaves very little time for exploration. Accommodation may be obtained on all the three islands, preferably by previous arrangement. Aran is quite one of the most interesting spots in western Europe, even apart from its remarkable flora, on account of its wealth of pagan and early Christian remains, and the primitive character of the life and customs of its people. The enormous and complicated prehistoric cliff-forts, built of loose stone, alone are worth a visit to Aran, being among the finest known structures of the kind.

H. C. HART: ''A list of Plants found in the Islands of Aran, Galway Bay.'' Dublin, 1875. J. E. NOWERS and J. G. WELLS: ''The Plants of the Aran Islands, Galway Bay.'' Journ. Bot. 1892, 180–183. N. COLGAN: ''Notes on the Flora of the Aran Islands.'' Irish Nat. 1893, 75–78, 106–111. R. LL. PRAEGER: ''Notes on the Flora of Aranmore.'' Irish Nat. 1895, 249–252.

LOUGH DERG.

353. Lough Derg has Tipperary for its eastern shore, and Clare and SE Galway for its western. In shape it preserves a sinuous, river-like form, with occasional bays of considerable size on either side, and many small islands. This great expansion of the Shannon has a length of 22 miles; the average breadth is about 2 miles. The greater part occupies a flat depression in the limestone, and has low, stony shores and no great depth; but in the S it fills a remarkable gorge between the Silurian masses of Slieve Bernagh and the Arra Mountains, and the depth increases to over 100 ft. The Shannon, which for more than a hundred miles above L. Derg has pursued a placid course across the Central Plain, breaks away after leaving the lake and its artificial extension, and dropping 97 ft. in 18 miles, reaches sea-level at Limerick.

On L. Derg botanical interest centres on the limestone shores and islands. Mr. Colgan's paper (*infra*) describes the flora remarkably well, and invites quotation.

''The most obvious characteristic of the Lough Derg flora is to be found in the great preponderance of a group of species, many of which are decidedly uncommon in various parts of Ireland. Omitting the very common plants, the chief members of this group are the following:—

PLATE 23.

ADIANTUM CAPILLUS-VENERIS, NEAR BLACK HEAD, CO. CLARE.

Filling joints in limestone pavements, accompanied by *Ceterach officinarum, Phyllitis Scolopendium, Rubia peregrina,* &c. May.

R. *Welch, Photo.*

352.

PLATE 24.

GENTIANA VERNA ON SAND-DUNES AT BALLYVAUGHAN, CO. CLARE.

The flora of the close sward in which the Gentian forms numerous patches includes much *Asperula cynanchica*, *Galium verum*, *Thymus Serpyllum*, with *Euphrasia salisburgensis* and *Neotinea intacta*. May.

R. Welch, Photo.

Hypericum perforatum	Cnicus pratensis
Geranium sanguineum	Lysimachia vulgaris
Rhamnus catharticus	Samolus Valerandi
Rubus cæsius	Chlora perfoliata
Rosa spinosissima	Erythræa Centaurium
Parnassia palustris	Gentiana Amarella
Viburnum Opulus	Lycopus europæus
Galium boreale	Teucrium Scordium
Eupatorium cannabinum	Littorella lacustris
Solidago Virgaurea	Juniperus communis
Antennaria dioica	Schœnus nigricans
Carlina vulgaris	Selaginella selaginoides

"Hardly an island or promontory was landed on, all down the lake, at least from Portumna to Mountshannon, about which point a change of rock takes place, that this group did not present itself in full development. Some of the lonely rocky islets rising a few feet above the lake surface were positively ablaze with the coral berries of the *Viburnum*, standing out against the sober ashen green of the Juniper; the Dewberry threw out its handsome bronzed streamers far over the naked limestone, and right in the wash of the waves the Water Germander (*Teucrium Scordium*) spread its matted roots through the shingle. However bare of novelty many of these desert islets may have been, and the uniformity of conditions was undoubtedly accompanied by uniformity of vegetation, not one of them lacked the charm of an absolutely unadulterated indigenous flora."

Where in the S the limestone gives way to slates of Upper Silurian age, this group ceases abruptly. "The change of flora with change of rock was very strikingly shown on touching at Freagh wood when running west from Dromineer to Mountshannon. Five minutes on the rough grits here gave *Cotyledon Umbilicus, Galium saxatile, Pyrus Aucuparia, Vaccinium Myrtillus, Digitalis purpurea, Scilla nutans, Luzula maxima, Lastrea dilatata,* not one of which had turned up on all the five days spent on the limestone further north. And similar results had been arrived at in the hour spent on the Woodford bogs two days before when sailing down from Portumna. Here on the pure peat *Gnaphalium uliginosum, Senecio sylvaticus, Calluna vulgaris, Erica Tetralix, Rumex Acetosella, Juncus supinus,* and *Lomaria Spicant* immediately presented themselves

when one passed from the limestone to the overlying peat''
(Colgan *l.c.*).

354. Lough Derg is famous as the only habitat in
Ireland for *Inula salicina* (**122**, plate 25), absent from
Great Britain; it is widely spread along the stony limestone
shores, occupying a belt at or a little below winter flood-
level (plate 26). Its far-creeping slender subterranean
stems form a loose mat. Whether it has been much
interfered with by the changes in the level of the lake
owing to the impounding of the waters for the Shannon
Electricity Works cannot yet be definitely stated, but
apparently it has not.

Another interesting L. Derg species is *Sisyrinchium
angustifolium* (**159**), which grows in abundance in rough,
damp ground about the Woodford River and northward,
and has more recently been found at Scarriff, further S,
and on the opposite Tipperary shore. The best water-
plant is *Chara tomentosa* (**207**), which in Britain is
unknown, and in Ireland is confined to the Shannon basin.
It is probably frequent; but, as the lake has not been
dredged, its recorded stations are few.

Other plants of the shores and islands include *Thalictrum
minus, Caltha radicans, Rhamnus Frangula, Geranium
columbinum, Agrimonia odorata, Galium uliginosum,
Orobanche Hederæ, Pinguicula vulgaris* var. *bicolor, Utricularia intermedia, Plantago maritima* (up to 25 miles
from the nearest sea), *Spiranthes spiralis, Epipactis
palustris, Ophrys apifera, Sagittaria, Butomus, Potamogeton coloratus, P. densus, Carex gracilis, C. lasiocarpa,
C. Pseudo-Cyperus, Osmunda, Chara aculeolata.* The
local *Vicia Orobus* is on Church Island.

A remarkable feature of L. Derg is the groves of *Taxus*
and luxuriant *Juniperus communis* found in many spots.
On the shore N of Bounla Islands for instance, they form
an open wood, with giant Junipers having a diameter of
up to 45 feet, and height up to 20 feet. *Sorbus porrigens*
also occurs, but much more sparingly.

Among introduced plants, *Valerianella rimosa, Crepis
biennis,* and *Verbena* occur about Dromineer; *Leucojum
æstivum* grows by the Little Brosna River in wild ground,
but there are gardens close by.

While the heather-clad hills rising over L. Derg impart
to its S portion a picturesqueness of a high order, the
lover of the beautiful will possibly even prefer the N part

PLATE 25.

INULA SALICINA ON THE SHORES OF LOUGH DERG.

Growing among boulders above flood-level near the edge of the wood-
land zone, in a coarse grassy vegetation including *Sesleria cærulea,
Schœnus nigricans, Plantago maritima, Campanula rotundifolia, Spiræa
Ulmaria.* July.

R. Welch, Photo.

PLATE 26.

ON THE SHORE OF LOUGH DERG.

Near Woodford, S.E. Galway. Below the woodland zone (*Fraxinus*, *Crataegus*, &c.) is a dry grassy zone, with *Rhamnus catharticus*, *Campanula rotundifolia*, *Galium boreale*, *Sesleria cærulea*, *Briza media*, &c., followed by a wet grassy zone, with *Spiræa Ulmaria* *Lathrum*

of the lake, where there is a noble breadth and expanse in the scenery that is impressive and exhilarating. Killaloe (**343**), at the S end, is a good centre. Hotels will also be found at Portumna (**355**) at the N end, Mountshannon on the W shore, Nenagh near the E side, etc.

N. COLGAN: ''On the Flora of the Shores of Lough Derg.'' Irish Nat. 1897, 189–197.

355. **Portumna** lies on the Shannon, where the river widens out suddenly into the expanse of Lough Derg, and it forms a good base for the exploration of the N part of the lake, having sufficient hotel and boat accommodation. The low river-banks E of the town yield *Lathyrus palustris, Sium latifolium, Galium uliginosum, Utricularia intermedia, Hydrocharis, Sagittaria,* etc. On the lake-shore and islets S of Portumna may be gathered *Inula salicina, Vicia Orobus, Teucrium Scordium, Lastrea Thelypteris, Chara tomentosa,* and to the SW *Sisyrinchium angustifolium;* not to mention other characteristic members of the L. Derg flora (**353**), and of the flora of the Central Plain (**240**). *Brachypodium pinnatum,* in Ireland often found in artificial habitats, such as railway banks, occupies twenty acres of old pasture land here, and is probably indigenous; the alien *Poa compressa* grows on wall-tops as usual.

THE COUNTY OF GALWAY.

356. This large county of 2452 sq. miles is heterogeneous. Lough Corrib, lying N of Galway town, divides a wide area of low-lying limestone on the E from the remarkable tract of bog and mountain which forms W Galway or Connemara. The former has much in common with its northward extension into E Mayo, and the whole area is best treated together. The western mountainous metamorphic area on the other hand has great affinities with W Mayo. Both the limestone and the older area to the W are so large and have so much of botanical interest that a number of sections (**357–415**) is devoted to their consideration.

SLIEVE AUGHTY.

357. In SE Galway, S. of Loughrea, the ground rises in a gentle slope out of the low limestone plain till it exceeds 1200 ft., and broad, flat hills stretch thence

SW far over the Clare border, overlooking Scarriff Bay in L. Derg on the E, and Crusheen on the W. These hills known as Slieve Aughty (so far at least as the higher Galway part is concerned), represent an anticline, and are formed of Old Red Sandstone with a core of Silurian rocks occupying about 250 sq. miles. On every side the slopes are gentle, and broad, heathery ridges, uninteresting to the botanist, form the higher grounds. A couple of lakes, of which the larger, L. Graney, is 3 miles in length, are embosomed in the hills. This area is not inviting to the explorer, and the most interesting ground is the wooded glens that here and there diversify the slopes. On the W, the valley of the Owendalulleegh River affords one of the few Connacht stations for *Euphorbia hiberna*,[25] and on the E side the Dalystown woods yield a group of plants, mainly upland in character, including *Thalictrum minus, Leucorchis albida, Milium, Festuca sylvatica, Equisetum hyemale, E. trachyodon*. Here also the rare *Lastrea remota* (**198**) has its only Irish station.[26] From Glendree and neighbouring places *Lastrea montana, Phegopteris polypodioides*, and both species of *Hymenophyllum* are reported. The flora of L. Graney itself is unrecorded, although it is a considerable lake. The little Lough Atorick yields *Calamagrostis epigejos*, one of the rarest of Irish grasses, and *Potamogeton prælongus*. The district is a remote one, and but little known. Gort, Woodford, Scarriff, and Mountshannon are the best bases for the visitor.

For the southern prolongation of this hill district, where, beyond a narrow band of limestone, stretching from Ennis to Scarriff, the ground rises to 1740 ft., see **343**.

WOODFORD.

358. The village of Woodford forms a good centre for the exploration of Slieve Aughty on the W and Lough Derg on the E; it lies beside a pleasant trout-stream, on the junction of the Old Red Sandstone of the hills with the limestone of the plain. Woodford is notable as the first-discovered Irish station of *Sisyrinchium angustifolium* (**159**), which will be found in meadows and rough ground

[25] H. C. HART in Journ. Bot., 1873, 338.

[26] R. LL. PRAEGER: ''*Lastrea remota* in Ireland.'' Irish Nat., 1909, 151.

from the village down to the lake, and thence nearly to
Portumna (plate 12). The hill-region (**357**) adjoining
Woodford on the W has been little explored. The eastern
environs of Woodford and especially the adjacent shores
of L. Derg (plate 26) yield a very interesting flora, which
includes, in addition to *Sisyrinchium*, *Inula salicina* (plate
25), *Chara tomentosa*, *Thalictrum minus*, *Prunus Padus*,
Drosera anglica, *D. longifolia*, *Rhamnus catharticus*, *R.
Frangula*, *Rubus ochrodermis*, *Galium boreale*, *G. uligino-
sum*, *Utricularia intermedia*, *Teucrium Scordium*, *Juniperus
communis*, *Ophrys apifera*, *Epipactis palustris*, *Rhynchospora fusca*, *Sesleria*, *Carex diandra*, etc. (see also **353**).
N of Woodford typical Central Plain country extends. At
Derryvunlam Wood, between Woodford and Loughrea,
Cephalanthera ensifolia is abundant in one of its rare
Irish stations. *Caltha radicans* is recorded from Loughrea.
Plantago maritima, which is widespread on the low limestones for some distance inland in E Galway, etc., and as
an alpine plant on non-calcareous rocks also, covers a
disused road on Old Red Sandstone near Woodford at
560 ft. Near Loughrea *Neotinea intacta* (**158**) has been
found.

THE GALWAY-MAYO LIMESTONES.

359. Northward of the Devonian uplands of Slieve
Aughty (**357**) and E and NE of Galway Bay there extends
a region of low-lying Carboniferous limestone, forming the
W edge of the Central Plain. Much of it is devoid of
covering, and lies open to the sky, its face weathered and
its joints widened by rain and wind. Eastward there is
no marked boundary; drift and peat begin to cover the
surface, and it passes imperceptibly into typical Central
Plain country. Most of the area which is bare limestone
(and consequently is the best botanical ground both in this
region and southward) lies W of the Ennis-Tuam line of
railway. On the W edge there is everywhere a marked
boundary—in the extreme S, as about Ardrahan, the low
ground gives way abruptly to the Burren hills, formed of
the same rock. N of that, Galway Bay adjoins. Further
N, away up to Castlebar, where its continuity is interrupted
by the Caledonian fold of the Ox Mountains (**421**), the
limestone laps against the ancient acid rocks of Connemara
and W Mayo, providing one of the most striking changes
of scenery and of vegetation to be found in Ireland.

Large lakes—Corrib (**367**), Mask (**371**), Carra (**372**)—lie on the limestone here close to the boundary, and add to the scenic and floristic contrast.

Though it has no sharp vegetational variations, the limestone area is so large that for purposes of botanical description it is best divided into sections.

360. The drainage of the western part of this district is of the most erratic description. The rock is a mere honeycomb, and streams disappear into the limestone continually, to re-appear perhaps several miles away. And while an impervious floor allows lakes to persist in some places, in many others "turloughs" abound—hollows sometimes as much as a mile across, and up to 30 ft. in depth, in which the water lies for periods depending on the amount and rate of inflow and outflow. The rapid rise and fall of the water produces a vegetation quite different from that of marshes or similar places where slower changes of the kind occur. In the turloughs the most conspicuous feature is the cessation of all arboreal or arbuscular growth below a certain contour. Blackish patches of the moss *Cinclidotus fontinaloides* provide another well-marked contour, generally several feet above the last, pointing like the shrub-limit to a certain period of flooding. Much lower down, *Fontinalis antipyretica* supplies a third datum level, but even this water-moss has an upper limit above the lower limit of such plants as *Rubus cæsius, R. saxatilis, Potentilla reptans, Salix repens*. A large number of plants, many of them distinctly zoned, descend below the level at which *Cratægus, Prunus spinosa*, and other prevailing shrubs are killed, producing right down to the floor of the turloughs a dense green sward, mostly closely nibbled. The plant *par excellence* of the turloughs is *Viola stagnina* (**91**), which haunts the deeper parts. *V. canina* forms a zone higher up, with *V. Rivinina* above it. *V. stagnina* hybridizes freely with the former and probably with the latter.

R. LL. PRAEGER: "The Flora of the Turloughs: A Preliminary Note." Proc. Roy. Irish Acad., **41**, B, 37–45. 1932.

THE ARDRAHAN LIMESTONES.

361. This name may be given to the stretch of low-lying bare limestones that occupy most of the district at the head of Galway Bay, and stretch up towards Craughwell, and down to Gort and Kinvarra and the foot of the hills of Burren. The greater part of this flat area is only from 100 to 200 ft. above sea-level, and a full "Burren flora" prevails. A certain amount of drift, thrown down here and there over the rocks, allows of patches of cultivation, mainly near the very irregular coast-line; inland of this, the grey rock often extends naked as far as eye can reach. Turloughs (**360**) are frequent; and in some of them the water is sufficiently permanent to merit the use of the term "lake."

The characteristic plant of the district is *Spiræa Filipendula*, which is very abundant towards the S, and is, indeed, in Ireland hardly found outside this area and its immediate neighbourhood. As one ascends from these low grounds towards the nearest of the Burren summits, this plant ceases, and others which are rare on or absent from the plain, such as *Dryas* (**106**) and *Epipactis atropurpurea* (**157**), become abundant. In other respects, the flora conforms to that of the Burren hills; and we walk amid a profusion of *Geranium sanguineum, Saxifraga hypnoides, Galium boreale, G. sylvestre, Asperula cynanchica, Rubia, Arctostaphylos Uva-ursi, Gentiana verna, Blackstonia, Euphrasia salisburgensis, Plantago maritima, Spiranthes spiralis, Sesleria*, often with *Neotinea intacta* and *Ophrys muscifera*. In wet ground *Caltha radicans* is frequent, and *Lastrea Thelypteris, Equisetum trachyodon*, and *E. variegatum* occur. *Monotropa* is at Castle Taylor and *Agrimonia odorata* at Ardrahan, where also are *Calamintha ascendens* and *Ophrys apifera*.

Two of the rarest members of the flora occur with some frequency about Gort — *Viola stagnina* and *Limosella*. *Adiantum* grows 4 miles NW of the same place. The stunted arbuscular vegetation of the limestones includes *Rhamnus catharticus, R. Frangula, Cornus, Juniperus communis, Taxus*. Other plants of the Gort area are *Papaver hybridum, Sorbus porrigens* and *S. Aria, Crepis biennis, C. taraxacifolia, Chenopodium rubrum, Epipactis palustris, Butomus, Potamogeton coloratus, Carex Hudsonii, C. lasiocarpa*; and *Ligustrum* looks really native here in chinks of the limestone.

At Castle Lough, 7 miles SW of Gort, there is an interesting outlier of *Teucrium Scordium* (**148**), a plant of the Shannon basin, and here we enter the main habitat in Ireland of *Potentilla fruticosa* (**107**). On the way thither the once important ecclesiastical settlement of Kilmacduagh (cathedral, round tower, etc.) may be visited.

The maritime vegetation has little variety, and the most conspicuous plant of the much indented and shallow head of Galway Bay is *Artemisia maritima.* *Cochlearia anglica* also occurs.

The district can be visited on a one-day trip from Galway by taking train or motor to Craughwell, Ardrahan, or Gort, or by staying at either of the two last places or at Kinvarra or Oranmore.

362. The characteristic flora may be excellently studied close to Gort, at **Garryland**—an exceedingly interesting place. One of the best examples of crag-land wood is found here. Though many trees have been introduced, the wood, as such, appears to be aboriginal. *Quercus Robur* (*pedunculata*) and *Fraxinus,* largely self-sown, are co-dominant in places; *Ulmus montana,* locally dominant; *Prunus spinosa* (abundant), *Cratægus monogyna, Pyrus Malus* var. *acerba, Ilex, Euonymus, Viburnum Opulus, Hedera,* and *Rubi. Fagus* and Conifers have been planted. Round the skirts of the wood are *Sorbus porrigens, Taxus,* and *Corylus,* with remarkably luxuriant *Juniperus communis.* "Although developed over limestone, the soil over considerable tracts gave no calcareous reaction, and the general character of the wood inclined to the oak-, rather than to the ash-type. It is, however, quite possible that it has been derived from a wood of the ash-type by continuous accumulation of soil and washing out of lime; and this view is supported by the abundance of ash in parts, and the occurrence of species like the Spindle-tree and the Helleborine [*E. atropurpurea*], as well as by analogy with various English woods.["][27]

In other parts of the wood, where the limestone lies practically bare, *Quercus* and *Corylus* are co-dominants, with a little *Fraxinus* and *Ulmus, Prunus spinosa, Rosa arvensis, Euonymus,* and amid a dense covering of mosses much *Hedera;* also *Anemone, Viola silvestris, Oxalis, Geum urbanum, Circœa lutetiana, Viburnum Opulus, Veronica Chamœdrys, Prunella, Neottia, Epipactis atropurpurea, Scilla non-scripta, Arum* (spotted-leaved form), *Carex sylvatica, Polystichum angulare, Lastrea Filix-mas.*

The country round the head of Galway Bay is the area of least rainfall in the W of Ireland; and the combination of this with a light limestone soil is probably the explanation of the weed flora, unusually diversified for the west, which

[27] [A. G. TANSLEY] in New Phytologist 1908, 260.

is found in the Kinvarra district and on the N coast of Clare. This includes three species of *Papaver* (*P. Argemone* absent), *Lepidium campestre,* and *Linaria Elatine.* The native flora of Kinvarra includes *Ranunculus circinatus* and the rare *Chara canescens.*

A. G. MORE: "Notes on the Flora of the neighbourhood of Castle Taylor, in the County of Galway." Proc. Bot. Soc. Edinb. 1855, 26–30, 60.

Lough Cooter, or Cutra, is a pretty lake, 3 miles in length, lying on the limestone SE of Gort, with the Old Red Sandstone hills (Slieve Aughty, **357**) rising close at hand. The flora of its shores includes *Ophrys muscifera, O. apifera,* and *Epipactis palustris. Euphorbia hiberna* (**154**), a representative of the adjoining calcifuge flora, descends the Owendalulleegh R. almost to the lake.

363. **Athenry,** an important railway junction, is a small town with a historic past, of which one is still reminded by imposing ruins. The botanist can spend a profitable time here. A fine esker-ridge passes close to the town on the N side. *Gentiana verna, Ophrys muscifera,* and *Juniperus communis* in abundance, as well as other species characteristic of the western limestone area, are there. *Cochlearia danica,* 15 miles from the sea, grows on old buildings. *Neotinea intacta* has been found near by. A few miles N and NW, crag country occurs about **Castle Lambert** and elsewhere, yielding *Vicia Orobus* in several stations, *Dryas, Arctostaphylos Uva-ursi, Gentiana verna, Euphrasia salisburgensis,* etc., often among *Erica cinerea.* Eastward, near Raford, *Sisyrinchium angustifolium* and *Equisetum variegatum* are found, and *Gentiana verna* has its most easterly Irish station on an esker 2 miles E of Athenry. To the southward, about Moyode (*Magh Fhóid,* sod plain), is more crag-land, with *Viola stagnina, Neotinea intacta,* and other characteristic species.

R. LL. PRAEGER: "Notes of a Western Ramble." Irish Nat. 1906, 257–266.

GALWAY TOWN.

364. This historic place is best known to botanists as the gate to Connemara, but it forms a very good botanical centre. Galway is situated at the mouth of the River Corrib, where it drops suddenly to enter the sea at the head of Galway Bay, and it is on the junction of the Carboniferous limestone and the more ancient metamorphic rocks. To the E and N are bare limestone pavements and grasslands; to the W, granites, covered with peat and heath. The sharply contrasting floras of these two areas may be studied close around the city. On the limestone are *Neotinea* (**158**), growing amid an abundance of *Geranium sanguineum*, *Rubia peregrina* (**121**), *Galium boreale*, *Asperula cynanchica*, *Carlina*, *Blackstonia*, *Gentiana verna* (**135**), *Euphrasia salisburgensis* (**141**), *Spiranthes spiralis*, *Sesleria* (**186**). *Adiantum* (**193**) has also been found. The limestone flora is very well developed towards **Menlo**, where, in addition to most of the foregoing, *Caltha radicans*, *Apium Moorei*, *Sorbus Aria* (type, very rare in Ireland), *Hieracium lasiophyllum*, *Allium Babingtonii*, *Lemna polyrrhiza*, *Carex Pseudo-Cyperus*, *Lastrea Thelypteris*, and *Juniperus communis* may be seen. To the W, on the other hand, stony heath prevails, with *Calluna* dominant. *Dabeocia* (**132**) may be gathered, and with it such plants as *Saxifraga spathularis* (**114**), the three species of *Drosera*, *Utricularia intermedia*, etc. **Gentian Hill**, on the coast 3 miles W of Galway, is a promontory of limestone drift resting on granite, colonized by an interesting outlier of the limestone flora, including *Cerastium arvense* var. *Andrewsii*, *Dryas*, *Asperula cynanchica*, *Gentiana verna*, *Blackstonia*, *Neotinea*, *Ophrys apifera*, *Anacamptis pyramidalis*, *Sesleria*, all within 50 ft. of sea-level. *Zostera nana* grows on mud on the west side of the promontory, and *Crambe* and *Cochlearia anglica*, with *Limonium humile*, sparingly on the shore. Barna Head, beyond Gentian Hill, is a similar deposit with a similar flora; *Artemisia maritima* has been found there. A little W of Salthill, *Helianthemum canum* has its only Galway station.

The gravelly shore E of Galway, and its vicinity, yield, among other plants, *Raphanus maritimus*, *Glaucium*, *Erodium moschatum*, *Dipsacus sylvestris*, *Solanum nigrum*, *Hyoscyamus*, *Chenopodium rubrum*, with *Chara canescens* in pools. Boulder-beaches there support a coarse vegetation of *Solanum Dulcamara*, *Scrophulularia aquatica*, *Glaucium flavum*, *Rumex crispus*, *Urtica dioica*—an inconsistent group. Ostenfeld's subsp. *celticum* of *Geranium Robertianum* has been collected near Galway. The E European *Lactuca tatarica* is naturalized at and above high water-mark at Cromwell's Fort ¾ mile E of the town

—its only Irish station. About the town itself colonists are frequent, and the roadside flora includes such plants as *Reseda lutea, Verbena, Malva rotundifolia. Matricaria occidentalis* is abundant all over the Galway neighbourhood, completely ousting the more widespread *M. suaveolens.* Bog-holes at Ballindooly, NE of Galway, form one of the few Irish habitats of *Nitella tenuissima* (**207**); *Lastrea Thelypteris* grows in wet meadows in the same neighbourhood.

THE MOYCULLEN LIMESTONES.

365. Moycullen (*Magh Cuilinn*, the plain cf holly), on the road and railway half-way between Galway and Oughterard, is the centre of the interesting strip of limestone which stretches along L. Corrib between these two places, with the metamorphic rocks of Connemara on its W side. In the village street one is on the acid rocks, looking down on the low-lying limestone country, which consists of poor tillage and pasture, with ridges and great flat areas of bare grey crags, limestone terraces clothed with dense *Corylus* scrub and groves of *Fraxinus,* esker-like rock-strewn mounds, and large desolate peat-bogs stretching away to L. Corrib. On flat levels many mushroom-shaped rocks give an exact indication of a former higher level of the lake water. Ballycuirke Lough, close to the village, lies on the junction of the two geological formations, and offers a striking contrast of vegetations, the W shore being heathery, with *Dabeocia* (**132**), the E side grassy, with *Neotinea* (**158**), the result of the abrupt change of rock. The limestones in this neighbourhood, down to the lake-shore, yield *Thalictrum minus, Rhamnus catharticus, Geranium sanguineum, Dryas, Rubia peregrina, Galium boreale, G. sylvestre, Asperula cynanchica, Arctostaphylos Uva-ursi, Gentiana verna, Euphrasia salisburgensis, Calamintha ascendens, Ophrys apifera, O. muscifera, Juniperus communis, Sesleria.* With them *Erica cinerea* grows abundantly on the bare limestone, with (bird-sown?) *Cotoneaster microphyllus* among it. *Pimpinella major* and *Verbena* are plentiful on roadsides. *Epipactis palustris* and *Potamogeton nitens* are at Ross Lake; *Carex Pseudo-Cyperus* is frequent.

R. Ll. Praeger: "Notes of a Western Ramble." Irish Nat. 1906, 257–66.

OUGHTERARD.

366. Oughterard forms an interesting centre for the botanist, inasmuch as, like Galway and Moycullen, it lies upon the junction of the Carboniferous limestone and the metamorphic and igneous rocks of W Connaught (**384–396**). The proximity of Lough Corrib (**367**) provides an additional attraction. The limestone extends SE from Oughterard towards Galway, between the railway and the lake. Grey stretches of bare terraced crags alternate with great peat bogs and farmland. Its plants include *Potentilla fruticosa* (shore of L. Corrib below Lemonfield, not found recently), abundance of *Gentiana verna, Euphrasia salisburgensis* and the commoner crag species, and more sparingly *Neotinea* and *Ophrys muscifera* (see **365**). N, W, and S the change of rock is shown by the heathery ridges that lie around the village, on which we at once find a typical Connemara flora, with *Saxifraga spathularis* and *Dabeocia,* on bogs *Rhynchospora fusca* and *Carex lasiocarpa,* and in the pools *Eriocaulon, Utricularia intermedia, Lobelia, Sparganium minimum,* and everywhere *Osmunda.* The tourist will find the limestone flora very well developed around Ross Lake and Ballycuirke Lake (see **365**), while for the opposite aspect of the flora, the beautiful walk along the north-western arm of Lough Corrib can be confidently recommended. *Rubus iricus* and *R. hesperius,* two characteristic western brambles, are abundant about Oughterard.

The **River Corrib** drains a large area of country stretching N and NE of Galway as far as Castlebar and Ballyhaunis, a distance of 30 miles, but the only part which can be called a river is the 5 miles extending from L. Corrib to the sea at Galway. Like the Erne, the river-basin is occupied by a chain of lakes—L. Carra (**372**), L. Mask (**371**), L. Corrib (**367**), and the river has no botany apart from that of these considerable expanses of water. Between L. Mask and L. Corrib the stream does not even appear above ground, but pursues a subterranean course into which one may descend near Cong (**370**).

LOUGH CORRIB.

367. L. Corrib lies on the W edge of the Limestone Plain, at 28 ft. above sea-level. In its S portion the limestone still prevails along its W edge for a mile or two from the water; but in the NW part the metamorphic and Silurian hills rise direct from the lake-shore. In shape this great lake is irregular, with a NW and SE trend, and a length of 27 miles. Galway town (**364**) lies close to its low-shored and shallow S end; Oughterard (**366**) is situated on its W bank; and its W termination is buried far away among the Connemara hills. Fretted limestone shores, or stony beaches, backed by bogs and extensive grey pavements, fringe the lake, except in the NW, from Oughterard to Cong (**370**), where the older rocks bring in a bolder and highly picturesque type of scenery. The shore vegetation is usually poor. The flat slope below Oughterard displayed the following succession :—about storm-level — *Briza, Gymnadenia conopsea, Anthyllis, Thymus Serpyllum, Galium verum, G. boreale, Rhin-anthus,* etc.; then *Schœnus* and *Scirpus pauciflorus* abundant to the water's edge, with *Euphrasia, Anagallis tenella, Leontodon taraxacoides, Ranunculus Flammula, Carex diversicolor, Samolus*; in the water no plants. A richer part of the shore is described in **367a.** The flora of the limestone will be found referred to in **365**; of the metamorphic rocks, in **366**. *Ranunculus scoticus* is plentiful at the narrow middle part (Kilbeg), and is probably widespread; *Pimpinella major* also grows there. The marshes on the edge of the lake are full of *Juncus subnodulosus, Cladium, Schœnus, Carex lasiocarpa, Epipactis palustris. Caltha radicans* is frequent. In hydrophytes the waters are not rich. *Chara aculeolata* is common. *Lobelia* and *Eriocaulon* occur in the NW, and *Potamogeton angustifolius. Potentilla fruticosa* (**107**), recorded long ago from Lemonfield below Oughterard, has not been found there recently, but it grows abundantly on the stony damp lake-shore for several miles N and S of Ballycurrin in the NE. The numerous islands have never been systematically explored. From Bilberry I., one of the many limestone knolls off Oughterard, *Vicia Orobus* and *Taxus* are recorded. Cannaver I. (miscalled Canova) at the entrance of the NW arm of the lake, and inside the metamorphic area, yields *Thalictrum minus, Rhamnus catharticus, Galium boreale, Dabeocia, Lastrea œmula.* The botanist visiting L. Corrib has an interesting problem before him in the re-discovery of *Potamogeton Babingtonii* (**169, 217**), which is known only from a single specimen found floæating by John Ball in 1835, and of *P. sparganiifolius* (**168**), apparently gathered there by David Moore in 1853,[28] and later (1858) by M. Norman (Bot. Exch. Club Report 1930, 394).

367a. The usual sharp contrast exists between the flora of the limestone parts of the lake-shores and those where non-calcareous rocks prevail. Thus, the flora of the islet of Roeillaun in the NW, formed of Gotlandian slates, may be compared with that of the flat limestone shore directly eastward.

Roeillaun (N of Inishdoorus).

Well-drained peaty and rocky small islet, never grazed or tilled. The following common :—

Ranunculus scoticus	Erica cinerea
Vicia Cracca	Tetralix
Sedum anglicum	Dabeocia polifolia
Galium boreale	Solanum Dulcamara
Leontodon autumnalis	Mentha aquatica
Campanula rotundifolia	Teucrium Scorodonia
Calluna vulgaris	Myrica Gale

Also present :—

Anemone nemorosa	Quercus sessiliflora
Viola canina	Alnus rotundifolia
Oxalis Acetosella	Salix aurita
Hypericum Androsæmum	Listera cordata[29]
pulchrum	Orchis elodes
Rhamnus catharticus	Platanthera chlorantha
Ulex europæus.	Luzula sylvatica
Rubus *sp.*	Schœnus nigricans
Cratægus Oxyacantha	Molinia cœrulea
Sorbus Aucuparia	Athyrium Filix-fœmina
Pyrus Malus (acerba)	Asplenium Trichomanes
Euonymus europæus	Ruta-muraria[30]
Fraxinus excelsior	Adiantum-nigrum
Lonicera Periclymenum	Lastrea aristata
Viburnum Opulus	Polypodium vulgare
Lythrum Salicaria	Osmunda regalis
Solidago Virgaurea	Equisetum arvense
Cnicus pratensis	litorale
Lysimachia vulgaris	Taxus baccata
Scutellaria galericulata	

[28] See Irish Nat. 1909, 83–85.
[29] Here only 35 ft. above sea-level.
[30] In crevices of storm-washed slate rocks.

Ballycurrin.

Flat limestone shore, very stony, with much grey limy deposit. No plants in the water, save starved *Chara*. Succession from water's edge to the incoming of the flora of the dry limestone pasture and rocks:—

Open vegetation.

Schœnus nigricans
Carex flava
Ranunculus Flammula
Salix repens

Molinia cœrulea
Lythrum Salicaria
Juncus sylvaticus

Closed vegetation begins.

Anagallis tenella
Scabiosa succisa
Spiræa Ulmaria
Euphràsia officinalis
Alnus rotundifolia
Salix aurita
   cinerea
Leontodon autumnalis
Mentha aquatica
Agrostis alba
Fraxinus excelsior
Potentilla fruticosa

Gymnadenia conopsea
Lotus corniculatus
Sieglingia decumbens
Carex fulva
Pinguicula vulgaris
Leontodon taraxacoides
Prunella vulgaris
Cratægus Oxyacantha
Carex panicea
   dioica
Hydrocotyle vulgaris

Flood level.

Thymus Serpyllum
Blackstonia perfoliata
Calluna vulgaris
Ulex europæus
Antennaria dioica

Bellis perennis
Galium boreale
Lastrea Filix-mas
Rubia peregrina
Sesleria cœrulea

(Potentilla fruticosa stops.)

Campanula rotundifolia
Thalictrum minus
Hypericum pulchrum

Galium verum
Solidago Virgaurea
Geranium sanguineum

Oughterard is the most central point for the exploration of L. Corrib by water: on account of the fishing, plenty of boats are obtainable there. A small steamer, running on uncertain days from Galway to Cong, calls at Kilbeg ferry, situated on the narrow waist of the lake; this is a strategic centre for the explorer, and modest accommodation may be obtained.

Cloghmoyne.—Three miles NW of Headford, in the townland of Cloghmoyne, a low ridge rises which furnishes the only Irish station of *Phegopteris Robertiana* (**199**). The fern is rather widespread in fissures of occasional patches of bare limestone. Mostly the rock here is buried under a stony sward in which *Geranium sanguineum* is enormously abundant, accompanied by *Calluna* (abundant), *Erica cinerea, Aquilegia, Vicia Orobus, Rubia peregrina, Euphrasia salisburgensis*, and a little *Gentiana verna, Neotinea, Juniperus communis*, and commoner plants. *Calamagrostis epigejos*, very rare in Ireland, also occurs sparingly.

R. LL. PRAEGER in Proc. Roy. Irish Acad., **41**, B, 122.

TUAM.

368. The old town of Tuam lies in the flat limestone country east of L. Corrib. Between the town and the lake, where the Clare River flows, there are great stretches of bog and level pasture, and several shallow loughs which in summer usually present flat expanses of white marl fringed by black peat. A few miles to the SE, at Castle Hacket, **Knockmae**, a limestone knob, rises to 543 ft., and is tenanted by an outpost of the Burren flora, including *Epipactis atropurpurea* (**157**), *Gentiana verna* (**135**), and *Euphrasia salisburgensis* (**141**). The last two, with others of the same group, become abundant further W towards L. Corrib, where crag country prevails. The best plant of the Tuam neighbourhood is *Potamogeton perpygmœus* (*lanceolatus* var. *hibernicus olim*) (**167**), which grows in the Grange River near the ruined mill at Barbersfort. On a bit of wet bog on the limestone near Killower L. *Eriocaulon* (**173**) and *Lobelia* flourish — their only station in NE Galway. On an esker 4 miles N of the town *Hieracium iricum* is abundant. Along the Clare R. *Apium Moorei* (**118**) grows; also *Sium latifolium* and the usually seaside *Carex distans*. Here *Alisma ranunculoides* × *Plantago-aquatica* has been found and described by Glück as a new hybrid: it is accompanied by the better-known *A. Plantago-aquatica* × *ranunculoides* (Irish Nat. 1913, 179). *Rhynchospora fusca* and all the species of *Drosera* are frequent on the bogs. *Blackstonia* is common. *Fumaria micrantha* is frequent in tilled ground. At Hacket Lough near Head-ford *Ceratophyllum demersum* grows. Hotels will be found at Tuam.

369. Ballinasloe is situated on the River Suck in the NE corner of SE Galway. The river is here sluggish, with marshy banks, and yields *Sium latifolium, Utricularia intermedia, Potamogeton angustifolius, Carex gracilis,* and similar plants. The surrounding district consists of typical Central Plain country, pasture and bog predominating, and is almost unknown botanically except around the demesne of **Clonbrock**, in NE Galway, 8 miles NW of the town. Here *Potamogeton perpygmæus* (**167**) grows in the river; and the plants of the neighbourhood include the three species of *Drosera, Sium latifolium, Utricularia intermedia, Epipactis palustris, Neottia, Ophrys apifera, Potamogeton coloratus, P. angustifolius, Rhynchospora fusca, Carex diandra, Osmunda,* etc. *Althæa officinalis* occurs as a denizen at Ballinasloe, and *Sedum dasyphyllum* on walls.

R. Ll. PRAEGER: "Flowering Plants and Vascular Cryptogams" [of Clonbrock]. Irish Nat. 1896, 239–244.

There is an extensive hinterland of County Galway, E of Loughrea, Athenry, and Tuam, extending with little change E over Roscommon and N across E Mayo, a land of flattish drift-covered limestone with occasional low ridges of Old Red Sandstone which, so far as known, offers very little to the botanist. It is, indeed, the least interesting region, and most monotonous, in the whole of Ireland—Central Plain country of a very dull type. It is drained in great measure by the R. Suck, a slow stream artificially deepened in many places for drainage purposes and seldom seen by the botanist except at Ballinasloe (**369**), where it passes under the Dublin-Galway railway.

370. Cong and Clonbur. — The narrow neck of land which separates L. Corrib (**367**) from L. Mask (**371**) is composed mainly of low-lying limestone, but the western part is hilly, and formed of the older non-calcareous rocks of Connemara. Clonbur or Fairhill lies near the dividing line; Cong more to the E on the limestone. Eastward stretches the Limestone Plain, with some areas of bare crags along its edge; westward, the surface soon rises into heathery hills (Partry Mountains, **400**). The flora of the Cong-Clonbur neighbourhood includes *Ranunculus scoticus* (**82**), *Draba incana, Viola stagnina* (**91**) (common, and hybridizing with *V. canina*, as it often does), *Rhamnus Frangula, R. catharticus, Geranium columbinum, Rubus iricus, R. hesperius, Rubia peregrina, Galium boreale, Carlina, Crepis paludosa, Lobelia, Dabeocia, Centunculus, Gentiana verna, Blackstonia, Euphrasia salisburgensis, Verbena, Calamintha ascendens, Epipactis atropurpurea, Neotinea* (on the non-calcareous rocks at Mount Gable as well as on the limestone), *Ophrys apifera, Allium vineale, A. Schœnoprasum* (claimed by the finder, E. S. Marshall, as indigenous here), *Potamogeton prælongus, P. filiformis, Eriocaulon, Carex Hudsonii, C. lasiocarpa, Sesleria, Cystopteris fragilis, Pilularia, Chara connivens.* Mount Gable and L. Coolin, to the W, add *Subularia, Hieracium iricum, Phegopteris polypodioides*; while the limestone SE of Cong yields *Potentilla fruticosa* (see **367**), etc. There is no better spot than this last to study the striking contrast of floras between the calcareous and the non-calcareous rocks. On the lake-shore, on the very last remnant of the slates, we find, growing in company, *Ranunculus scoticus, Rhamnus Frangula* and *R. catharticus, Potentilla fruticosa, Dabeocia, Scutellaria galericulata, S. minor.*

E. S. MARSHALL and W. A. SHOOLBRED: "Irish Plants observed in July, 1895." Journ. Bot. 1896, 250–8. E. S. MARSHALL: "Irish Plants collected in June, 1896." *Ibid.*, 496–500.

LOUGH MASK.

371. Lough Mask is the second largest of the chain of lakes lying along the edge of the Central Plain, with low limestones stretching from the E shore, and the more ancient Connacht highlands rising from the other, the rocks prevailing here being Silurian slates and Lower Carboniferous Sandstone. The lake is oblong, about 10 by 4 miles, with two narrow arms penetrating W into the mountains. The E shore of limestone is much fretted, with many promontories, bays, and islands; the W shore (mainly Silurian slates) much less so. Height above the sea, 62 ft. In places, especially towards the S, the shores are well wooded, but the general aspect of the lake is open and the landscape bare. This is especially true of the W shore, where the miles of flowering *Fuchsia* hedges add a welcome note of colour. On the limestone the woodland takes the form of scrub, chiefly *Corylus* and *Cratægus*, with *Fraxinus, Betula pubescens, Euonymus, Rhamnus cath-*

articus, R. Frangula, and much *Rubus cæsius,* while on the foreshore the characteristic plants are *Viola canina, Hypericum perforatum, Parnassia, Galium boreale, Carlina, Anagallis tenella, Blackstonia, Samolus, Thymus Serpyllum.* On the SE shore, on broad flattish bare limestone beds resembling floors, *Ceterach, Asplenium Trichomanes,* and *A. Ruta-muraria* grow 1 ft. below the fringe of flood-rubbish, among mosses, on mixed humus and limestone pebbles 2 inches deep, fully exposed to the sky. The flora has been worked most on the S side, where limestone prevails, and it there includes *Euphrasia salisburgensis, Neotinea,* and other characteristic pavement species; but the calcicole flora is not nearly so rich as on the shores of L. Carra adjoining. *Sesleria,* for instance, has been seen at only one place. The numerous islands are mostly denuded whale-backs of drift. Some are densely wooded with native trees, chiefly *Fraxinus, Quercus, Betula pubescens, Ilex, Alnus,* which testify by their general outline the force of

Fig. 17.—Wind-moulding of wood on islands in Lough Mask, as seen from the north. The tallest trees are about 30 feet high.

the westerly winds (fig. 17); in open places *Rhamnus Frangula* is abundant. Much *Pinus sylvestris* has been planted, and it is spreading freely. On the island shores *Galium boreale* is conspicuous; *Ranunculus scoticus* and *Dabeocia* occur among a mixture of calcicole and calcifuge plants, due to the mixed nature of the substratum—non-calcareous or slightly calcareous drift resting on limestone. The W side of the lake yields a calcifuge flora including *Dabeocia, Listera cordata* (at 70 ft. above sea-level), *Agrimonia odorata, Carex strigosa, Osmunda,* etc. In sandy bays grow *Lobelia, Potamogeton filiformis, Isoetes lacustris.* The two *Rhamni* continue abundant on the Silurian rocks, on islands. *Plantago maritima,* plentiful around the lake as it is round L. Corrib and L. Carra, continues as far inland as Claremorris, Tuam, and Athenry. *Cuscuta Epithymum* (aggr.) is on *Ulex* near the bridge over the "Narrow Lake." On the western mainland there is a great deal of yellow or reddish sand, with *Filago minima,*

Anthemis nobilis, etc. Between the two western arms there is a considerable area of *Quercus-Betula* wood, with much *Alnus* in the damper parts and some *Corylus* and *Ilex.* The ground vegetation is very poor, no doubt on account of long-continued grazing, and consists mainly of *Pteris* and grasses: near by, on a cliff over Maamtrasna Bay, *Taxus* grows with the Himalayan *Cotoneaster microphyllus, Phegopteris polypodioides* (100 ft. above sea-level), *Lastrea æmula, L. montana.* An interesting feature of L. Mask is the occurrence round its shores of an old terrace and scarp about 15 ft. above present summer level (9 ft. above storm-mark), showing a former higher level of the lake (fig. 18). This lowering of the water-level was due, at least in part, to the construction of the futile Mask-Corrib canal, through the floor of which the water pours as through a sieve. Water-passages through the limestone now carry the whole drainage of L. Mask S into the adjoining lake. The flood-beach, often broad, is tenanted by dense growth of *Alnus,* under which are *Rubus cæsius, Lythrum Salicaria, Mentha aquatica.*

Fig. 18.—East and west section across Devenish, Lough Mask, showing of Boulder-clay, with ancient church, subtended by old beach, now wood

The contrast between the flora of the limestone and of the non-calcareous rocks is very marked. Half a mile W of Partry, where the Carboniferous Sandstone comes in, on one side of a narrow shallow depression rise green grassy limestone bluffs clothed with *Sesleria, Blackstonia, Carlina.* On the other side are dark-brown heathy knolls covered with *Erica cinerea, Calluna,* and *Dabeocia* in profusion, with *Listera cordata* sheltering under them. See also under **370.**

Ballinrobe is the best centre for exploring L. Mask. *Geranium pusillum* occurs there. At Tourmakeady on the W, accommodation will also be found.

R. Ll. Praeger: ''Botanical Notes, chiefly from Lough Mask and Kilkee.'' Irish Nat. 1909, 32–40.

LOUGH CARRA.

372. Lough Carra is a pretty and interesting lake lying on the limestone near the E margin of the great metamorphic area, and close to the N end of L. Mask. It is of very irregular shape, with numerous peninsulas and islands. Length 6 miles, breadth 1 to 2 miles. Height above sea, 69 ft. The shores and islands are frequently wooded, the islands with native trees, including *Betula pubescens, Rhamnus catharticus, Fraxinus, Salix aurita, S. cinerea, Juniperus communis*; less of *Cratægus, Ilex, Euonymus, Sorbus Aucuparia, S. porrigens* (**111**), *Viburnum Opulus, Ulmus montana, Corylus, Populus tremula*; and a little *Prunus avium, Sambucus, Alnus, Salix caprea, Quercus, Taxus.* The shores are mostly stony. The water is of a wonderful pale pellucid green, owing partly to its purity (being derived mainly from springs) and partly to a curious soft whitish calcareous deposit which envelops the whole bottom and reflects the light. Even on the boulders just below water-level this is an inch or two in thickness, a soft crust, with pinkish or greyish blotches due to algal growth. Deeper down the deposit is softer and more soapy in feel, and white or cream in colour. The incrustation is most dense in the S section of the lake, and has there a very deleterious effect on plant life. Hydrophytes are rare, and a few starved beds of *Chara*, heavily incrusted,. and some spindly stems of *Potamogeton perfoliatus* rising from deep water, represent the aquatic vegetation where the incrustation is at its maximum. The surrounding country is low. In some places on the W side drift is absent, and limestone crags are developed. The N part of the lake is deeper than the S, and the shores are rather higher. The margin, especially on the islands, shows four distinct zones of vegetation :—

(1) HYDROPHYTE ZONE.—A very poor flora, *Potamogeton filiformis*, which is frequent, being the most interesting plant. In the upper less incrusted part of the lake *Chara hispida* (aggregate) forms great beds in 10–20 ft. of water, mixed with *C. aculeolata* and *Potamogeton perfoliatus*.

(2) SCHŒNUS ZONE.—A wet-soil calcicole group dominated by *Schœnus*, with abundance of *Epipactis palustris, Galium boreale, Cnicus pratensis, Parnassia, Pinguicula vulgaris, Gymnadenia conopsea, Orchis* several spp., *Selaginella*; more locally *Ophrys apifera, O. muscifera,* and *Lastrea Thelypteris.*

(3) SESLERIA ZONE.—A dry-soil calcicole association, with much *Sesleria, Thalictrum minus, Viola canina, Rhamnus Frangula, Rubus saxatilis, Rubia peregrina, Carlina, Leontodon hispidus, Campanula rotundifolia, Blackstonia, Plantago maritima, Anacamptis pyramidalis, Ophrys apifera,* and rarely *Gentiana verna* (N extremity of Irish range).

(4) WOODLAND ZONE.—With the trees above-mentioned, and a shade flora including *Orobanche Hederæ, Neottia,* and *Epipactis latifolia.*

The limestone pavements are interesting as forming the N limit of the great tract of bare limestone which, beginning in Co. Limerick, attains its maximum development on the Burren hills in Clare, and continues along the chain of lakes on the W edge of the Limestone Plain northward to L. Carra. The very remarkable flora which characterizes this formation in Clare and Galway is much reduced by the time L. Carra is reached; but its most interesting member, *Neotinea intacta,* remains, and is frequent on the crag-lands. On the bare limestone *Erica cinerea* flourishes; also *Aquilegia, Rubia peregrina, Hieracium iricum, Spiranthes spiralis* and *Equisetum trachyodon,* with most of the plants of the *Sesleria* zone of the shores. On a bog near the lake *Listera cordata* grows at 100 ft. above sea-level. The lake shores are remarkably rich in orchids. Out of some 26 species in Ireland, 19 may be seen here.

Ballinrobe is the best place to stay for the exploration of L. Carra.

Northward, **Castlebar** lies amid a group of small lakes, which present no special interest.

R. LL. PRAEGER: "The Botany of Lough Carra." Irish Nat. 1906, 207–214. IBID.: "The Calcareous Deposit in Lough Carra." *Ibid.,* 232–3.

373. The River **Moy** rises high up on the Ox mountains near Lough Easky (**421**). After the manner of Irish rivers, it flows first SE, the direction opposite to that in which its mouth lies, and after nearly boxing the compass becomes tidal at Ballina (**375**), and enters the broad shallow sandy Killala Bay. Above Ballina (**375**) it yields *Caltha radicans* and *Potamogeton densus.* Its seven miles of estuary, with *Cochlearia anglica* and *Zannichellia* on tidal mud, is incompletely known from a botanical standpoint. The Moy is a featureless stream, draining much of E. Mayo. Above

Foxford it passes but does not enter Lough Conn (**374**), receiving the drainage of that considerable lake by a short tributary from Lough Cullin.

LOUGH CONN.

374. Lough Conn lies near Ballina, in NW Mayo. It is 9 miles long by 2 to 3 miles wide, and 41 ft. above the sea. L. Cullin, 2 miles by 2 miles, which adjoins on the S, is really portion of the same depression, cut off from the main lake by a narrow rocky ridge, and may be treated as portion of L. Conn. A stream only a few yards in length joins the two, and the difference in water-level is normally about a foot. L. Cullin drains into the Moy, which adjoins on the E. When the Moy is flooded, it pours back into L. Cullin, which then rises and discharges back into the larger lake. The two lie in the low limestone valley of the Moy, with the metamorphic highlands towering close on the W side; but old gneissic rocks surround the S end of L. Conn and the whole of L. Cullin— the W end of the great "Caledonian" fold which formed the Ox Mountains (**421**).

The limestone shores of L. Conn are low and mostly very stony, with a number of outlying islands and reefs. The characteristic scene on the NE shore consists of a slope like a badly macadamized road gay with *Poterium officinale, Thalictrum minus, Rubus saxatilis, Galium boreale, Plantago maritima,* all in profusion, set in sheets of *Thymus,* with clumps of *Lithospermum officinale. Poterium* is *par excellence* the characteristic plant of L. Conn. It occurs on the shores (and only there), right round both Conn and Cullin, being elsewhere in Ireland found sparingly only in a few stations in the NE. On either side of the middle of the lake a boggy patch over the limestone yields *Erica mediterranea* (**131**), elsewhere known only in the wild metamorphic country nearer the western seaboard in Mayo and Galway; also the three species of *Drosera. Ranunculus scoticus* (**82**) is on the lake-shore, and by Levally Lough to the SW.

Once the limestone is left, the landscape gets bolder, and rocky bluffs with an ericaceous flora, *Osmunda,* and so on, overhang the lake, so that the scenery around **Pontoon**, on the neck separating L. Conn from L. Cullin, is very picturesque, resembling that of Connemara lakes, but without their most characteristic plants, and differing also in fine *Quercus* and *Quercus-Betula* woods, which fringe the water and clothe low rocky hills on either side. *Sorbus rupicola, Juniperus communis* and *J. sibirica, Taxus* and *Populus tremula* grow on the granite rocks, with *Hieracium orarium* (var. *fulvum*), *H. iricum*, etc.; and *Lobelia* abundantly in sandy bays. The last, with *Potamogeton filiformis* (bays in SE and SW corners) and *Chara aculeolata*, are the only interesting hydrophytes yet found in the lake. *Pyrola media* and *Cephalanthera ensifolia* are near Pontoon, and *Scutellaria minor* is frequent.

Listera cordata grows on Illandooish, less than 50 ft. above sea-level, and a good colony of *Alisma Plantago* × *ranunculoides* in a sandy bay near by. The curious form *anomalum* of *Blechnum*, with all fronds similar and partially fertile, is frequent: elsewhere in Ireland it is known about Slieve Donard, Co. Down.

On Freaghillaun, an islet of an acre or two, never grazed, all but the exposed NW edge is occupied by *Quercus* wood, with here or there a *Betula pubescens* or *Corylus*. Undergrowth dense, about 2 ft. high; *Vaccinium Myrtillus* dominant; much *Luzula sylvatica, Rubus, Melampyrum pratense*; also *Calluna, Blechnum, Primula vulgaris, Hedera, Ilex, Sorbus Aucuparia, Lonicera, Pteris, Prunella, Asperula odorata, Oxalis, Anthoxanthum, Agrostis tenuis, Stellaria Holostea, Teucrium Scorodonia, Hypericum Androsæmum, Vicia sepium, Lastrea aristata*. Only two grasses (*supra*) seen! *Ilex* and *Sorbus Aucuparia* appear frequently as young trees in the undergrowth, apparently recent bird-brought arrivals, whereas no seedlings of *Quercus* are to be found. Where *Quercus* fails on the exposed side of the island *Calluna* supervenes, with a little *Pyrola media* among it. The association described above may usefully be contrasted with that of similar wood on the mainland a few hundred yards away, where, presumably owing to grazing, ferns (*Lastrea aristata* and *L. æmula*) have become dominant, with *Vaccinium* subordinate among *Blechnum, Melampyrum, Saxifraga spathularis*, and both species of *Hymenophyllum*.

L. Cullin, surrounded by metamorphic rocks, has a flora resembling that of L. Conn with the calcicole species omitted. *Poterium officinale* still prevails, also *Galium boreale* and *Ranunculus scoticus; Rhamnus catharticus* and *Crepis paludosa* are present; *Equisetum trachyodon* (**204**) grows on the shores and in Drumman Wood, and *E. litorale* (**201**) on wave-washed peat banks in several places. *Potamogeton filiformis* is at the outlet and near Pontoon, *Utricularia intermedia* and *Apium Moorei* on the S shore. Drumman Wood, mainly of *Quercus,* has a wonderful undergrowth of ferns, chiefly *Lastrea æmula, L. spinulosa* and *L. aristata,* all 2 to 3 ft. high.

The shores of L. Cullin near Pontoon yield some very ine old *Juniperus communis,* the largest being a dead uprooted tree measuring 18 ft. in height, with a trunk over 2 ft. in circumference at 3 ft. above ground. Some of the Junipers, varying from one species towards the other, are difficult to name.

The visitor to L. Conn will find accommodation at Crossmolina, on the Errew peninsula in the middle of the lake, and at Pontoon. The towering mass of Nephin (2646 ft., **411**) rising near the W shore imparts an impressive character to its beautiful scenery.

R. LL. PRAEGER: "Round Lough Conn." Irish Nat. 1900, 224–9.

BALLINA.

375. Ballina is situated on the River Moy (**373**), where t becomes tidal, in a wide limestone valley. Northward, he river broadens to the sands of Killala Bay. To the SW, L. Conn (**374**) lies at a distance of 4 miles, with the huge mass of Nephin (**411**) towering behind it. To the SE, Slieve Gamph, the W end of the Ox Mountains (**421**), presents a series of rather sterile hills. About Ballina tself few interesting plants have as yet been found (but little work has been done)—*Poa compressa* on the railway, *Epipactis palustris* and *Pinguicula lusitanica* in marshy ground, *Caltha radicans* and *Potamogeton densus* in the iver, *Cochlearia anglica* in the estuary. Good hotel accommodation.

Killala is a village with a fine round tower, on the shore 8 miles NW of Ballina. *Draba incana* and *Gentiana Amarella* grow on sandhills here, and *Scirpus rufus* in salt marshes. About the picturesque ruins of Moyne Abbey, between Killala and Ballina, *Inula Helenium* and *Carum Petroselinum* still tell of the old monkish kitchen-garden.

At Inischrone, a watering place opposite Killala, on the Sligo side of the river-mouth, one may stay, and find an area but little explored round about.

Botanical interest in E Mayo centres along the line of lakes which have just been described (**371, 372, 374**). East of these, around Ballyhaunis, Claremorris, Swineford, lies a wide stretch occupied mostly by poor pasture and peat bog, of few attractions.

THE COUNTY OF ROSCOMMON.

376. Roscommon would be dull botanically were it not for the River Shannon and its lake-expanses, which form the whole E boundary, a length of 60 miles. Another river, the Suck, an uninteresting stream, margins the county on the W. Area, 850 sq. miles, consisting, except in the N and NE, of flattish undulating drift-covered limestone country, mostly under grass. An anticline near L. Boderg exposes the Silurian slates in the ridge of Slieve Bane (857 ft.), and a more pronounced fold in the NW produces the Curlew Mountains (**381**), whose heathery slopes, formed of Old Red Sandstone and Silurian rocks rise to 863 ft. In the extreme N, by Lough Allen (**427**) a small part of the Connacht coal-field lies within Roscommon, the ground rising there to 1377 ft. on the border of the county (see **427**). In the E, the whole W shore of L. Ree (**379**) belongs to Roscommon, and other smaller Shannon lakes lie further N. The Curlew Hills are associated with several beautiful lakes, of which L. Key belongs wholly, and L. Arrow partly, to Roscommon (see **381**). Grass-land, as is usual in these Central Plain areas claims nearly two-thirds of the surface of the county; turf bog about one-eighth, crops one-fifth. In the S is much poor stony land, with bog and esker, and some rather desolate and little-known lakes.

377. **Roscommon,** the county town, offers no special attraction to the botanist, lying as it does amid great expanses of flattish grass-land. In Carrowroe Park *Monotropa* grows plentifully. *Rhamnus Frangula* and *Carex strigosa* are at Kilteevan. *Geranium pratense* is naturalized in several places, and *Campanula Trachelium* at Lough Glynn. An excursion E to Lough Ree (**379**) will allow the peculiar flora of the limestone lake-shores to be studied *Lathyrus palustris* (**103**) is found in abundance by the Clooneigh River. Westward, the banks of the R. Suck

are in some places, as at Donamon, good ground for the
botanist.

ATHLONE.

378. Athlone is a busy town situated on the Shannon.
This was always an important place, for it formed the
only spot where the river was fordable for a long distance
N and S, and, situated in the middle of Ireland, it was
the point at which traffic between E and W converged.
Hence the old fortifications, of various dates, which guard
the river on either side. Sir Henry Sydney built the first
bridge in 1566. On account of the proximity of Lough
Ree (**379**), with its interesting flora, a very pleasant day
may be spent here by the botanist, either on foot or by
boat. Furthermore, he has the very fine early Christian
monuments of Clonmacnoise or the Seven Churches within
a day's excursion, where, when surfeited with antiquities,
he may study bogs and eskers on an imposing scale. In
wet meadows on the W side of the Shannon above the town,
where it widens into L. Ree, *Lathyrus palustris* (**103**) and
Lastrea Thelypteris grow in delightful profusion; with
them are *Galium boreale, Teucrium Scordium* (**148**),
*Epipactis palustris, Carex lasiocarpa, C. Hudsonii, C.
diandra, Apium Moorei* (**118**), *Callitriche autumnalis*, and
in the river the rare *Chara tomentosa* (**207**) — all
characteristic L. Ree species. Coosan L., on the Westmeath
side of the lake, furnishes an excellent sample of the
L. Ree flora. Several esker-ridges lie close to the town,
and in consequence we get a good gravel flora, which
includes *Papaver Argemone, Arabis hirsuta, Cerastium
tetrandrum, Arenaria tenuifolia, Crepis taraxacifolia,
Blackstonia, Linaria minor, Hordeum murinum. Sesleria*
grows near the Shannon, presumably on limestone gravel,
its most inland Irish station. Bogs on the Roscommon side
yield *Rhynchospora fusca, Pinguicula lusitanica*, and so
on, and drains *Carex Pseudo-Cyperus, C. diandra*, etc.

There is an old record of *Geranium sanguineum* from
Brideswell, 7 miles NW of Athlone, of which confirmation
is desirable. Another on the same authority (Dr. Patrick
Browne, c. 1785), which appears certainly erroneous (see
Proc. R. Irish Acad., **41**, B, 116–7, 1932), attributes
Euphorbia hiberna (**154**) to Slieve Bane. This is a ridge
of over 800 ft., harbouring *Lastrea montana* and a calcifuge
flora.

LOUGH REE.

379. Lough Ree is the more northern of the two great lake-like expansions of the Shannon. The upper end is at Lanesborough, and thence it stretches S for 17 miles nearly to Athlone (**378**). The lake lies entirely in the great Central Plain of limestone. The shores are irregular (fig. 19). Twice it narrows to one mile, and further S throws out an arm which gives it its greatest width, namely, 7 miles. Islands are numerous. Height above the sea 122 ft. in summer. Average depth only 20 to 25 ft. Woods are rare; the shores and islands are largely under pasture, and the scenery in general rather bare.

The aquatic flora is remarkable chiefly for *Chara tomentosa* (**207**), unknown in Britain, and in Ireland confined to the Shannon basin (in which it has a wide range) and the lakes of Westmeath; it is of frequent occurrence throughout L. Ree. Other interesting aquatics are *Utricularia intermedia, Potamogeton angustifolius, P. filiformis* (frequent), *Isoetes lacustris, Chara aculeolata*, with *Hydrocharis* and *Lemna polyrrhiza* in ditches. The rough limestone shores have a characteristic flora, which includes, mostly in quantity, *Teucrium Scordium* (**148**) (confined to the Shannon basin save for a station in Clare) *Thalictrum minus, Papaver Argemone, Galium boreale Carlina, Campanula rotundifolia, Lysimachia Nummularia Blackstonia, Calamintha ascendens, C. Acinos, Ophrys apifera.* On the islands, thickets of *Rhamnus catharticus Populus tremula,* and *Sorbus porrigens* are frequent, while among the rarer plants of their undergrowth are *Rhamnus Frangula, Crepis taraxacifolia, Hieracium umbellatum Orobanche Hederæ, Cephalanthera ensifolia.* In marshy places *Stellaria glauca, Cicuta, Sium latifolium, Galium uliginosum, Cladium, Carex diandra, C. lasiocarpa, C. Pseudo-Cyperus* are characteristic plants. *Galium erectum* and *Equisetum variegatum* are at Coosan Lough, *E. litorale* at Lanesborough; *Lathyrus palustris* and *Lastrea Thelypteris* have a wide but local distribution. In bogs near the lake *Andromeda* and *Rhynchospora fusca* are abundant while on the E arm *Stachys officinalis*, rare in Ireland, is found, and *Barbarea intermedia* at Lanesborough.

A typical piece of shore (near Rathcline in the NE corner) shows a grassy slope dropping about 6 ft. in 100 yards from a bushy fringe of *Ulex europæus, Cratægus*

and *Rubus* (with some *Rhamnus catharticus, Euonymus,*
etc., occasionally) to the water. Soil, mostly a mixture of
limy gravel with peat or loam. Mixed vegetation of *Briza,
Schœnus, Lotus corniculatus, Galium boreale, Orchis* spp.,
*Trifolium pratense, Lysimachia Nummularia, Lythrum
Salicaria, Achillea Ptarmica, Mentha aquatica, Ranunculus
acris, R. Flammula, Spiræa, Hydrocotyle, Scabiosa succisa,*

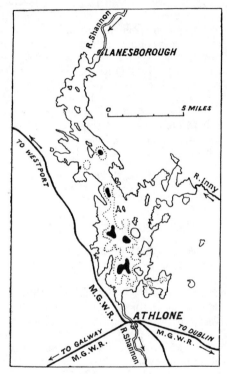

Fig. 19.—Lough Ree.
Dotted line = 5 fathoms.   Black area = over 10 fathoms.

*Juncus inflexus, Rhinanthus minor, Vicia Cracca, Sela-
ginella, Pinguicula vulgaris, Scirpus pauciflorus, Carex
hirta, C. fulva, C. panicea, C. Goodenowii.* Finally a more
frequently washed, more gravelly zone with *Carex diversicolor, Teucrium Scordium, Mentha aquatica, Potentilla
Anserina, Carex Œderi, Littorella.* Then summer water-
level, with mostly no aquatic vegetation below it. Winter

storms cast their debris about half-way up the slope, and the edge of the zone of bushes above represents maximum winter flooding.

Lough Ree is best explored by boat from Athlone (**378**). A day spent in working the wet meadows above the town (*Lastrea Thelypteris* and *Lathyrus palustris* in abundance, also *Apium Moorei*, *Callitriche autumnalis*, *Chara tomentosa*, *C. aculeolata*, etc.), and on to the finely wooded Hare Island (*Cephalanthera ensifolia*, etc.) with perhaps a bit of Coosan L. (good marsh flora and plenty of *Chara tomentosa*) on the way home, will give a good idea of the vegetation and scenery of the lake, and provide a profitable day's botanizing. Ballymahon, on the River Inny near the E shore of the lake, is also a good centre, not well worked. *Lathyrus palustris* is by the river there, and *Mercurialis perennis*, local in Ireland, occurs. Two uncommon hybrid pondweeds, *Potamogeton gracilis* (*alpinus* × *gramineus*) and *P. Tiselii* (*natans* × *obtusifolius*) are in the Inny (Journ. Bot. 1907, 173).

R. M. BARRINGTON and R. P. VOWELL: "Report on the Flora of the Shores of Lough Ree." Proc. Roy. Irish Acad., ser. 2 (Science). **4**, 693–708. 1887. R. LL. PRAEGER in Irish Nat. 1901, 39–41.

THE DRUMOD REGION.

380. Above L. Ree the Shannon has not been adequately examined. The river maintains its characteristic form as a very broad slow-flowing stream, with reed-beds and marshes mostly submerged throughout the winter, and occasional lake-like expanses. W of Drumod there is a remarkable area of this kind, half land and half water with *Carex axillaris* among other things. *Nasturtium barbaroedes* (*amphibium* × *sylvestre*) has been found at **Carrick-on-Shannon**, where also *Potamogeton filiformis* grows (at Annaghearly Lake). E of Drumod is a curious region consisting of low parallel ridges and hollows, the former tilled, the latter filled with bog or lake (**426**). *Viola stagnina* (**91**) and *Teucrium Scordium* (**148**) grow on the Shannon bank below Roosky. Many of the L. Ree aquatic and paludal species are widely spread in this area.

One of the most remarkable sights in the district is the great colonies of *Sarracenia purpurea* in a bog lying 1½ to 2 miles WSW of Termonbarry, which have arisen from a few plants introduced there in 1906. When in full flower in June they form a very beautiful sight. *S. flava* and *S. Drummondii* planted in the same bog (a little further S) at the same time, have not succeeded (see PRAEGER in Proc. R. Irish Acad., **41**, B, 104–5, 1932). *S. purpurea* has been introduced by the same agency into Westmeath (**487**). *Apium Moorei* and *Equisetum litorale* are on the Shannon bank close by, with *Lysimachia Nummularia*, etc.

About **Castlereagh**, in the NW, the Old Red Sandstone breaks through the limestone and forms low hills, with a consequent change of flora; but plant-records from that district, a wide rather dull area extending to Elphin, Strokestown, and Frenchpark, are very few.

THE CURLEW MOUNTAINS AND BOYLE LAKES.

381. A fold in the rocks of the Central Plain, running from the S end of L. Allen (**428**), SW to Kilkelly in E Mayo, reveals the older rocks lying below the limestone, and forms in parts a low heathery range of Old Red Sandstone hills, contrasting with the green pastures of the plain. L. Key and L. Gara lie in breaks in the ridge, and L. Arrow on the limestone on its N edge; the whole forms a picturesque group of hills and lakes around the prettily situated town of **Boyle**, in Roscommon. L. Gara (*infra*), 222 ft. above the sea, a very sinuous, shallow lake, about 4 miles by 3, with bare shores and islands, and a poor vegetation, drains NE into L. Key (about 3 by 3 miles, 139 ft. above the sea), a beautiful and interesting sheet of water, embellished with woods and hills. L. Key in turn drains SE, by means of a slow, winding, lake-like river, into the headwaters of the Shannon, some miles below L. Allen. L. Arrow (4 by 1½ miles, 181 ft. above the sea), also a delightful lake, adjoins L. Key on the other side of the Curlew Mountains, as the Old Red Sandstone ridge in the neighbourhood of Boyle is called. It drains NW into Ballysadare Bay.

L. Gara is desolate, with shallow water and a sandy bottom, its shores fringed with vast bogs. *Sorbus porrigens* and *Galium boreale* are the only interesting plants seen on its margins, with *Oxycoccus, Carex diandra*, and *C. limosa* in quantity on the bogs. The flora of **L. Key** is richer, with *Ranunculus scoticus, Lathyrus palustris* (N end); there are also *Thalictrum flavum, Myriophyllum verticillatum, Utricularia intermedia, Cladium, Carex gracilis,*

C. Hudsonii, C. lasiocarpa, C. acutiformis, Osmunda, etc. *Ranunculus Lingua, Cicuta, Sagittaria,* and *Butomus* are also found about the lake. From the fine demesne of Rockingham, on the S shore, *Neottia, Lastrea Thelypteris,* and *Veronica peregrina* are recorded. **L. Arrow,** the water of which is wonderfully clear (being supplied mostly by springs), like L. Key needs further examination. *Saxifraga hypnoides,* growing on its shore, is one of the few plants recorded from it; *Carex diandra* also occurs. Rising over the W side, Carrowkeel Mountain (1062 ft.) (**382**), an outlier of the Ben Bulben limestone hills, yields some of the plants of that famous range.

The **Curlew Mountains** (highest point 863 ft., NW of Boyle) call for no remark, save for the sharp contrast between their calcifuge flora and that of the surrounding plain; *Lycopodium clavatum* and *Lastrea montana* are the only uncommon plants recorded from them.

Boyle forms a convenient centre for the exploration of this district.

R. LL. PRAEGER: ''A Botanist in the Central Plain.'' Irish Nat. 1899, 89–90.

KESHCORRAN AND CARROWKEEL.

382. These two hills, lying W of L. Arrow (**381**), form an important outlier of the limestone range of Sligo. They are twenty miles S of Ben Bulben (**422**), but Knocknarea (**419**) forms a connecting link, and all their plants are familiar Ben Bulben species. **Keshcorran** (1163 ft.), which rises boldly from the plain near Ballymote, is a dome with a conspicuous low cliff on its SW side, pierced by a row of caves in which many bones of Bears and other extinct animals have been found.[31] *Cornus sanguinea* is on the cliff with native *Ulmus montana*: about the summit are *Draba incana, Cerastium tetrandrum, Saxifraga hypnoides, Cystopteris fragilis, Asplenium viride.* On Glacial drumlins along the S base of the hill, formed of limestone debris and resting on limestone, an unexpected calcifuge flora prevails, including *Hypericum humifusum, Cytisus* (abundant), *Galium saxatile, Digitalis, Blechnum, Athyrium.* **Carrow-**

[31] R. F. SCHARFF, etc.: ''The Exploration of the Caves of Kesh, County Sligo.'' Trans. Roy. Irish Acad., 32, B, 171–214, pl. ix–xi. 1903.

keel, which adjoins on the E and looks down on L. Arrow,
is a curious hill, with a flat top traversed by parallel cliff-
walled rifts, and dotted with Bronze Age carns. It has
yielded *Meconopsis, Draba incana, Saxifraga hypnoides,
Epilobium angustifolium, Hieracium stenolepis.* The
visitor to this region can stay at Ballymote or Boyle.

THE COUNTY OF MAYO.

383. Like Galway, Mayo (Irish *Muigheo*, the Plain of
the Yew-tree) offers a sharp contrast in its geology and
consequently in its topography and botany due to the
substitution, along the line of L. Mask and L. Conn, of
ancient metamorphic rocks for the low limestones which
prevail in the E, and which have already been dealt with
(**359, 370–5**).

The E–W political boundary between Galway and Mayo
is political only. The topographical and botanical
boundary—and a very striking one—runs N and S along
the chain of lakes, separating the low limestone grassy
country of E Galway and E Mayo from the bog-covered
and largely mountainous metamorphic part of the two
counties. The limestone tract has been dealt with above;
the western part remains to be described.

THE GALWAY-MAYO HIGHLANDS.

384. A profound and dramatic change of scenery and
of vegetation is obtained by venturing only a mile westward from the edge of the N portion of the Limestone
Plain with its fringe of bare "crags," into the metamorphic
region which immediately adjoins. W of a line drawn
(roughly) from Galway N by L. Corrib, L. Mask, and L.
Conn to Killala, the Carboniferous limestone gives place
to acid rocks, mostly very ancient, and brought up by
folding, of the "Caledonian" epoch. In the S, granite
occupies much of Connemara. N of that, quartzite, gneiss,
Silurian schists and slates predominate, and in the NE,
from Nephinbeg to Killala, the slates and sandstones
which underlie the Carboniferous limestone cover a
considerable area. The deep indentation of Clew Bay,
where alone the limestone reaches W to the Atlantic, breaks
the area into almost equal parts. The great metamorphic

rock-mass of W Galway and W Mayo, stretching N and S
for 75 miles, has resisted marine denudation; its fretted
coast stands out boldly, sentinelled with brown mountains,
protecting the low grassy limestone plain in its rear. This
is a rugged region, with a complex of high hills in both
of the counties just mentioned; for the rest, vast stretches
of peat-bog, low and undulating, often lying direct on
glaciated rock without the intervention of drift. Here in
Connemara, Erris, and the adjoining tracts, despite the
result of centuries of farming and reclamation, grass-land
of all kinds occupies only about one-third of the total
area of over 2000 sq. miles, while bog and mountain land,
generally dominated by an ericaceous vegetation, claim
nearly one-half. The S part of the region, Connemara,
has uninterrupted boglands in the S, sloping gently to
the sea along a highly dissected coast-line; and towards
the N, bold groups of bare quartzite mountains. Around
the remarkable long narrow fiord of Killery and on N to

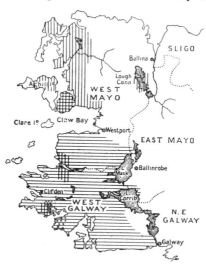

Fig. 20.—Distribution of *Dabeocia polifolia* (horizontal lines),
Erica mediterranea (vertical lines), and *E. Mackaii* (dots).

Note.—L. Beltra in W Mayo, half way between Clew Bay and
L. Conn, has now to be added to the range of *E. mediterranea*.

Clew Bay, rocks of Silurian age prevail, and there are
similar rocky hills, and, in the N, boglands of great extent.
Lakelets are numerous, especially in Connemara, and the

country is exceedingly wild and beautiful. The sunken coast-line, with outlying islands and peninsulas and innumerable inlets, is almost as complicated and varied as that of western Scotland. In the E, the low limestone of the Central Plain laps up against the older rocks, producing a startling change of vegetation. Crossing the Galway-Oughterard road westward, for instance, one passes abruptly from the calcicole flora—grass-land and *Corylus* scrub, with *Gentiana verna, Euphrasia salisburgensis, Ophrys muscifera, Sesleria,* to peat-bog and heath, with great stools of *Osmunda, Dabeocia* nodding its big purple bells on the wayside banks, and *Saxifraga spathularis* clinging to the rocks.

The flora is especially interesting as containing several Lusitanian plants not found elsewhere in Ireland, nor in Great Britain. Such are the three heaths (fig. 20), *Erica mediterranea* (**131**), *E. Mackaii* (**130**), *Dabeocia* (**132**). The presence of these compensates for the loss of *Arbutus* and *Pinguicula grandiflora,* which are absent from this region. *Saxifraga spathularis* (**114**) is everywhere; *S. Geum* and *Euphorbia hiberna* appear to be vanishing species here : each is confined to one island. Equally interesting is the occurrence of American plants—*Sisyrinchium angustifolium* (**159**), *Naias flexilis* (**172**), and *Eriocaulon* (**173**). These three range along the W coast of Ireland from Kerry to Donegal, and the second and third re-appear sparingly in northern Britain.

H. C. HART: ''Report on the Flora of the Mountains of Mayo and Galway.'' Proc. R. Irish Acad., 2nd ser., 3 (Science), 694–768. 1883.

SOUTH-EASTERN CONNEMARA OR IAR CONNACHT.

385. If one follows the Clifden road from Galway NW through Oughterard to Maam Cross, a distance of nearly 30 miles, and then turns S to Cashla, and back E along the coast to Galway, one encircles an area of some 400 sq. miles formed almost entirely of lake-strewn bog spreading in broad undulations, rising towards the N but not attaining 1000 ft. Here one can tramp for ten miles without encountering a single road or house — only the soaking brown bog and little lakes of that indescribable blue that is given by the Connemara sky reflected in brown peat-water. This is no place for the botanist intent merely on collecting rare species — *Saxifraga spathularis* on the few rocks, and *Lobelia* and *Eriocaulon* in the water, are the chief of the plants which will reward him—but to the ecologist and plant-geographer the area, absolutely untouched by man, is full of interest.

From lake through bog to rocky heath shows a transition in which *Phragmites, Cladium, Carex lasiocarpa, Schœnus, Eriophorum, Rhynchospora, Calluna, Molinia* may in turn be dominant, and particularly the last: the vegetation consists mainly of combinations of these with much *Scirpus cœspitosus* and the usual concomitants. *Calluna* and *Erica Tetralix* occur throughout, but are often quite subordinate, as are *Rhynchospora alba, Scirpus cœspitosus, Narthecium, Eriophorum* spp., and *Drosera* spp. There are wetter areas, which probably represent rock-basins not yet quite filled up. Here *Rhynchospora alba* is dominant, with much *Schœnus* and *Sphagnum,* some *Rhynchospora fusca, Menyanthes,* and the other species mentioned above; *Ericaceœ* occurring only as odd, starved plants. Patches of *Phragmites* and *Cladium* growing here and there suggest the last survivals of the flora of lakelets now overwhelmed by the bog. The *Molinia* moor extends down to sea-level, and up to about 1000 ft. on the hills, where, with better drainage, it is replaced by the familiar *Calluna* formation, with patches of *Scirpus cœspitosus* moor on high-level flats. The impression conveyed by the vast mantle of bog in Connemara is that peat-formation is still proceeding steadily over the great driftless rock-floor. While innumerable lakelets fill the hollows yet, every gradation may be traced from deep water, through *Cladium* and *Phragmites* swamp, and *Rhynchospora* bog, to the prevailing *Molinia* moor. *Molinia* is everywhere especially conspicuous, usually in the small form *depauperata,* and *Eriophorum* moor also abundant. The pH is very high, ranging from 4 to 5. *Schœnus* tends to get reduced, as it is extensively cut for thatching.[32]

[32] G. CONNOLLY: ''The Vegetation of South Connemara.'' Proc. Roy. Irish. Acad., **39**, B, 203–231. 1930. See also the British Vegetation Committee's Excursion to the West of Ireland [by A. G. TANSLEY and others]. New Phytol., 1908, 253–260.

SOUTH-WESTERN CONNEMARA.

386. The district lying between Clifden (**390**), Roundstone (**389**), and Slyne Head possesses high botanical interest, and differs in some important respects from the rest of Connemara. Stretching from the back of Urrisbeg, the dark serrated hill which rises behind Roundstone, an extraordinary maze of lakelets and rocky heath extends W to Clifden, a type of ground in Great Britain only matched in the Outer Hebrides. Along the shore, limy sands, formed mainly of foraminifera and nullipores, have been widely scattered by westerly gales; and the coastal region harbours in consequence a calcicole vegetation very different from the bog and lake flora of the hinterland.

The rarer plants of the great bogland which stretches from Roundstone to the western ocean includes quantities of *Dabeocia* (plate 27, **132**), *Drosera anglica, D. longifolia, Arctostaphylos Uva-ursi, Rhynchospora fusca* (**177**), *Deschampsia setacea*; the lake-margins are fringed with *Cladium, Carex lasiocarpa, C. limosa,* and their bottoms covered with *Lobelia, Utricularia intermedia* (**144**), *Eriocaulon* (**173**), *Littorella,* but very few other plants.

The occurrence in and around the lakelets of abundance of *Schœnus* and *Cladium* and occasionally of *Juncus subnodulosus* is interesting in view of the highly acid nature of the water, since in other places these may often be reckoned as distinctly calcicole.

The country in general is treeless, and only where rising rocky ground provides relief from the all-prevailing soaking peat is there any arboreal or arbuscular vegetation—as for instance a mile W of Craigga-more. But the rocky islets in the lakes, well-drained and protected from grazing animals, are frequently densely clothed with a wind-shorn woodland vegetation in which *Taxus* often predominates, with *Sorbus Aucuparia, Ilex, Juniperus sibirica, Ulex Gallii, Myrica, Erica cinerea, Calluna, Arctostaphylos Uva-ursi; Hedera, Rubus* spp., with *Osmunda* often form a fringe, while *Lastrea aristata, Polypodium vulgare* and sometimes *Listera cordata* and both species of *Hymenophyllum* grow as shade plants. *Sorbus rupicola* is near Recess.

R. LL. PRAEGER: ''The Island-flora of the Connemara Lakes.'' Irish Nat. 1896, 292–3.

387. Analysis of a typical sq. yard of flat soaking bog near Craigga-more showed the following composition :—

Molinia cœrulea, *dominant.*
Schœnus nigricans, *c.*
Rhynchospora alba, *c.*
Campylopus *sp., c.*
Sphagnum *spp., c.*
Narthecium ossifragum, *f.*
Erica Tetralix, *r.*
Calluna vulgaris, *r.*

Cladonia sylvatica, *r.*
    uncialis, *r.*
? Pleurozia *sp., r.*
Dicranum *sp., r.*
Potentilla erecta, *v.r.*
Drosera rotundifolia, *v.r.*
    anglica, *v.r.*
Pedicularis sylvatica, *v.r.*

with patches of *Eriophorum vaginatum* and *Sphagnum magellanicum.* Small low mounds rose a few inches above the general surface. On them were :—

Calluna vulgaris,
        dominant.
Molinia cœrulea, *v.c.*
Cladonia sylvatica, *v.c.*
Erica Tetralix, *f.*
Schœnus nigricans, *f.*
Leucobryum glaucum, *f.*

Cladonia uncialis, *f.*
Potentialla erecta, *r.*
Myrica Gale, *r.*
Eriophorum vaginatum, *r.*
    angustifolium, *r.*
Racomitrium uliginosum. *r*
Sphagnum *spp.*

Where the bog is protected from grazing, as inside the fence along the railway, *Molinia* becomes very conspicuous forming a dense waving shining meadow over a foot high. The *Molinia* heath extends up the hills (*e.g.* Ben Lettery) to about 1000 ft., where it gives way to *Calluna* moor.

388. The maritime calcicole flora referred to is best developed at **Dog's Bay** near Roundstone (**389**), and over an extensive tract of sandy and rocky grassy pasture in the **Slyne Head** district, and resembles (in a reduced form) that which characterizes the limestone pavements. The most interesting plant is *Neotinea intacta* (**158**), which grows sparingly at both ends of Dog's Bay, and more frequently on the Bunowen peninsula beyond Bally-conneely, where it has been observed in three stations. *Euphrasia salisburgensis* (plate 28) is abundant both at Dog's Bay (a corrupted translation of Portnafedog, *Port na Feadóige,* the plover's stream) and in the Ballyconneely and Bunowen district, accompanied by much *Asperula cynanchica, Centaurea Scabiosa, Carlina, Blackstonia, Spiranthes spiralis, Anacamptis pyramidalis, Sesleria* (**186**), and less of *Arabis Brownii* (**85**), *Geranium colum-binum, Saxifraga tridactylites,* the last two growing in

DABEOCIA POLIFOLIA AT ROUNDSTONE, CONNEMARA.
Growing through *Ulex Gallii*, among *Rubus*, *Pteris*, and the commoner *Ericaceæ*.
July.

R. Welch, Photo.

PLATE 28.

EUPHRASIA SALISBURGENSIS AT DOG'S BAY, CONNEMARA.

Growing on a small steep bank beside a granite rock, accómpanied by
   Thymus Serpyllum, *Galium saxatile*, *Asperula cynanchica*, *Arabis Brownii*, &c.   July.

R. *Welch*, *Photo*.

crevices of gneissose rocks. *Bartsia viscosa* (**142**) is at Dog's Bay also. *Ajuga pyramidalis* (**149**), elsewhere in Ireland known only in Burren and Aran, grows on Bunowen Hill, a little volcanic neck that rises abruptly from the lowland. The lakes of this maritime fringe have likewise a flora different from that of the bogland, more calcicole and slightly halophile in character. *Lobelia* and *Eriocaulon* are rare; and instead we get quantities of Charophyta (*C. aculeolata, C. hispida, C. delicatula*), and Pondweeds (*P. coloratus, P. nitens, P. angustifolius*), *Ranunculus Baudotii, Œnanthe Lachenalii.*

The sand-surrounded marshes yield *Sium erectum, Œnanthe Lachenalii, Rumex Hydrolapathum, Epipactis palustris, Carex diandra*—like the other plant-groups, a series quite at variance with the typical Connemara flora. (See Praeger in Irish Nat., 1907, 241–3).

At Clifden and Roundstone hotels will be found from which the botanist can visit all parts of this region.

389. Roundstone itself, apart from its interesting calcicole outlier-flora (**388**), is an excellent botanical centre. This neighbourhood is the only home outside the Pyrenean region of *Erica Mackaii* (plate 39, **130**), which has its head-quarters at **Craigga-more Lough,** four miles NNW of Roundstone. From that place it has been traced for a mile E and W, and S across the wet bogs to Letterdife and up to 600 ft. on the hill of Urrisbeg, where it may be seen mingling with *E. mediterranea* (**131**). It occurs again in some quantity about Carna, 7 miles to the SE. With it at Craigga-more *E. Praegeri* (**130**), a cross between *Mackaii* and *Tetralix*, is not uncommon. Another form, found by Charles Stuart of Edinburgh in 1890, has been determined by E. F. Linton (Ann. Scott. Nat. Hist., 1902, 176–7, and Irish Nat., 1902, 177–8) as *Mackaii × mediterranea* and named *E. Stuartii* (**130**). One clump was found growing with *Mackaii*; the other parent occurs 2½ miles S. A double-flowered form of *Mackaii* was collected in 1869 by A. G. More, and on being refound by F. C. Crawford of Edinburgh in 1891, has been called by nurserymen *E. Crawfordii* (**130**). *E. Mackaii* has a rather straggling habit, recalling that of *E. Tetralix* among which it grows, and from which it is at once picked out by its broader ciliate leaves, and flowers rose-coloured on the sunny side, almost white on the other. On the margin of Craigga-more L. its blossoms may be seen mixing with those of *Dabeocia*, with *Eriocaulon* and *Lobelia* rising out of one to three ft. of water to meet them, forming a charming and unique picture.

Craigga-more is also the reputed habitat of *Erica ciliaris* (**210**), a plant twice definitely recorded from this place and even supported by a specimen, but which has defied every effort to refind it, and is not believed to have ever occurred here (for a digest of the facts see "Cybele Hibernica," ed. 2, 498). It would be highly interesting if one of the most characteristic of the English Lusitanian species were found to grow among its Hibernian congeners.

Another rare Roundstone plant is *Naias flexilis* (**172**), long known from Cregduff Lough (plate 30), one mile SW of the village, where it grows accompanied by *Elatine hexandra*, and by *Nymphœa occidentalis*, which up to the present is unknown elsewhere in Ireland. This water-lily was first recorded as *N. candida* (see *Irish Nat.*, 1911, 198, 1912, 136). *Erica mediterranea* also occurs, occupying with *E. Mackaii* wet boggy valleys on the NW slope of Urrisbeg, the dark serrated ridge (987 ft.) which rises immediately behind Roundstone. This hill yields also *Sagina subulata, Juniperus sibirica, Juncus subnodulosus* (in intensely acid soil), *Eriophorum latifolium.* Other Roundstone plants are *Agrimonia odorata, Bartsia viscosa, Lamium molucellifolium, Allium Babingtonii, Sparganium minimum, Carex lasiocarpa,* and the critical *Carex trinervis* (found by E. F. Linton in a seaside field "near Roundstone" in 1885, not collected in Ireland since). *C. Hudsonii* and *Scutellaria minor* are widespread.

Several little hotels offer accommodation to the visitor to Roundstone.

R. LL. PRAEGER: "Notes of a Western Ramble." Irish Nat. 1906, 257–66; *ibid.*: "Recent Extensions of the Range of some rare Western Plants." *Ibid.* 1907, 241–3.

390. Clifden, the railway terminus, lies in a sheltered nook, and has actually trees about it. The tongues of land which run towards the ocean furnish excellent studies in the response of the vegetation to exposure. Of trees, *Fraxinus* is the one which penetrates furthest west.

A very exposed rocky heath facing the Atlantic, 60–70 ft. above the sea, showed a composition as follows:—

PLATE 29.

ERICA MACKAII ON THE EDGE OF CRAIGGA-MORE LOUGH, CONNEMARA.

The accompanying plants include *E. cinerea, Calluna vulgaris, Dabeocia polifolia, Myrica Gale, Molinia cærulea, Osmunda regalis.* July.

R. Welch, Photo.

PLATE 30.

CREGDUFF LOUGH AND URRISBEG, CONNEMARA.

In the deeper water *Naias flexilis, Scirpus lacustris, Nymphaea alba.* Round the margin a
fringe of *Eriocaulon septangulare, Lobelia Dortmanna, Littorella lacustris, Mentha aquatica.*
At flood-level a fringe of *Osmunda regalis,* and above *Dabeocia polifolia* and the commoner
heaths, *Juniperus sibirica, Ulex Gallii.* July.

R. Welch, *Photo.*

Calluna vulgaris,
        dominant.
Ulex Gallii, *v.c.*
Erica cinerea, *c.*
Molinia cœrulea, *f.*
Potentilla erecta
Scabiosa Succisa
Solidago Virgaurea

Dabeocia polifolia
Salix repens
Eriophorum angustifolium
Carex diversicolor
        binervis
Sieglingia decumbens
Polygala serpyllacea
Cladonia sylvatica

Where the ground was slightly wetter, also

Ranunculus Flammula
Cnicus pratensis
Erica Tetralix
Anagallis tenella
Pedicularis sylvatica
Prunella vulgaris
Juncus bulbosus

Narthecium ossifragum
Carex panicea
Holcus lanatus
Agrostis tenuis
Nardus stricta
Poa annua

Still wetter ground added

Hydrocotyle vulgaris
Lythrum Salicaria
Potamogeton polygoni-
    folius

Carex echinata
Blechnum Spicant

The immediate neighbourhood of Clifden is poor botanically, but it makes a good centre for exploration in all directions. *Saxifraga spathularis* is the only one of the notable plants of Connemara which approaches the town. *Centunculus* occurs. *Cotoneaster Simonsii* is naturalized and spreading in wild ground. In view of the occurrence of a *Saxifraga Geum* hybrid on Clare Island to the N (the species is confined to Cork and Kerry), it is worth noting that a specimen of *S. Geum* purporting to have been gathered near Clifden by J. H. Balfour in 1835 is in the Edinburgh Herbarium (Irish Nat., 1912, 245); it has been usually attributed to an error in labelling.

The railway colonist *Diplotaxis muralis* has penetrated to this remote terminus, and *Erodium moschatum* is recorded.

RECESS.

391. Recess, where modest accommodation may be obtained, is extremely well situated for the exploration of the mountains that rise to E and W (**392, 393**), with lake and rocky moor on every side. **Lisoughter,** the isolated hill rising close to Recess, yields, in addition to "Connemara marble" (serpentine) *Dryas* (**106**), *Saxifraga oppositifolia, Juniperus sibirica, Asplenium viride,* and of course *Saxifraga spathularis, Dabeocia,* etc. Beautiful lakes abound : Glendalough L. lying close by, while Ballynahinch L., Derryclare L., and L. Inagh (*Loch Eidhneach,* ivy lake), sweep in a semicircle round the base of the Twelve Bens. These lakes yield *Subularia, Utricularia ochroleuca* in three of its very few Irish stations, *Pilularia* (growing a foot long in deep water), and plenty of *Eriocaulon* and *Lobelia.* The bogs are full of the characteristic Connemara plants (see **385–7**), and *Eriophorum latifolium* is at Oorid L. *Erica mediterranea, E. Mackaii,* and *Naias* lie within a day excursion in the Roundstone direction (**388–9**).

THE TWELVE BENS.

392. The beautiful Twelve Bens (corrupted into Twelve Pins) or Bennabeola and the adjoining mountains of Maam Turk (**393**) consist of masses of quartzite flanked with schists, and separated from each other by a low bog-filled valley with lakes. So many of the summits range a little over 2000 ft. that the whole suggests the wreck of an old plain of denudation. The Twelve Bens occupy an approximately circular area in W Connemara, stretching from Ballynahinch to Kylemore, and fringed on the S and E with a chain of lovely island-studded lakes (**391**). The mountainsides are steep, and to a great extent devoid of any covering of peat or vegetation ; much of the ground is sufficiently precipitous to give the climber ample scope for the enjoyment of his hobby. Some of the valleys are filled with deep, rough bog, and the ridges give better walking. The Twelve Bens form a compact and very beautiful mountaingroup, with rugged quartzite points everywhere standing up against the sky (plate 31). Towards the W, mica schist comes in, and forms a more hospitable substratum for alpine plants.

The vegetation in general is sparse and stunted; its most interesting features are the occurrence of some of the well-known lowland plants of Connemara, such as *Dabeocia,* which creep up into the high grounds, and the comparative richness of the alpine flora which, especially

PLATE 31.

AMONG THE TWELVE BENS, CONNEMARA.
Quartzite slopes with sparse ericaceous vegetation.

R. Welch, Photo.

PLATE 32.

CROAGHMORE, CLARE ISLAND,
1520 feet, from the NE.

R. Welch, Photo.

on one or two mica-schist peaks, attains a development seldom met with on Irish mountains. Peat supporting *Molinia* heath ascends the hills to 1000 ft. or so, when it is replaced by a drier *Calluna* association. Of the plants which occur on the high grounds the following may be mentioned :—*Thalictrum alpinum, T. minus, Cochlearia alpina, Meconopsis, Silene maritima, Sagina subulata, Rubus saxatilis, Saxifraga oppositifolia, S. incurvifolia, S. rosacea, S. spathularis, Sedum roseum, Saussurea, Crepis paludosa, Hieracium argenteum, Arctostaphylos Uva-ursi, Dabeocia, Armeria maritima, Oxyria, Salix herbacea, Juniperus sibirica, Carex rigida, Hymenophyllum peltatum, Asplenium viride, Cystopteris fragilis, Polystichum Lonchitis, Lastrea œmula, L. montana, Lycopodium alpinum.*

The best ground for the botanist is Muckanaght, in the middle of the group, where an "oasis of schist in a Sahara of quartz" brings in a very pretty colony of alpine plants. Ben Lettery, which overlooks Ballynahinch Lake and the Galway-Clifden railway and road, also yields a good many interesting species, and is much more easy of access. Lisoughter (**391**), above Recess, though only 1313 ft. in height, also holds a fair mountain flora. At the head of Glen Inagh a curious form of *Saxifraga stellaris* grows with flowering stems covered with innumerable small leaf-rosettes, approaching the arctic var. *comosa* Poiret (*S. foliolosa* R.Br.).

The area of which the Twelve Bens form the centre is a famous tourist district, and for such not too abundantly supplied with hotels; but one can stay comfortably at Oughterard (**366**), or Clifden (**390**) or Letterfrack in the W; and, in the N, Leenane (**399**) or Renvyle (**396**).

H. C. HART: "Report on the Flora of the Mayo and Galway Mountains." Proc. Roy. Irish Acad., 2nd ser., 3 (Science), 736–744. 1883. N. COLGAN: "Botanical Notes on the Galway and Mayo Highlands." Irish Nat. 1900, 111–114.

MAAM TURK.

393. The Maam Turk mountains form a well-defined ridge of quartzite stretching SE and then E for a distance of about 20 miles from Killery Harbour to the NW corner of L. Corrib near Maam (**394**). On the W side the hills rise very boldly like "a zigzag series of beehives" from the valley which separates them from the Twelve Bens—a row of rounded, massive, bare quartzite knobs. As H. C. Hart has concisely said, the Maam Turk range is the tangled cluster of the Twelve Bens stretched out in a curved line with their tops rubbed off. For climbing and botanizing they are neither so picturesque nor so interesting as their neighbours. The N end of the ridge is formed of Silurian slates, but does not ascend high enough to allow this more hospitable rock to influence the alpine vegetation. The highest point of the group is 2307 ft.; but the best ground for the botanist is Maumeen, more to the SE, and not far from Recess. As in the case of the Twelve Bens, there is an absence of tarns, although innumerable lakes fringe the base of the hills.

The rarer plants recorded from the higher grounds of Maam Turk are:—*Thalictrum alpinum, Sagina subulata, Saxifraga oppositifolia, S. spathularis, Sedum roseum, Crepis paludosa, Hieracium argenteum, Lobelia, Dabeocia, Salix herbacea, Juniperus sibirica, Cystopteris fragilis, Lycopodium alpinum*; and the majority of the other characteristic plants of Connemara occur about the hills. On a bed of crystalline limestone at Corco Gap, in dry crevices devoid of peaty soil, *Saxifraga spathularis, Athyrium* and *Blechnum,* three typically calcifuge species, were observed growing well.

H. C. HART: "Report on the Flora of the Mayo and Galway Mountains." Proc. Roy. Irish Acad., 2nd ser., 3 (Science), 732–735. 1883.

394. Maam.—The village (if it may be dignified by this name) of Maam lies in the fine mountain valley which descends to the NW arm of L. Corrib — a lonely and picturesque spot. Hither come botanical pilgrims to see *Potamogeton sparganiifolius* (= *Kirkii*) (**168**) in its original habitat. It grows in the Bealanabrack River near the bridge, close by which is the little hotel. Along the banks, *Carex aquatilis* (**181**) may be gathered, with abundance of three characteristic western Brambles—*Rubus iricus, R. hesperius,* and *R. Borreri. Lycopodium inundatum* has one of the few Irish stations on wet heath on the W side of Maam. The Maam valley is reached by a mountain road from Maam Cross on the Galway-Clifden road and railway. Its W side is formed by the imposing quartzite ridge of Maam Turk (**393**). On the E side grassy

Silurian hills rise to 1500–2000 ft., over which one crosses by a steep road to Lough Nafooey and the "Narrow Lake" on Lough Mask (**371**).

E. S. MARSHALL and W. A. SHOOLBRED: "Irish Plants observed in July, 1895." Journ. Bot. 1896, 250–258.

KYLEMORE AND BENCHOONA.

395. Stretching N from the Twelve Bens to the Killery fiord, and cut off from the former by the beautiful pass of Kylemore (*An Choill Mhór,* the great wood), with its lakes and woods, a group of hills rises, of which Benchoona (1975 ft.) is the highest. Several rugged shoulders extend from the central mass; and a number of lakes lie among the hills. At the higher levels *Thalictrum alpinum, Saxifraga oppositifolia, Sagina subulata, Salix herbacea, Isoetes lacustris* occur. *Erica mediterranea* grows not infrequently in this area. *Rubia peregrina* is found at Salruck on the Killery—a delightful spot well worth a visit. The lakes yield *Lobelia, Utricularia ochroleuca, Eriocaulon,* etc. The steep scarp behind Kylemore Castle is clothed with native scrub (probably the remains of the "great wood"), which deserves study.

396. Renvyle is a beautiful lonely spot situated on the N shore of Connemara, in the extreme NW of the district. *Arabis Brownii* (**85**) was first recorded (as *A. ciliata*) from the sandhills near by. The neighbourhood is not rich botanically, and is rather far removed to be a convenient centre for the Twelve Bens, etc.; but the Killery fiord (**399**) and Kylemore (**395**) are within striking distance; and boat excursions may be made to Inishbofin (**405**) and Inishturk (**406**).

THE MWEELREA MOUNTAIN-GROUP.

397. Mweelrea (2688 ft.), the highest mountain in the West of Ireland, is one of a very fine group lying on the N side of the Killery fiord (**399**). Gotlandian and Ordivician rocks prevail, and slate is the usual material of which the mountains are built. The higher hills are grouped close together, the mass being cut into three portions by the deep Doo Lough valley running S into the Killery, and the tributary valley Glenummera descending into Doo L. from the E. Mweelrea, with its neighbours Benbury and Benlugmore, the W portion of the group, forms a huge flat-topped mass edged with spurs, cliffs, and ravines of thoroughly alpine character, and superbly picturesque. The SE section is dominated by Bengorm (2303 ft.), and the NE by Sheffry, also called Loughty or Tievummera (2504 ft.). A 4-mile radius from the S end of Doo L., which lies only 109 ft. above the sea, will enclose the whole mass, which forms one of the finest mountain-groups to be found in Ireland, intersected by valleys of great beauty.

The flora of the lofty precipices and cirques is not so alpine as might be expected; the Mweelrea section and the Sheffry ridge, lying north of Glenummera, are about equally rich as regards the flora of the high grounds. The following list shows what the botanist may expect in his exploration :—*Thalictrum minus, Subularia, Saxifraga oppositifolia, S. spathularis, Sedum roseum, Rubia peregrina, Lobelia, Vaccinium Vitis-Idæa, Arctostaphylos Uva-ursi, Dabeocia, Erica mediterranea, Oxyria, Plantago maritima* (to 2600 ft.), *Salix phylicifolia, S. herbacea, Juniperus sibirica, Carex rigida, C. lasiocarpa, Deschampsia alpina, Cystopteris fragilis, Lastrea montana, Lycopodium alpinum, L. clavatum, Isoetes lacustris.* While a few of the more alpine of these descend into the lower grounds (below 1000 ft.), *Rubia* and *Erica mediterranea* are the only ones whose range is quite lowland. To this list must of course be added many of the characteristic western plants whose presence may be taken for granted when dealing with the flora of the Connaught uplands and boglands, such as *Drosera anglica, D. longifolia, Utricularia intermedia, Rhynchospora fusca, Eriocaulon, Cladium, Carex limosa, C. lasiocarpa, Lastrea æmula, Osmunda, Hymenophyllum tunbridgense, H. peltatum,* etc.

The visitor can stay conveniently at Leenane.

H. C. HART: ''Report on the Flora of the Mayo and Galway Mountains.'' Proc. Roy. Irish Acad., 2nd ser., 3 (Science), 716–730. 1883. N. COLGAN: ''Botanical Notes on the Galway and Mayo Highlands.'' Irish Nat. 1900, 114–116.

398. Dooaghtry.—At the W base of Mweelrea (**397**), and forming the N side of the entrance to Killery Harbour (**399**), is a curious low-lying area, sandy beaches being backed by a plain of flat sand barely raised above sea-level, and strewn with shallow lakelets. The place is utterly windswept. The sandy waste supports a close sward of *Œnanthe Lachenalii, Sagina nodosa, Leontodon taraxacoides, Parnassia, Epipactis palustris, Gymnadenia conopsea, Scirpus pauciflorus, Carex dioica, Selaginella,* etc. As the ground gets wetter, *Sium erectum* and *Carex diandra* come in, with *Bidens cernua* f. *radiata* and *Eleocharis acicularis.* Rocky knobs rise, yielding *Agrimonia odorata, Antennaria, Carlina, Campanula rotundifolia, Thymus glaber, Juniperus sibirica.* Another bluff, rising from a sandy tidal inlet, supports stunted *Populus tremula, Quercus sessiliflora, Betula, Ilex, Corylus,* with an undergrowth of *Thalictrum minus, Erica mediterranea, Dabeocia, Osmunda, Saxifraga stellaris.* The sands at the base of this bluff yield *Scirpus Tabernœmontani,* S. *rufus, Salicornia, Œnanthe crocata, Ruppia rostellata.* It will be seen that the place presents a very curious mixture of floras, and an incongruous assemblage of plants, many of them rare in the surrounding region; it is well worth a visit.

R. LL. PRAEGER : ''A Note on Dooaghtry.'' Irish Nat. 1911, 103–4.

399. Killery Harbour.—The remarkable sea-inlet of Killery or The Killeries, 10 miles long, $\frac{1}{2}$ mile wide, and very deep, is the best example Ireland affords of the true fiord—a drowned valley much deepened by glacial action. It runs far in among high hills, and its flora is the flora of the steep ground on either side. On its N edge Mweelrea (**397**) descends in a grand slope into its calm dark water. Near its head stands Leenane, surrounded by huge glacial terraces.

Leenane is well situated as a centre for exploration, both by land and sea, of the Killery fiord and the adjoining mountain-groups (**392, 393, 395, 397**). The hill which rises immediately S of the hotel yields *Thalictrum minus, Saxifraga stellaris, Sedum roseum, Leucorchis albida, Carex rigida, Lastrea montana, Lycopodium clavatum. Cochlearia anglica* at the head of the bay is the only uncommon halophile species.

THE MAAMTRASNA OR PARTRY MOUNTAINS.

400. From Leenane, at the head of the Killery fiord (**399**), a range of high hills, the Partry Mountains, runs NE for about 14 miles. The summit forms a tableland, or

series of tablelands, dropping NE from about 2000 to
1000 ft. The highest point is Maamstrasna, 2207 ft. On
the NW side the hills descend in a bold slope into the
valley of the Erriff River, through which the Westport-
Leenane road runs. The SE side, which faces L. Mask
(**371**), is irregular and most picturesque, with high cliffs
and cirques embosoming lakes. But this aspect is
unfavourable for alpine vegetation, and the flora of the
range is very poor. *Hieracium anglicum, Dabeocia, Salix
herbacea, Juniperus sibirica,* and of course *Saxifraga
spathularis* are recorded between 1350 ft. and the summit
of Maamtrasna, while about Loughnadirkmore grow *Ranunculus scoticus, Saxifraga stellaris, Lobelia, Oxyria, Eriocaulon, Lycopodium inundatum, Isoetes lacustris. Lastrea
montana* is abundant, and of course *L. æmula.* In spite
of the rarity of botanical riches, a tramp from the beautiful
Lough Nafooey NE over the hills to the Owenbrin River
(near the mouth of which a very fine fossil forest of
Pinus has been exposed by turf-cutting and subsequent
weathering) can be confidently recommended. The
Maamtrasna range forms part of the mass of Silurian
slates that extends thence W along the Killery to Mweelrea.

See also Lough Mask, **371**.

CLEW BAY, WESTPORT, AND NEWPORT.

401. It is a remarkable experience to drop down from
the wild metamorphic highlands which lie N and S, to
the shores of **Clew Bay**, where nestle the little towns of
Westport and Newport. Bog and heather give way
abruptly to grass and tillage, and gnarled rock is replaced
by trees. The drumlins (whale-backed Glacial ridges),
formed of calcareous drift covering the limestone, here dip
down below sea-level, producing a labyrinth of grassy islands
almost shutting out the view of the bay. The limestone
area is too much farmed to be attractive to the botanist,
and in any case few of the characteristic calcicole plants
push into this remote corner.

About **Westport** *Nasturtium sylvestre, Crepis paludosa,
Rumex Hydrolapathum,* and *Potamogeton angustifolius*
are recorded. Around Westport Quay *Limonium humile*
grows remarkably luxuriantly, also *Cochlearia anglica,* and
Lepidium Draba is well established. Extensive plantings
of *Spartina Townsendii* have been made during 1929–32

by the Marquess of Sligo, and are growing well (see Proc. Roy. Irish Acad., **41**, B, 120–1, 1932). Westport is important to the botanist chiefly as forming the base for the exploration of Croaghpatrick (**403**) and the wild country beyond it. The vicinity of the smaller town of **Newport**, lying a couple of miles N, with hills and lakes close at hand, is richer; and the botanist may find *Cochlearia anglica, Rubus iricus, Eleocharis uniglumis, Carex Hudsonii, C. diandra, Isoetes lacustris, I. echinospora, Nitella translucens; Erica mediterranea* (**131**), which is abundant further W, grows close to the town. E of Westport, where the Aille River, coming from the Partry Mountains, disappears under a limestone arch, *Ulmus montana* is native. Newport lies within striking distance of the Nephinbeg range of hills (**412**), and of the wild lakes L. Feeagh and L. Beltra (**402**).

E. S. MARSHALL: "Plants observed in West Mayo, June, 1899." Journ. Bot., 38, 184–188. 1900.

402. Lough Beltra.—This little-known lake is remarkable for the fine fringe of *Erica mediterranea* which, as at Carrowmore Lake out to the NW, covers the broad stony storm-beach of the eastern side, half of it being submerged during average floods. *Thalictrum minus, Potamogeton filiformis* (SE corner), and *Plantago maritima* also occur along the shore.

CROAGHPATRICK.

403. One of the noblest mountains in Ireland, Croaghpatrick (*Cruach Phádraig,* St. Patrick's rick) rises in a beautiful cone from the island-studded waters of Clew Bay to a height of 2510 ft. The peak is flanked by rugged shoulders for several miles, and the group thus formed stands in conspicuous isolation from the other mountains of Mayo. These shoulders run east and west in a narrow line, so that from either of these directions the hill appears as a perfect cone for its full height. "The Reek" (rick), as it is locally called, is formed mainly of quartzite, with some gneiss and schist. A precipitous scarp on the N face yields *Thalictrum alpinum, Saussurea, Oxyria, Salix herbacea, Asplenium viride.* Other high-level plants are *Sedum roseum, Saxifraga spathularis, Hieracium anglicum, Juniperus sibirica.* In the middle zone occur *Rubus saxatilis, Crepis paludosa, Dabeocia.*

Croaghpatrick is easily visited from Westport (**401**). A little chapel, erected on the summit some years ago, is approached by a bridle-path from the base near Murrisk, which gives easy access to the higher grounds. An annual pilgrimage, very largely attended, accounts for the presence of this well-beaten stony track. Additional interest may be added to a day's work on Croaghpatrick by a visit to the ruins of Murrisk Abbey. *Allium Babingtonii* and *Iris fœtidissima* are two plants of the low grounds, the latter probably an escape.

H. C. HART: "Report on the Flora of the Mountains of Mayo and Galway." Proc. Roy. Irish Acad., 2nd ser., 3 (Science), 702–705. **1883.**

LOUISBURGH.

404. Louisburgh, on the S shore of Clew Bay, is not so close to the Mweelrea mountain-group (**397**) as Leenane (**399**), being separated by several miles of wild bog-land, but it allows of direct attack upon the N slopes, which are but slightly explored. Above the little Lough Brawn, *Thalictrum alpinum* and *Polystichum Lonchitis* grow among a number of more widespread alpine plants, and there are further promising cliffs and corries. *Utricularia ochroleuca* is at Lugaloughan, and *Elatine hexandra* is frequent in the area.

INISHBOFIN.

405. Inishbofin lies off the Connemara coast, S of Inishturk (**406**), 5½ miles from the nearest point of the mainland. Close by is the smaller island of Inishark. The two may be classed together conveniently. Their combined area is 4½ sq. miles; and over the whole, Silurian slates prevail. The highest point is 292 ft. The greater part of the surface consists of undulating, rough moorland, now entirely stripped of the peaty covering which once clothed it, leaving a barren waste only relieved by the abundance of *Helianthemum guttatum* (**89**). There is a fair extent of cultivated land with potatoes, oats, barley, rye, etc., and of pasture. Five small lakes occur, and a few pools. The shores are almost everywhere cliffy, with *Cochlearia grœnlandica* (**87**), *Spergularia rupicola, Sedum roseum, Crithmum*; there is sand at the E end of Bofin, yielding *Cakile, Raphanus maritimus, Eryngium, Centaurea Scabiosa,*

Polygonum Raii; and blown sand helps to make a light calcareous soil N of Church Lake, providing a habitat for *Arabis Brownii, Ophrys apifera,* and (on a dry-built wall) *Ceterach.* Arbuscular vegetation is at a minimum, but is represented by *Prunus spinosa, Sorbus Aucuparia, Populus tremula,* and a few *Salices, Rubi,* etc. The lake flora is more interesting, and includes *Ranunculus Baudotii, Elatine hexandra* (common), *Lobelia, Sparganium angustifolium, Eriocaulon, Isoetes echinospora.* Looking down through the clear water of L. Gowlanagower, one sees a lovely continuous green sward, formed of about two parts of *Isoetes echinospora* to one part of *Lobelia,* every interstice between these plants being filled with *Elatine,* with here and there a cloudy mass of *Apium inundatum.*

On the heaths, *Helianthemum guttatum* (**89**), one of the rarest of Irish (as of British) plants, is abundant, both on Bofin and Shark. *Sagina subulata, Radiola, Saxifraga spathularis* (**114**), *Centunculus, Bartsia viscosa* (**142**), *Juniperus sibirica, Hymenophyllum peltatum, Lastrea æmula* may be mentioned also. *Calamagrostis epigejos,* very rare in Ireland, grows among rocks in two places. *Osmunda* is frequent in peaty spots, with a little *Drosera anglica, Utricularia intermedia, Allium Babingtonii, Carex limosa, Selaginella,* and at one spot *Lycopodium inundatum*; also *Inula Helenium,* the ubiquitous *Matricaria suaveolens,* and *Lamium molucellifolium.* There is a little hotel on the island. Cleggan, N of Clifden, is the usual point of embarkation.

A. G. MORE: "Report on the Flora of Inishbofin, Galway." Proc. Roy. Irish Acad., 2nd ser., 2, 553–578. 1876. R. LL. PRAEGER: "Notes on the Flora of Inishbofin." Irish Nat. 1911, 165–172.

INISHTURK.

406. Inishturk rises on all sides boldly from the sea, with very fine cliff-scenery in the W. The greatest elevation is 629 ft. Only a small part, in the most sheltered situations, is under cultivation, the remainder consisting of grassy heath and maritime sward, with the Silurian slate forming small sharp E-W folds, and constantly breaking through, so that the surface often resembles a sheet of corrugated iron on a large scale. Area $2\frac{1}{4}$ sq. miles. The island lies 7 miles out from the nearest land, and supports some twenty families, who live by farming and fishing. No more than five per cent. of the surface is under cultivation. The few native trees of the island, such as *Populus tremula, Betula pubescens, Ilex, Pyrus Malus,* grow only 3 to 4 ft. high, but sometimes twice that breadth. The herbaceous vegetation is usually correspondingly stunted. The flora is remarkably large in comparison with that of the adjoining islands, numbering 329 species on a conservative estimate. Rocks over the sea yield *Cochlearia grœnlandica, Sagina subulata,* and quantities of *Sedum roseum.* The *Plantago* formation, here often composed entirely of *P. maritima* and *P. Coronopus* without any other ingredient, occupies a large area at the W of the island, as close and smooth as if shaved with a razor. Tiny lakelets in the W are tenanted by *Lobelia Dortmanna* and *Isoetes lacustris. Juncus effusus* f. *spiralis,* with spreading loosely spiral stems, occurs in wet places, with *Pinguicula lusitanica.* The spiral rush just mentioned would appear to be an Atlantic form: it is common on many of the Irish western islands, and is stated to be abundant in Orkney (Journ. Bot. 1919, 69). *Euphorbia hiberna* grows on slopes over the small harbour. *Rubus Borreri* and *R. iricus* are among the most abundant of the half-dozen brambles. *Saxifraga spathularis* is of course everywhere, even on the walls of houses. The rarest Irish plant on the island is *Helianthemum guttatum* (**89**), which has its Hibernian head-quarters on Inishbofin near by. *Athyrium Filix-fœmina* is present in remarkable profusion. *Asplenium marinum* grows on rock-faces over the whole island. *Centunculus* is not infrequent, and *Allium Babingtonii* occurs. Renvyle is the most convenient point of embarkation for Inishturk, or the adjoining Tully Pier; but there is no accommodation on the island for visitors, though a bed may be obtained.

R. Ll. Praeger: "The Flora of Inishturk." Irish Nat. 1907, 113–125.

CLARE ISLAND.

407. Clare Island (6⅛ sq. miles), lying across the entrance of Clew Bay, is one of the most interesting of Irish islands. It is composed mainly of Gotlandian slates, but Ordovician slates, Old Red Sandstone conglomerates, and Carboniferous sandstones occupy the N part. The surface is diversified, Knocknaveen rising to 749 ft. in the

E, while the NW is dominated by Croaghmore (1520 ft.), which drops in a magnificent cliff into the Atlantic (plate 32 and fig. 21). Interest in the flora centres in the group of alpine and other rare plants which occupy this cliff at about 1000–1300 ft. About one-third of the island is farmed (tillage and pasture): this area extends along the S and E coasts, and is coincident with the distribution of Glacial drift. Most of the remainder is thin *Calluna*-covered bog, giving way to grass along the high

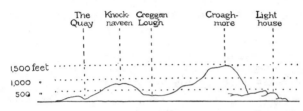

Fɪɢ. 21.—Outline of Clare Island from the North-east.

exposed N and W sides. In the extreme W this is replaced by an almost pure sward of *Plantago maritima* and *P. Coronopus.* Analysis of a typical example of this association shows:—

Plantago maritima	Anagallis tenella,
(*about 80 per cent.*),	Plantago Coronopus,
Radiola linoides,	lanceolata,
Potentilla erecta,	Luzula campestris,
Sedum anglicum,	Aira præcox,
Galium saxatile,	Festuca ovina,
Jasione montana,	Mnium hornum.

These form a dense shining sward ½ inch high, with flower stems rising to 1 or occasionally 2 inches. The rosettes of *Plantago maritima* and *P. Coronopus* measure ½ to ¾ inch across.

For the rest, lakelets are few and small, with a poor flora; a limited area of salt-marsh and another of blown sand occur at the E end near the little harbour. In the E also are a few patches of low scrub 4 to 8 ft. high— *Corylus, Betula pubescens, Ilex, Sorbus Aucuparia, Quercus, Salix aurita, S. cinerea, Myrica, Lonicera,* and *Rubi* (including *R. iricus*): these shelter a good shade-flora, and *Osmunda* 6 ft. high.

The great scarp of Croaghmore, which is very interesting botanically, may be traversed by any one with a good head by means of sheep-paths starting from its E end at about 1000 and 1200 ft. respectively. The plants of the cliff include *Silene acaulis* (abundant), *Hypericum pulchrum* var. *procumbens, Saxifraga oppositifolia, S. spathularis* (**114**) (common all over the island), *S. Geum* (**113**) (hybridized with the last and apparently now extinct in the pure state), *S. rosacea* (widespread along the northern cliffs), *Sedum roseum, Saussurea, Hieracium anglicum, H. hypochæroides* var. *saxorum, Oxyria, Salix herbacea, Cystopteris fragilis, Polystichum Lonchitis, Asplenium viride, Lastrea æmula, Hymenophyllum peltatum.* Among the other rarer plants of the island are *Erica mediterranea* (**131**) and *Cephalanthera ensifolia* (on boulder-clay banks sloping to the sea in the N), *Orobanche rubra* (**143**) on cliffs at the lighthouse, *Cochlearia grœnlandica* (**87**) on rocks at W end, *Drosera anglica.* Near the sea, *Sagina subulata, Spergularia rupicola, Crithmum, Scirpus filiformis* are frequent; *Juniperus sibirica* grows on one sea-stack.

The effect of maritime influence and humid air in raising or depressing the vertical limits of plants is seen in the occurrence of *Saxifraga rosacea* and *Hymenophyllum* at the lowest limit of terrestrial vegetation, of *Silene acaulis* at 400 ft., *Sedum roseum* at 150 ft., *Oxyria* at 600 ft.; and on the other hand, *Plantago Coronopus,* 1200 ft., *Glaux,* 400 ft., *Asplenium marinum,* 500 ft.

The few rarer bog and water plants include *Drosera anglica, Pinguicula lusitanica, Utricularia intermedia, Carex limosa, Nitella translucens.* The introduced flora, which is tolerably large, is of the usual type. *Matricaria suaveolens* has spread long since to the island. Examination of a very small quantity of mud from my boots on landing at the quay showed four seeds of this plant, which abounds on roadsides on the adjoining mainland; so its mode of introduction is clear. The flora numbers, on a conservative estimate, 393 species, a total smaller by only 21 than that of the adjoining island of Achill, which has an area nine times as large. Apart from botanical features, Clare Island is interesting on account of the magnificent scenery of the N coast, where the high cliffs are in summer tenanted by great colonies of breeding seabirds. On the S coast stands a little ruined abbey with quaint archaic frescoes in colour, mainly of animals : these are now suffering from damp. There is a small inn by the harbour at the E end. The island may be reached from Roonah quay near Louisburgh (3 miles, post boat 3 times weekly), or from Achill Sound (railway terminus), 11 miles, mostly sheltered water.

Clare Island was in 1909–1911 the subject of an intensive study of the whole of its natural history—animal, vegetable, and mineral, the results occupying three large 8vo volumes (Proc. Roy. Irish Acad., **31**, 1911–15). In Part 10 (112 pp., 6 plates) will be found a full account of the flowering plants and their allies, with discussion of the problem of their immigration and kindred subjects.

R. LL. PRAEGER : ''The Flora of Clare Island.'' Irish Nat. 1903, 227–294. *Ibid.*: ''Phanerogamia and Pteridophyta'' (Clare Island Survey, Part 10) in Proc. R. I. Acad., 31 (1911). See also parts 2 to 16 of same series, which deal with botany and allied subjects.

ACHILL ISLAND.

408. Achill is the largest island on the coast of Ireland, its area being 57 sq. miles. It is formed of gneiss, mica-schist, and quartzite. The island may be described as consisting of wild undulating bog-land, relieved by small patches of cultivation (only 15 per cent. of the surface reclaimed, fig. 22), and dominated by three ridge-like mountains—Croaghaun (2192 ft.) in the W; Slievemore (2204 ft.) on the N shore; the Meenaun (1530 ft.) towards the S. Croaghaun has suffered erosion by the Atlantic, and drops into the ocean in a glorious precipice of the full height of the mountain (plate 33). Lakelets are numerous, and blown sand occurs. The coast is usually precipitous in the W, low in the E, where a winding channel (Achill Sound) severs the island from the mainland. Here peat often forms the shore-line, and *Fucus* grows on tree-stumps *in situ* from which the over-lying peat has been washed away. In circumstances of maximum exposure up to 300 or 400 ft., the *Plantago* association is well developed. At the extreme point of Achill Head, an extraordinarily wild spot, this vegetation consists of a dense smooth mat of *P. maritima* and *P. Coronopus,* dotted with *Armeria, Spergularia rupicola, Sagina maritima, S. procumbens, Cerastium tetrandrum, Radiola, Festuca ovina, Aira præcox, Agrostis tenuis.* The greatest exposure is found

not on the summits of the higher hills, but on the ridge o₁
Meenaun (1530 ft.). Here a patchy skin of peaty soi
alone covers the disintegrated quartzite. The vegetatior
consists of a close mat of *Calluna*, with *Empetrum*
Vaccinium Myrtillus, *Erica cinerea*, *Salix herbacea*, and
a little *Arctostaphylos Uva-ursi* and *Juniperus sibirica*

The whole island may be described as a wild wind-
swept bogland with occasional towering hills, a coast-line
which vies in savage grandeur with anything in Ireland
or Britain, and a surface with numerous brown lakes and
tarns. Rainfall is high, and humidity likewise. On the

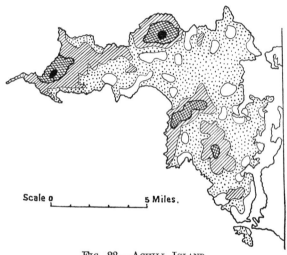

FIG. 22.—ACHILL ISLAND.
Cultivated areas, white; bogland up to 500 ft., dotted; to 1000 ft.,
hatched; 2000 ft., cross-hatched; over 2000 ft., black.

N face of Slievemore, *Hymenophyllum peltatum* is a
constant component of the turf which covers the slope,
growing among the short *Calluna* and *Sphagnum* quite
unprotected and unshaded.

The flora of Achill numbers 420 species and subspecies
(Irish Top. Bot. standard). The striking features of the
vegetation are its treeless and windswept character, and
the almost unbroken continuity of bog and heath associations. Inside the narrow fringe occupied by the halophile
groups, the moorland flora holds undisputed sway, save
where, in sheltered places, man has reclaimed some acres

PLATE 33.

CROAGHAUN, ACHILL ISLAND,
Showing the seaward face, 2192 feet in height.

James Grierson, Photo.

408.

PLATE 34.

R. Welch, Photo.

SAXIFRAGA SPATHULARIS AT MALLARANNY, CO. MAYO.
June.

from the bog. The wettest swamps support an abundance
of *Carex limosa, Hypericum elodes, Menyanthes, Potamogeton polygonifolius,* and the three species of *Drosera,*
forming together a close floating felt among which
Nymphœa grows as a semi-terrestrial plant, with short-
stalked aerial leaves. Over large areas of lowland bog the
surface is composed of almost bare peat, thickly studded
with little stools of vegetation 1 to 2 ft. high; the floor
here is thinly colonized by *Eriophorum angustifolium*
and *E. vaginatum,* the stools being occupied by *Calluna,
Molinia,* and other common bog plants. Calcicole and
light soil plants are almost absent. Rare and interesting
species are fewer than the position of the island and the
nature of the ground might lead one to expect. Lake-
shores in the E are crowded with *Eriocaulon* (**173**), and
especially in the W with *Ranunculus scoticus* (**82**). *Lobelia
Dortmanna* is common in most of the low-level lakes;
Isoetes lacustris is rarer, and *I. echinosposa* grows in
Sraheens Lough. Keel L. is full of *Potamogeton nitens,
P. filiformis,* with *Nitella translucens* and *N. batrachosperma*; *Mimulus Langsdorffii* has found its way here.
Wet bogs yield abundance of *Hypericum elodes, Drosera
longifolia, D. anglica, Utricularia intermedia, Carex limosa,
C. lasiocarpa. Lycopodium inundatum* occurs near
Sraheens L. *Saxifraga spathularis* (**114**) and *Lastrea
œmula* are all over the island at all elevations. *L. montana*
is rare. *Cochlearia anglica* grows along the Sound. *Erica
mediterranea* (**131**) is widely but locally distributed, chiefly
along the edges of streamlets and lakes. A curious bushy
form of *Erica Tetralix* occurs in the west. As one ascends,
Arctostaphylos, Juniperus sibirica, J. communis, and
the two species of *Hymenophyllum* come in—the last-
named chiefly in holes among the rocks and boulders.
Higher up on Slievemore and Croaghaun are *Saxifraga
stellaris, Salix herbacea* (descends to 1200 ft.), *Listera
cordata, Carex rigida.* The scarp below the summit of
Slievemore yields in addition *Epilobium angustifolium,*
and abundance of *Oxyria* and *Cystopteris fragilis.* The
grand precipice of Croaghaun is quite poor in its flora, but
the interesting Shetland var. *procumbens* of *Hypericum
pulchrum* is frequent; and another northern plant,
Euphrasia frigida (**140**) is on the summit. On Achill
Head and elsewhere *Cochlearia grœnlandica* (**87**) grows;
near by *Adiantum* (**193**) grows in an almost inaccessible

station. The low sea-cliffs in the north are hung with *Sedum roseum.* Among Brambles, the local western *Rubus altiarcuatus* is abundant at Dugort; *R. iricus* is at the same place. The form of *Juncus effusus* with spreading spiral stems (**406**) occurs; *Hieracium umbellatum* and *H. anglicum* are the only Hawkweeds; *Ammophila arenaria* is found in peaty soil on a cliff-edge 70–80 ft. above the sea. (See also **448**.) The introduced flora is quite limited. *Geranium pratense* has naturalized itself in a meadow at Dugort. There is an old record of *Erodium moschatum.* Near the Sound *Rhododendron ponticum* is naturalized and spreading widely, as in many exposed peaty areas in Ireland.

The hotels are mostly at Keel, at Dugort at the base of Slievemore, and at the Sound, which narrows to a few hundred yards, and is crossed by a causeway and swinging bridge close to the railway station; but one can stay also at other places.

R. LL. PRAEGER: ''The Flora of Achill Island.'' Irish Nat. 1904, 265–289.

CURRAUN ACHILL.

409. Curraun Achill is a wild mountainous sea-girt peninsula—very nearly an island—forming for the most part a heathery tableland of 1300–1700 ft., edged with cliffs in many places. Old Red Sandstone prevails over the S part, but soon gives way to the ancient metamorphic series. On the lower grounds *Erica mediterranea* (**131**) is abundant, *Dabeocia* (**132**) has its most northerly Irish and European station, and the two species of *Hymenophyllum, Saxifraga spathularis* (**114**), and other western plants occur. On the plateau, *Arctostaphylos Uva-ursi, Empetrum,* and *Juniperus sibirica* form a thick carpet; and *Salix herbacea, Carex rigida,* and *Lycopodium alpinum* grow high up. Hotels at Mallaranny and Achill Sound provide the visitor with convenient starting-points for the working of this treeless, almost houseless and roadless area.

H. C. HART: Mayo and Galway Mountains, *cit.*

MALLARANNY.

410. Mallaranny or Molreny (*Mullach Raithnighe,* summit of fern), where the railway company has erected

PLATE 35.

ERICA MEDITERRANEA AT MALLARANNY, CO. MAYO,
Growing six feet high around the head of Bellacragher Bay.   June.
R. Welch, Photo.   410.

PLATE 36.

ENTRANCE of GLENCAR, Co. SLIGO.
Cliffs of Carboniferous Limestone.

R. Welch, Photo.

a large tourist hotel, overlooking Clew Bay from the N, is the best centre for the exploration of the Nephinbeg group of hills (**411**), and of Curraun Achill (*supra*). It is worth while going to Mallaranny to see one plant only, namely *Erica mediterranea* (**131**, plate 35), in its glory. It greets the traveller from the edge of the platform as he descends from the train, and grows with *Saxifraga spathularis* (**114**, plate 34) in the hotel grounds; and around the head of Bellacragher Bay close by, it rises in masses, 6 feet and more in height, a sheet of pale purple blossoms before winter is fairly over. At this spot it descends to sea-level, and bushes of it may be seen hung with sea-weed thrown up during storms. Close by the remarkable var. *muscoides* of *Fucus vesiculosus* (see COTTON: "Marine Algæ," in Survey of Clare Island (Proc. Roy. Irish Acad., vol. **31**, no. 15, pp. 80, 127, pl. 5)) grows above high-water mark mixed with *Plantago maritima*; *Cochlearia anglica* (**86**) occurs also; and in spring, groves of *Iris Pseudacorus* burst through the fringe of *Fucus* left by winter gales. The plants recorded from the immediate neighbourhood of Mallaranny include *Rubus iricus, R. mucronatoides, Rosa mayoensis (mollis × spinosissima), Centunculus, Lamium molucellifolium, Utricularia intermedia, Juniperus sibirica, Listera cordata, Carex diandra, C. limosa, Phleum arenarium.* The maritime flora is limited, as is usual in W Ireland. Botanists visiting Mallaranny should watch for *Orchis Traunsteineri,* which E. S. Marshall believed he saw there.

E. S. MARSHALL: "Plants observed in West Mayo, June, 1899." Journ. Bot. 1900, 184–188.

NEPHIN AND NEPHINBEG.

411. Nephin forms a huge hump of quartzite, 2646 ft., rising from the W shore of L. Conn (**374**), and close to the edge of the limestone. In height it is exceeded in the West of Ireland only by Mweelrea (**397**) : but the even weathering of the dry intractable quartzite furnishes no suitable habitat for the alpine plants which one might hope to find around so lofty a summit. *Saxifraga stellaris, Hieracium anglicum, Arctostaphylos Uva-ursi, Vaccinium Vitis-Idœa, Salix herbacea,* and *Carex rigida* occur, and *Saxifraga spathularis* is as abundant on the summit as at the base. *Armeria* and *Silene maritima* also haunt the high grounds.

A few miles W of Nephin, separated from it by a deep valley, Birreencorragh and Buckoogh form an irregular mass of quartzites and schists rising to 2295 ft., dropping on the W into L. Feeagh. These hills are uninteresting botanically.

412. West again of L. Feeagh, **Nephinbeg** and its neighbours form a lofty ridge, stretching northward from Clew Bay (**401**) to Bangor, with an expanded S end. The principal points are Cushcamcarragh (2343 ft.), Glennamong (2067), Nephinbeg (2065), and Slieve Cor (2369). Here again quartzites mainly prevail. These hills rise from the vast bogs of Erris with flanks clothed with peat. Along the crest there is some fine rock-sculpture, but nevertheless the alpine flora of the range is poor, embracing *Thalictrum alpinum, Saxifraga oppositifolia, S. stellaris, S. spathularis, Sedum roseum, Saussurea, Arctostaphylos Uva-ursi, Oxyria, Salix herbacea, Carex rigida, Lycopodium alpinum, Isoetes lacustris.* Descending the hills S, W, or N we get into the domain of *Erica mediterranea* (**131**), with the two species of *Hymenophyllum, Lobelia,* etc. A curious instance of bird-dispersal is recorded in the occurrence of *Cotoneaster microphyllus* growing over a rock by a tarn in these mountains, far from any road, house, or sign of cultivation (Irish Nat. 1902, 7. See also **73**).

The district may be explored from Crossmolina or Pontoon (**374**) in the E, or from Mallaranny (**410**), where there is a large tourist hotel, or Newport (**401**) in the S.

H. C. HART: ''Report on the Flora of the Mountains of Mayo and Galway.'' Proc. Roy. Irish Acad., 2nd ser., 3, 705–709. 1883. N. COLGAN: ''Botanical Notes on the Galway and Mayo Highlands.'' Irish Nat. 1900, 116–118.

ERRIS.

413. The Barony of Erris comprises W Mayo as far S as (but not including) Achill and Mallaranny. It presents a more unbroken expanse of peat-bog, both lowland and mountain, than can be found elsewhere in Ireland. The mountains are dealt with above under the heading Nephin (**411**) and Nephinbeg (**412**). The large peninsula of the Mullet is described separately (**414**). Elsewhere the ground is undulating, forming a vast moorland, with a low indented coast-line on the W, while the N coast is magnificently precipitous. On this treeless, trackless waste

Erica mediterranea (**131**) has its Irish headquarters, and is widely spread. "To find," wrote Dr. David Moore, "a district of at least a quarter of a million acres in extent covered with this lovely heath, in full bloom, during the second week in April, forms one of the most remarkable botanical features the British Islands can afford." *Saxifraga Hirculus* (**116**) has been found in the N, near Ballycastle. Wherever the continuity of the peat is interrupted, the characteristic flora of the Mayo-Galway uplands comes in— *Osmunda* fringing the streams, *Saxifraga spathularis* decking the rocks, *Lobelia* and occasionally *Eriocaulon* in the lakelets, *Rhynchospora fusca,* etc., in the bog-pools. The peat itself harbours a very limited flora. *Wahlenbergia* has its only known W Ireland station by the Owenduff River about Lagduff Lodge.

Carrowmore Lake is an extensive sheet of water, about 4 miles in length, open and desolate. It has not been exhaustively examined, but appears to yield the characteristic local flora—*Erica mediterranea* (**131**) forming a broad belt on the shore above storm-level, with *Lobelia Dortmanna* in the water. Belmullet (**414**) in the W, and Mallaranny (**410**) in the S, provide hotels for the visitor; and primitive accommodation may be obtained at Bangor in the centre of the area.

C. C. Babington: "On the Botany of Erris." Mag. of Zool. & Bot. 1838, 119–124. D. Moore: "Observations on the prevailing and rarer Plants of Erris, and of some other portions of the County of Mayo." Nat. Hist. Review, 7 (Proc.), 414–417. 1860.

THE MULLET.

414. The Mullet is a wild narrow peninsula, with a surface divided between peat and blown sand, lying N of Achill on the W Mayo shore. Its longer axis (15 miles) lies parallel to the coast, with Blacksod Bay between. The isthmus which connects it with the mainland at the little town of **Belmullet** narrows to 300 yards. Most of the area is quite low, and the greatest hill rises only to 434 ft. Nowhere on the Irish coast can the effects on the flora of exposure and continually soaked soil be studied better than on this treeless tract. No bush raises its head more than a few feet above the ground. Stunted *Senecio aquaticus* brightens the pastures instead of *S. Jacobœa*; *Lythrum Salicaria* is through the fields everywhere; and no bank is too dry for *Hydrocotyle* and *Anagallis tenella.* The total flora (350 species) of this 45 sq. miles of land, possessing a large variety of features—cliffs, moors, cultivation, sandy and muddy shores, sand-dunes, lakes, and rocks—is actually smaller than that of Clare Island (6 sq. miles), and only a little larger than that of Inishturk (2¼ sq. miles). The flora of the dunes, as is usual on the W coast, is extremely limited in variety. *Eryngium maritimum* is characteristic of the drier parts, as *Pinguicula vulgaris* and *Selaginella* are of the damp flats. Several lakes, some of them brackish, occur, and yield some interesting species —*Lobelia Dortmanna, Ceratophyllum demersum, Potamogeton filiformis* (**170**), *Eriocaulon* (**173**), all within 50 ft. of sea-level. In the Portnafranka inlet, which fills at high water through a narrow channel and is slightly brackish, *Bidens cernua, Utricularia intermedia* (**144**), *Chara aspera* grow amicably with *Triglochin maritimum, Scirpus maritimus, S. rufus,* and other halophytes. The very shallow Ardmore Lough goes dry in summer, and then presents the appearance of a plain of red sand, much of it covered with a green mantle of *Littorella,* thickly strewn over its whole area with the graceful flower-stems of *Lobelia,* and more sparingly with those of *Eriocaulon.* The boggy areas in the N yield stunted *Erica mediterranea* (**131**), and abundance of *Arctostaphylos Uva-ursi.* The field and roadside flora includes *Centunculus, Lamium molucellifolium*; and in the lee of the peat banks are *Rubus iricus, R. dumnoniensis,* etc., with *Osmunda* and *Lastrea æmula.* The cliffs which fringe the N coast are hung with *Sedum roseum* and *Spergularia rupicola*; *Epipactis palustris, Juncus subnodulosus,* and the three species of *Drosera* occur. Small hotels at Belmullet. Steamer service from Sligo, along the magnificent cliff-bound coast of N Mayo, or bus from Ballina.

415. Inishkea consists of two reefs of gneiss, each with an area of about 400 acres. They had a population of about 150, which was evacuated in 1932 on account of the miserable conditions prevailing there. No botanist has as yet reported on the changes brought about by the cessation of agricultural and other human activities. The present account dates from 1905. The S island rises to 230 ft. Both are absolutely wind-swept, a few prostrate *Rubi* being the only shrubby vegetation. The *Plantago* association is well developed, and also a close sand-sward with *Daucus Carota*, etc., growing only an inch high (fig. 5). In storms the ocean pours across the island in various places, and so great is the force of the waves that the highest point of the north island (over 100 ft.) is actually the crest of a storm-beach of huge fragments of rock. *Cakile* grows as a weed in the potato fields, and *Agropyron junceum* on thatched roofs. *Drosera longifolia*, *Carlina* and *Centunculus* occur. Flora, 185 species and subspecies. Two plants, *Hyoscyamus* and *Convolvulus arvensis,* are not known elsewhere in W Mayo. Each island possesses a safe little sandy bay on its leeward side.

R. LL. PRAEGER: ''The Flora of the Mullet and Inishkea.'' Irish Nat. 1905, 229–244.

THE COUNTY OF SLIGO.

416. Co. Sligo is a region of high botanical interest, and contains much beautiful scenery. Area 721 sq. miles. Three types of ground and of vegetation are conspicuous. The S part, from L. Gara to Collooney and Tobercurry, is typical Central Plain country — flattish drift-covered limestone ground, consisting mainly of pasture, with some bog, marsh, and esker-ridges, and a good many lakelets. In the NW the limestone rises into a high cliff-walled plateau from 1500 to over 2000 ft. in elevation (Ben Bulben plateau, **422**); and these uplands are continued SW in the form of isolated terraced limestone hills of over 1000 ft. in height (**Keshcorran and Carrowkeel, 382**). While the flat tops of the hills are covered with bog, supporting a calcifuge heath vegetation of ordinary type, their wall-like sides and talus-slopes are tenanted by a remarkable alpine and calcicole vegetation. Lastly, in the W, one of the great ''Caledonian'' folds throws up the ancient metamorphic floor of the country as a wide range of bare bog-smothered hills, rising to over 1700 ft. (**Ox Mountains, 421**), with a few alpine plants included in their calcifuge vegetation. This fold continues across the county in the range of low rocky hills that passes Collooney, and overlooks the S shore of L. Gill (**420**). In the SW a portion of the coal-field of Connacht lies within the county, and forms the long, bare sandstone ridge of Bralieve (1498 ft.). There are several considerable lakes. The lovely L. Gill (**420**) lies close to the town of Sligo, between the metamorphic and the limestone ranges. L. Arrow (**381**), also a beautiful lake, and the bare L. Gara (**381**) lie in the SE. The extensive coast-line is irregular, generally rocky, with some sandy beaches, and several large, almost land-locked, sandy inlets. The only important island is Inishmurray (**424**), a mile in length, lying well out in the Atlantic, and notable for its early Christian remains and its primitive customs.

SLIGO TOWN.

417. The flourishing town of Sligo is surrounded by a district unsurpassed for beauty in Ireland, and of high botanical interest. It is chiefly known to the botanist as a base for the exploration of the Ben Bulben limestone hills (**422**), with their profuse alpine flora. L. Gill (**420**), which lies close to the town, is equally attractive to the lover of botany and of the picturesque. Knocknarea (**419**) and Rosses Point (**418**), referred to below, also lie close at hand. The knobby metamorphic hills which form the continuation of the Ox Mountains (**421**) pass within a few miles of the town. The botany of the immediate vicinity of Sligo calls for little remark : *Potagometon angustifolius* and *Apium Moorei* (in the Garvogue River), *Cochlearia anglica* and *Limonium humile* (in the estuary), *Rubus iricus* and *Ophrys apifera* are the best plants found within the municipal boundary. *Diplotaxis muralis, Lepidium latifolium,* and *Impatiens glandulifera* are spreading colonists which are now naturalized.

The neighbourhood of Sligo is particularly rich in megalithic monuments, burial cairns, and other early relics. On the way to botanize on the fine limestone hill of Knocknarea (**419**), the traveller may visit the famous series of dolmens and stone circles at Carrowmore. Other monuments lie between Sligo and L. Gill.

ROSSES POINT.

418. Rosses Point is a pleasant seaside resort, with several little hotels, on a small low peninsula 5 miles NW of Sligo. The sands, knolls, and sea-rocks yield within 50 ft. of sea-level a flora in which plants of various types of distribution are mixed, as the following list will show :— *Papaver hybridum, Draba incana, Saxifraga aizoides,*

Parnassia palustris var. *condensata, Crithmum, Black-stonia, Orobanche rubra, O. Hederæ, Euphrasia Pseudo-Kerneri, Spiranthes spiralis, Ophrys apifera, Juncus subnodulosus, Carex contigua, Sesleria, Juniperus communis, Adiantum, Equisetum variegatum* var. *arenarium* (also at Mullaghmore to the northward). Frequent buses from Sligo.

419. Knocknarea, the conspicuous flat-topped hill which rises over the sea 4 miles W of Sligo, is an outlier of the Ben Bulben limestone plateau (**422**): its cliffs yield a reduced alpine flora, including *Draba incana, Saxifraga hypnoides, Asplenium viride, Sesleria.* The great carn on the summit, 34 ft. high and 80 ft. across its flat top, the reputed burial-place of Queen Maev of Connacht (the English Queen Mab), is one of the most impressive pre-historic monuments in Ireland. The extensive sand-dunes at Strandhill, facing the Atlantic at the W base of Knocknarea, are barren as regards rare plants, like most of the western dunes; but *Cochleria grœnlandica* (**87**) has been found on rocks here, *Gentiana Amarella* on the sands, and *Fœniculum* is abundant by the shore southward. Knocknarea Glen, situated at the S base of the hill, a curious, wooded, cliff-walled rift in the limestone running across, not parallel to, the slope, is well worth seeing for the remarkable luxuriance of its fern vegetation — chiefly *Phyllitis Scolopendrium*, growing in immense profusion, 3 to 4 ft. high, with *Festuca sylvatica.* Around the muddy head of Ballysadare Bay, which adjoins on the S, one may find *Scirpus rufus, Potamogeton interruptus, Ruppia rostellata*; the last named covers acres of tidal mud-flats, usurping the usual habitat of *Zostera.*

LOUGH GILL.

420. Lough Gill is an exquisite lake lying between the limestone mountains of the Ben Bulben range (**422**) and a ridge of heathery hills formed of metamorphic rocks— one of the old "Caledonian" foldings (**421**). Length 5 miles, breadth 1 to 2 miles; elevation, only 20 ft. above low-water mark. The shores and islands are richly wooded, and it is chiefly in the woods that the botanist will reap his harvest. The most interesting plant of Lough Gill is *Arbutus Unedo.* I have endeavoured to show that it is certainly indigenous there (Proc. R.I. Acad., **41**, B, 105–113, 1932). It occurs on rocks and rough ground mostly close to the water's edge on the shores and islands in the E half of the lake in many places, accompanied especially by *Taxus* and *Sorbus rupicola,* exactly as on the Lower Lake at Killarney. See also **129.**

The flora includes *Ranunculus scoticus* (**82**), *Circœa alpina* (abundant, and accompanied, as it frequently is in Ireland, by *C. intermedia*), *Galium boreale, Hieracium repandum* (Goat Island), *H. stenolepis* (E end), *Pyrola media* (Hazlewood), *Monotropa* and *Neottia* in several places, *Orobanche Hederæ, Carex Hudsonii, Sesleria, Polypodium vulgare* var. *cambricum* (Goat I. and shore at Doonee Rock (Irish Nat., 1905, 39, ? not there now); and on the gneiss *Festuca sylvatica, Lastrea æmula,* and both species of *Hymenophyllum.* At the SW corner two interesting hybrids grow, *Apium Moorei* (**118**) and *Equisetum litorale* (**201**), the last-named being rather frequent about the lake. *Tamus* occurs in woods, hedges, and bushy ground, and has good claim to be considered native, although it is unknown elsewhere in Ireland.[33] *Lysimachia Nummularia* is plentiful. *Scutellaria minor* has here its most N Irish station, and *Ranunculus scoticus* and *Polygonum laxiflorum* are on the shores. *Linaria Cymbalaria* has forsaken its usual mural habitat, and is abundant on the rough limestone rocks of promontories and islands, amid a native flora. The S European *Calystegia sylvestris* is naturalized and frequent in hedges and rocky places round the E part of the lake. Other aliens are *Polygonum Bistorta* (Church I.) and *Saxifraga umbrosa.*

An interesting *Corylus-Fraxinus-Quercus* wood with a delightfully flowery ground-vegetation covering almost bare moss-grown limestone is found on the W side of the mouth of the Bonet River. *Anemone, Ranunculus auricomus, Oxalis, Conopodium, Primula vulgaris, Scilla non-scripta* are all most abundant here. With them are *Viola Riviniana, Sanicula, Solidago, Veronica Chamædrys, V. montana, Neottia, Orchis mascula, Arum, Luzula sylvatica.* Deeper humus brings in groves of *Lastrea æmula + aristata* and *aristata + spinulosa.* The edge of the adjoining metamorphic rocks is shown by the sudden incoming of *Vaccinium Myrtillus* and *Blechnum,* and the

---

[33] See PRAEGER in Proc. R. I. Acad., **41,** B, 117–9. 1932.

cessation of almost all the above flowers except *Oxalis* and *Scilla.* Simultaneously the wood changes to *Quercus-Betula-Ilex* (*Betula, Ilex,* and *Cratægus* being very rare in the limestone wood).

Slish Wood or Rockwood, which for two miles occupies the S shore, is a very fine native oak-forest. Under or among the dominating *Quercus, Betula pubescens, Ilex, Corylus* are common, and a little *Arbutus* occurs along the stony lake-margin. The luxuriance of the ferns and mosses which, with *Vaccinium Myrtillus, Luzula sylvatica,* and *Melampyrum pratense,* form the ground-vegetation, is remarkable. The ferns are mainly *Lastrea æmula* and *Blechnum,* the former attaining 3 ft. in height. At the E end of the lake, in Co. Leitrim, limestone cliffs overhanging the water yield *Sorbus rupicola, Taxus, Cornus, Sesleria,* and the woods *Agrimonia odorata, Monotropa,* and *Orobanche Hederae. Montbretia Pottsii,* often seen in wild places in Ireland from corms thrown out or carried down by streams, is self-sown and naturalized in chinks of limestone rocks close to water-level along the S side of Sriff Bay.

L. Gill lies close to Sligo, and from the town one rows up the River Garvogue into the lake. At the opposite (SE) end, on the Bonet River, is Dromahair, with its interesting ecclesiastical ruins and little hotel (with *Circæa intermedia* in the garden), whence one can conveniently explore the E half of the lake.

Along the N side, between L. Gill and the mountains, a belt of interesting tumbled limestone country extends, with little cliff-walled plateaux, small lakes, and ferny glens. At "The Duns," two fine limestone bluffs behind Newton Manor, *Draba incana, Saxifraga hypnoides, Asplenium viride,* etc., come down to 700 ft. Caps of peat, with *Calluna* and *Erica cinerea,* on 6 ft. high limestone blocks on the talus, show the former greater extension of the peat. *Orobanche rubra* (**143**) occurs half a mile further N. At Colgagh Lough an extensive white deposit forming a kind of sand consists of washed-up fragments of *Chara* much encrusted with lime; *Potamogeton coloratus* grows in a starved condition in the water.

THE OX MOUNTAINS AND SLIEVE GAMPH.

421. One of the ancient "Caledonian" folds which have been referred to elsewhere. In the centre of this great anticline, from L. Cullin in Mayo to Collooney in Sligo, the crumpled metamorphic rocks rise as a dark forbidding ridge, the highest point being Knockachree (1780 ft.), near the E end. The ridge is mostly broad and smooth, about 1600 ft. elevation, and for miles the summit is occupied with deep flat spongy bog. Steep slopes are rare, and high cliffs do not occur. Lakes, mostly quite small, are numerous. Native wood is almost absent. The scenery about L. Talt is picturesque; elsewhere it is rather monotonous. The flora is poor, but a few very rare plants are present. *Trichomanes radicans* (**192**) has recently been found. *Rubus saxatilis, Saxifraga aizoides, Crepis paludosa, Vaccinium Vitis-Idœa, Salix herbacea* occur, mostly about Knockachree. *Phegopteris polypodioides* is at Knockachree and near the E end of the hills, and *P. Dryopteris* has one of its few recorded Irish stations near L. Talt, but has not been found lately. *Equisetum variegatum, E. litorale, Lycopodium clavatum, Plantago maritima, Epipactis palustris,* and *Eriophorum latifolium* grow by L. Talt, *Ranunculus scoticus* and *Prunus Padus* by L. Achree. The W part of the range is known as Slieve Gamph, and lies in Mayo. Here the hills are lower. A fine cliff overlooks the valley and lakelets of Glendaduff, and furnishes the only E Mayo station for *Saxifraga spathularis* (**114**). By the sea-shore at the foot of the Ox Mountains, near Dromore West, *Adiantum* (**193**) grows on limestone rocks.

At the E end the hills are lower but very rocky. They drop down to the conspicuous gap at **Collooney** which allows the Ballysadare River and also the main lines of communication to pass from the Central Plain N to the ocean and Sligo : the ancient ridge then rises again to form the heathery hills on the S side of L. Gill (**420**). At Collooney *Erinus alpinus* and *Lactuca muralis,* naturalized on the demesne wall of Markee Castle, are the only plants of note.

N. COLGAN: "On the Flora of the Ox Mountains, Co. Sligo." Irish Nat., 1896, 301–308.

The Ox Mountains occupy a tract of somewhat remote and primitive character, and are best worked from Dromore West, where a little hotel will be found, from Ballina (**375**) near the W extremity of the range, or from Collooney or Sligo (**617**) at the E end.

THE BEN BULBEN PLATEAU.

422. One of the most interesting and beautiful districts in Ireland. It may be defined as stretching from L. Gill (**420**) to L. Melvin (**431**), and from near the ocean to

Fig. 23.—Ben Bulben from the North-West.

Upper Carboniferous Limestone above shaly limestone.
Bogland of Grange in foreground.

Manorhamilton (**429**), where it adjoins the L. Allen hills or Connacht coal-field described below (**427**). It takes its name, Ben Bulben, from the prominent western spur which stands out grandly N of Sligo (fig. 23). The district lies half in Sligo and half in Leitrim, and covers about 200 sq.

miles. The Carboniferous limestone here forms a plateau
1500 to 2000 ft. above the surrounding country (fig. 24).
In places, as on the highest point, Truskmore, 2113 ft.,
the limestone is still capped by the wreck of the Upper
Carboniferous. In the N, denudation has proceeded apace,
and sloping hill-sides are common; but the main mass

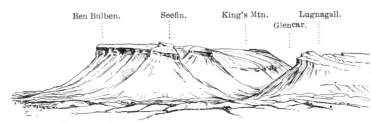

Ben Bulben.   Seefin.   King's Mtn.   Lugnagall.
Glencar.

FIG. 24.—BEN BULBEN RANGE FROM THE SOUTH.
After a Sketch by A. B. Wynne.

remains, intersected by two splendid cliff-walled valleys—
Glencar, plate 36, running E and W, and **Glenade (429)**,
running SE and NW. Each of these valleys embosoms a
pretty lake. To the western ocean also the plateau presents
a grandly precipitous front (fig. 23), subtended by a
stretch of arable land and bog. The typical succession in
the district (in ascending order) is :—

(1) Drift-covered arable land with bog, to 400–700 ft. ;

(2) Very steep grassy talus, variable in height, top at about
900–1200 ft. ;

(3) Vertical limestone cliff, very variable in height, top at
about 1600 ft. ;

(4) Flattish plateau covered by thick peat, highest point
(outlier of Yoredale sandstone) 2113 ft. The flora
of the peat is normal, fully calcifuge, with calcicole
and alpine plants wherever a rift or block of limestone betrays the underlying rock. Where it rises
to Truskmore, the highest point, the flora is :—

Potentilla erecta	Euphrasia officinalis
Galium saxatile	Empetrum nigrum
Solidago Virgaurea	Orchis maculata
Vaccinium Myrtillus	Juncus squarrosus
Calluna vulgaris	Luzula sylvatica
Melampyrum pratense	multiflora
var. montanum	Scirpus cæspitosus

Eriophorum vaginatum angustifolium	Deschampsia flexuosa
	Anthoxanthum odoratum
Carex rigida	Lastrea aristata
binervis	Blechnum Spicant
diversicolor?	Lycopodium Selago.

423. The remarkable feature of the vegetation is the alpine character of the flora of the limestone escarpments. *Arenaria ciliata* (**95**), which has here its only station in Ireland and does not occur in Britain, is a plant with a high northern and alpine distribution. It grows on the W part of the range at 1000 to 1950 ft.; a line drawn N from Glencar Lake defines its E limit. *Saxifraga nivalis* (on Annacoona), and also *Epilobium alsinefolium*, which grows on the Leitrim part of the range (along half a mile of cliffs in Glenade, 700–1200 ft.), are alpine plants which have here their only Irish station. The former was recently reported as being on the increase. On the escarpments occur in profusion—*Cochlearia alpina, Draba incana, Dryas, Saxifraga aizoides, S. hypnoides, Sedum roseum, Crepis_paludosa, Cystopteris fragilis, Asplenium viride.* More locally abundant are—*Thalictrum minus, Silene acaulis, Polygala vulgaris* var. *Ballii* (= *grandiflora*), *Saxifraga oppositifolia, Galium sylvestre, Circæa alpina, Euphrasia salisburgensis, Oxyria, Salix phylicifolia, Sesleria, Polystichum Lonchitis, Phegopteris polypodioides*; while the following may be classed as rare—*Thalictrum calcareum,*[34] *Meconopsis, Vicia sylvatica, Hieracium hypochæroides, H. repandum, H. britannicum, H. cymbifolium, Vaccinium Vitis-Idæa, Thymus glaber, Polygonum viviparum, Salix herbacea, Juniperus communis, J. sibirica, Carex rigida, Poa alpina* (second Irish station), *Adiantum.*

Two of the alpines, *Thalictrum alpinum* and *Saxifraga nivalis* (**112**), are confined to Annacoona, and two, *Arabis petræa* (second Irish station) and *Epilobium alsinefolium*, to Glenade. The most interesting alpine ground will be found on the cliffs of Annacoona which may be reached from the S by crossing the plateau from Glencar Lake, or from the N by road to Gleniff via Ballaghnatrillick; and the cliffs which lie W of Glenade Lake are also rich (**429**). The alpine flora extends from about 800 to 2000 ft.,

[34] Identified by Jordan as his *T. calcareum* (Ball in Bot. Gazette, **1**, 312–3), which appears to have no other station in British Isles.

its average being 1300 ft. In the arboreal flora of the cliffs *Sorbus rupicola, Ulmus montana* and *Taxus* are conspicuous.

The valleys and skirts of the mountains also yield some interesting plants. *Caltha radicans, Equisetum trachyodon* (**204**) and *Epipactis palustris* grow about Glencar L., which shelters great quantities of *Tolypella glomerata. Sisyrinchium angustifolium* (**159**) is in several places in Glencar; *Pyrola minor, Equisetum hyemale* and a rayless form of *Senecio aquaticus* near Lurganboy; *Drosera anglica, D. longifolia, Rhynchospora fusca, Carex diandra* in bogs in the N. On low limestone cliffs on the coast in several places *Adiantum* (**193**) is found. Other seaside plants of the area are *Trifolium fragiferum, Cuscuta Trifolii, Chenopodium rubrum, Elymus.* The district shows the mixture of alpine and southern plants characteristic of the west. The climate is very mild; in Glencar the Mexican *Sedum præaltum* is hardy and perennial on a cottage roof—a plant which never survives a Dublin winter.

The Ben Bulben area may be worked from Sligo (**417**) or from Bundoran, both well-provided tourist centres, or from Manorhamilton. The visitor to Glencar should not miss the fine waterfall there.

The Leitrim portion of the area, with Glenade, is mentioned more particularly in **429**.

T. H. CORRY: "On the Heights attained by Plants on Ben Bulben." Proc. R. Irish Acad., 2nd ser., 4 (Science), 73–77. 1884. R. M. BARRINGTON and R. P. VOWELL: "Report on the Flora of Ben Bulben and the Adjoining Mountain Range in Sligo and Leitrim." Proc. R. Irish Acad., 2nd ser., 4 (Science), 493–517. 1885. R. LL. PRAEGER: "Phanerogams and Vascular Cryptogams" (of the Sligo Field Club Conference). Irish Nat. 1904, 204–8.

424. INISHMURRAY.—Inishmurray, well-known for its early Christian remains, offers no special attraction to the botanist. The island lies 3 miles off the N Sligo coast, and is about $\frac{1}{2}$ sq. mile in area. It consists of a low reef of Carboniferous sandstone, with a thin boggy covering. *Lythrum Salicaria* is the most conspicuous species, occurring in vast profusion; *Osmunda* forms large masses; *Platanthera bifolia* and *Orchis maculata* are abundant. *Asplenium marinum* grows among boulders. *Bassica alba* is the commonest weed of tillage. *Radiola* and *Peplis* are among the many calcifuge species. A form of *Juncus effusus* with stems spreading in all directions and often spiral (**406**) occurs. An hour's work yielded a list of 145 species.

R. Ll. Praeger: ''The Plants of Inismurray, Co. Sligo.'' Irish Nat. 1896, 177–8.

THE COUNTY OF LEITRIM.

425. Leitrim is a long narrow area of 613 sq. miles, extending from the Atlantic into the centre of Ireland, and offering three distinct types of rock, of scenery, and of vegetation. In the N, half of the Ben Bulben range (**422**, **429**) belongs to the county. Here we have imposing cliff-walled limestone mountains, their summit forming a heathery bog-covered plateau, their sides clothed with alpine and calcicole plants. In the middle of the county, where L. Allen (**428**) cuts it nearly in two, high hills rise (Slieveanieran, 1922 ft., **427**) formed of Upper Carboniferous sandstones and shales, with seams of coal— a region of dark heather and clay-lands, with a rather poor and calcifuge flora. The S part of Leitrim (**426**) is low-lying; while the aquatic vegetation is here tolerably rich, the flora is on the whole uninteresting.

In the N the beautiful valley of Glenade (**429**) provides quite a botanical paradise; and the county boundary near by crosses the summit of Truskmore (2113 ft.), the highest point of the Ben Bulben range (**422**). L. Allen (**428**) in the centre, the uppermost of the Shannon lakes, is rather bare, like the mountains which surround it. The Shannon (**428**), rising on the hills a few miles outside the county boundary, flows through L. Allen (**428**), and almost at once assumes the lake-like sluggish course which characterizes it; and in this guise it flows along the W edge of the area. Portions of two beautiful Sligo lakes, L. Gill (**420**) and Glencar L. (**422**), infringe the W boundary of the county, and in the N, L. Melvin (**431**), 8 miles in length, belongs mainly to Leitrim. About three-fifths of the county is under grass, and one-fifth under crops. The sea-coast, in the extreme NW, consists of three miles of low rocky shore, with very few maritime plants.

SOUTH LEITRIM.

426. South of Slieveanieran and L. Allen, the S half of Co. Leitrim extends, a low, undulating area, composed mainly of tilled ridges formed of drift overlying impure limestone, with intervening hollows filled with bog or occupied by lakelets. In the extreme S, the narrow, alternating parallel tracts of farm-land and of bog or lake give the country a quite peculiar aspect. The Shannon, more like a winding lake than a river, pursues a sinuous course along the W edge of the district, fringed with great beds of reeds, and marshy meadows; and innumerable other sheets of water, of every size and shape, are found throughout the area. *Cicuta, Stellaria glauca,* and many commoner marsh plants are frequent here, and the two species of *Sium* are recorded from Carrick-on-Shannon, near which place also *Potamogeton filiformis* (Annaghearly Lough) and *Hydrocharis* occur—the last also with *Carex diandra* and *C. Pseudo-Cyperus* at Rinn Lough. *Viola stagnina* and *Teucrium Scordium* have their most northern Irish stations on the Shannon banks at Roosky. *Ranunculus scoticus,* doubtfully recorded from L. Bofin (Journ. Bot., 1892, 377) has recently been refound there. In the bogs *Rhynchospora fusca, Drosera longifolia,* and other characteristic species are common. *Polygonum minus* is abundant by Garadice L. See also **380**.

Towns are few and small in this area. Carrick-on-Shannon forms the best centre for the visitor, or he can stay at Ballinamore.

CUILCAGH AND SLIEVEANIERAN.

427. Around L. Allen (**428**), and stretching thence to L. Macnean, a mass of shales and sandstones of the Yoredale series, capped in the south by Coal-measures, rests on the limestone, and forms an imposing hill-mass. The main ridge runs NE from the E shore of L. Allen, from Slieveanieran (1922 ft.), overlooking that lake, to Cuilcagh (2188 ft.) on the borders of Cavan and Fermanagh; other mountains of less elevation are grouped W and N of L. Allen. Cuilcagh (pronounced Kulk-yach) presents a fine scarp of sandstone facing NNE for two miles, but is singularly poor in mountain plants, as are the other hills

within this area. *Cryptogramme* has been found on
Cuilcagh; also *Saxifraga stellaris, Vaccinium Vitis-Idœa,
Phegopteris polypodioides, Lycopodium clavatum* and
Hymenophyllum peltatum, the last being frequent in this
region. The Slieveanieran district is gloomy, with wet
pastures and boggy heaths, giving way, as one ascends, to
shaggy heather. *Listera cordata* occurs. The Arigna
hills, on the W shore of L. Allen, yield bands of hæmatite
and coal (part of the Connacht coal-field), which were
extensively worked formerly.

428. Lough Allen, lying between Slieveanieran (**427**)
and Arigna, is a large and rather desolate sheet of water,
pear-shaped, 9 miles in length, covering 14 sq. miles, and
elevated 160 ft. above the sea. The River Shannon, rising
to the E of Cuilcagh (**427**), falls with a rapid course into
the upper end of the lake, and leaves the lower end near
Drumshanbo as a deep slow-flowing stream. The accepted
source of the Shannon is a round pool, called the Shannon
Pot; this is the exit of an underground stream, coming
probably from Eden Lough, a mile to the E, which receives
a stream from Tiltinbane (1949 ft., the W end of the
Cuilcagh ridge), but has no visible outlet. The flora of
L. Allen is poor, but includes *Cicuta, Polygonum minus,
Carex aquatilis, Isoetes lacustris, Chara contraria, Nitella
flexilis* var. *nidifica.* The two species of *Hymenophyllum*
grow on Kilronan Mountain by Arigna.

All parts of this region are rather inaccessible. Belcoo
is the best base for Cuilcagh, and Carrick-on-Shannon for
L. Allen.

S. A. STEWART: ''Report of the Botany of Lough Allen, and the
Slieveanieran Mountains.'' Proc. Roy. Irish Acad., 2nd ser., 4
(Science) 426–442. 1885.

MANORHAMILTON AND GLENADE.

429. Manorhamilton forms a good base for the
exploration of Glenade and Glencar (**422**), the two beautiful
cliff-walled valleys that intersect the limestone plateau
of Ben Bulben. These valleys are somewhat similar—each
about 1000 ft. deep and a mile wide, each with a pretty lake
lying between its grey rock-walls, with patches of wood
to relieve the sternness of the massive cliffs. The low
grounds in **Glenade** have yielded *Pyrola minor, Polygonum*

*laxiflorum, Carex Hudsonii, Equisetum hyemale; Salix
pentandra* is abundant; but interest centres in the flora of
the cliffs which face N beyond Glenade Lake. Here *Arabis
petræa* has its second station in the country, and *Epilobium
alsinefolium* is found in its only Irish habitat. It grows
in considerable abundance on dripping rocks among
Cochlearia alpina and *Chrysosplenium oppositifolium*, from
700 to 1200 ft. Most of the following are frequent or
common—*Draba incana, Meconopsis, Polygala vulgaris* var
Ballii (**93**), *Silene acaulis, Dryas, Circæa alpina, Saxifraga
oppositifolia, S. aizoides, S. hypnoides, Epilobium angustifolium, Euphrasia salisburgensis* (**141**), *Oxyria, Polygonum
viviparum, Asplenium viride, Polystichum Lonchitis,
Lastrea montana, L. æmula.*

The talus under the Glenade cliffs, facing N, presents a
curious jumble of plants—alpine and lowland, calcicole
and calcifuge, sun-loving and shade-loving, dry-soil and
wet-soil. Limestone scree is mixed with the remains of a
former covering of peat. Ferns are unusually abundant and
luxuriant—*Polystichum Lonchitis, P. aculeatum, P.
angulare, Lastrea Filix-mas, L. montana, L. aristata, L.
spinulosa* (to 3½ ft. high), *L. æmula, Athyrium, Phyllitis
Scolopendrium, Blechnum, Cystopteris fragilis, Asplenium
Trichomanes, A. viride, A. Adiantum-nigrum, A. Ruta-
muraria, Ceterach, Polypodium vulgare, Pteris, Hymenophyllum peltatum.* These twenty species all grow among
or close to a profusion of many of the alpines enumerated
above, which are mixed with *Geranium Robertianum,
Oxalis, Epilobium palustre, Campanula rotundifolia,
Primula vulgaris* (all in quantity), *Viola Riviniana, Spiræa
Ulmaria, Geum rivale, Alchemilla vulgaris, Valeriana,
Solidago, Bellis, Vaccinium Myrtillus, Veronica Chamædrys,
V. officinalis, Pinguicula vulgaris, Prunella, Urtica dioica,
Empetrum, Scilla non-scripta, Luzula sylvatica*—a very
queer assemblage!

Phegopteris Dryopteris has been found on Benbo.
Glenade may be easily approached from Manorhamilton
or by its N end, from Bundoran or Kinlough. The hills
lying on other sides of Manorhamilton (S and E) are not
well known, but they are less attractive as regards flora
and scenery than the region about Glenade and Glencar.
The country all about is hilly and interesting; and L. Gill
(**420**), L. Melvin (**431**), and L. Macnean (**430**) all lie within
striking distance (6 to 8 miles), and invite exploration.

LOUGH MACNEAN.

430. Upper and Lower Lough Macnean lie close together, with the village of Belcoo between them. The Upper Lake is 5 miles long, the Lower 3 miles. Both are picturesquely situated among hills. They drain eastward into the Erne. Very few plants are recorded from their shores or waters : *Rosa Nicholsonii, Potamogeton nitens,* and *Ranunculus scoticus* have been found at the Lower Lake, and *Galium boreale* by the Upper.

Florencecourt, the seat of the Earl of Enniskillen, is a fine demesne on the S side of the Lower Lake. Its remarkable limestone caves (the best known called the Marble Arch[35]) have made it long a place of resort for tourists, and a good many rare plants have been found. *Hieracium tridentatum* has one of its few Irish stations. The woods yield *Carex strigosa, Epipactis latifolia.* On the hills above grow *Listera cordata, Lastrea montana, Asplenium viride,* and both species of *Hymenophyllum.* Bogs on the low grounds contain *Drosera anglica, D. longifolia, Rhynchospora fusca.* It was here that the "Irish Yew," *Taxus baccata* f. *fastigiata,* was originally found (**191**).

The neighbouring villages of Blacklion and Belcoo offer hotel accommodation to the visitor.

LOUGH MELVIN.

431. Lough Melvin, famous for its fishing, has not been thoroughly explored botanically. It is a pretty lake, 8 miles long by about 1½ wide, lying on the Carboniferous limestone mainly in Co. Leitrim, partly in Fermanagh, and containing a number of islands, mostly planted with trees. The interesting Ben Bulben area of limestone mountains (**422**) adjoins immediately on the SW. The E end of the lake is the most interesting part. A small islet here (Gorminish of one-inch map) is the only Irish station outside Donegal for the Globeflower, *Trollius* (Irish Nat., 1901, 33–4). *Caltha radicans* is on the shores. At **Garrison**, where there is a little fishing hotel, *Potamogeton*

---

[35] See H. BRODRICK: "The Marble Arch Caves, County Fermanagh." Proc. Roy. Irish Acad., 27, B, 183–192, pl. 12. 1909.

juliformis (**170**) is abundant on a shallow sandy bottom, and again at the mouth of the Glenaniff River; *P. angustifolius* also occurs, and a rare form of *Tolypella glomerata* (var. *erythrocarpa*, see Journ. Bot., 1919, 224). The lake is within easy reach of Bundoran.

THE COUNTY OF FERMANAGH.

432. Fermanagh might be defined as occupying the lower two-thirds of the basin of the Erne (**433**). This river enters the county at its SE end, spreads out into first one and then a second great island-filled expanse of water (Upper and Lower Lough Erne, **435–6**), and at the NW end of the county plunges down to meet the Atlantic a few miles further on, below Ballyshannon. L. Macnean (**430**) and L. Melvin (**431**), on the SW edge of the county, are the only other important lakes. All of these occupy troughs in the limestone; but the surrounding country is formed largely of other materials. Heathery hills of metamorphic rocks fringe Lower L. Erne in the NW, and of Old Red Sandstone in the NE; while SW of Lower L. Erne, between it and L. Macnean, Yoredale sandstones overlying the limestone form a high moorland, with heathery hills which stretch S, culminating in Cuilcagh (**427**) (2188 ft.) on the borders of Fermanagh and Cavan, and continuing across Leitrim to L. Allen (**428**) and Slieveanieran (**427**). Fringing the sandstone hills are others of limestone, with a characteristic flora.

The county approaches to within a few miles of the Atlantic in the west. Area 714 sq. miles, of which three-fifths are under grass, and a quarter under crops. Low-level bogs are very few, and, except in the hill-regions, the aspect of the country is rich and prosperous. Enniskillen (**437**), built on an island in the Erne between the two great lakes, forms an excellent centre for the visitor, whether he explore on land or (preferably) by boat.

THE RIVER ERNE.

433. This considerable river is unique in Ireland on account of its frequent mazy lake-expansions (fig. 25). It rises in Lough Gowna (**434**), a tortuous island-studded sheet of water lying mostly in Longford, among Silurian rocks. Thence the Erne flows N from the slates to the

imestone country to enter Lough Oughter (**434**), a maze
of land and water some 50 sq. miles in area, with similar
country to the west.   This region can be worked from
Cavan or Belturbet, as L. Gowna can from Granard.   At
Belturbet the Erne is for a few miles river-like, but it
soon opens into Lough Erne.   This great lake occupies the
centre of Co. Fermanagh for a length of nearly 40 miles.
At Belleek, at the W end of Lower Lough Erne, the river
plunges into a limestone gorge, to reach sea-level at

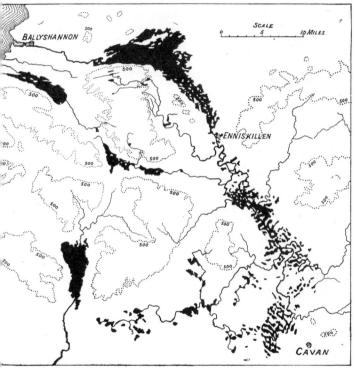

FIG. 25.—THE RIVER ERNE.

Ballyshannon. Hotels at Enniskillen, Rossclare (5 m. NNW
of the last), Pettigo and Belleek.

434. **Lough Gowna** is not at all well known botanically.
*Typha angustifolia, Lastrea montana, L. æmula, Nitella
flexilis* are on record.   Out to the E, at L. Sillan, the rare
var. *nidifica* of the last species is found.   *Phegopteris*

polypodioides is reported from Bruce Hall. **Lough Oughter** also invites further work. From its vicinity *Cardamine amara, Rhamnus catharticus, Callitriche autumnalis, Sium latifolium, Cicuta, Crepis paludosa, Utricularia intermedia, Polygonum laxiflorum, P. minus, Hydrocharis, Juncus diffusus, Lemna gibba, Potamogeton angustifolius, Cladium, Carex Hudsonii, C. gracilis, C. Pseudo-Cyperus, Lastrea Thelypteris, L. montana, Isoetes lacustris, Chara aculeolata* are recorded. *Melittis Melissophyllum* appears naturalized by a lake at Farnham.

The **Belturbet** area, between L. Oughter and L. Erne, is again a mere maze of land and water, with *Juncus subnodulosus, Potamogeton obtusifolius, Sparganium minimum, Butomus, Carex lasiocarpa, C. Pseudo-Cyperus,* and many other plants of the lakes further up and down the river. (Praeger in Irish Nat., 1905, 260.)

Between Cavan and Crossdoney *Andromeda* and *Drosera anglica* are in the bogs, and *Ophrys apifera* grows close to the edge of the limestone. For the exploration of this archipelagic area the visitor can stay at Cavan or Belturbet.

LOUGH ERNE.

435. This large lake is constricted near the middle for a distance of 8 miles into a river-like form, so that two tolerably distinct areas of water, Upper and Lower Lough Erne result. The **Upper Lake** is 151 ft. above the sea, filled with islands, points, and bays till it becomes a maze of land and water, like L. Oughter (**434**) and L. Gowna (**434**). The basin is formed in Carboniferous limestone, and the mazy nature of L. Erne and L. Oughter is due to the irregular solution of the limestone basin in which they lie. This explanation, however, does not account for the very similar form of L. Gowna, for that lake lies on the Silurian slates. L. Gowna is in fact a piece of submerged drumlin country (drumlins being parallel whale-backed ridges of drift which are often well developed in the Silurian area). The land around L. Erne is everywhere drift-covered, mostly tilled or grazed, sometimes well wooded as at Belleisle and Crum Castle. The flora as a whole possesses no special feature; but some very interesting plants occur. On the Upper Lake the best are *Sisyrinchium angustifolium* (**159**) on the promontory of Derryvore opposite Crum Castle and also on the W shore, and *Carex elongata* at Crum, where *Stratiotes* has an old-established head-quarters. *Cicuta, Neottia, Epipactis palustris, Ophrys apifera* flourish. *Nasturtium sylvestre* has one of its few Ulster stations at Belleisle, where *Lathyrus palustris* (**103**) is also found. *Viola Curtisii* has one of its rare inland habitats by the lake at Lisnaskea; *Typha angustifolia* is at L. Killygreen near Newtownbutler. The visitor to the Upper Lake may stay at Enniskillen or at Lisnaskea.

436. The **Lower Lake** is larger, much more open, and in places high ground adjoins. It lies 149 ft. above sea-level, is nearly 20 miles long by up to 5 miles wide, full of wooded islands in the upper part, broader and quite open below, where on the S shore the limestone cliffs of Shean North (the N end of the very interesting uplands described in **438**) rise to over 1000 ft. above the lake. Its basin is composed mainly of limestone, but in the upper part the E shore is occupied by sandstones, partly of Yoredale, but chiefly of Old Red age. The ground is largely drift-covered. The scenery is more diversified and picturesque than that of the Upper Lake, and attains a high order of beauty, and the flora is more varied and interesting.

Many of the islands are characterized by the abundance of some one species often rare over the rest of the area. Thus Heron Island is in possession of *Rhamnus catharticus*; Dacharne I. of *Solidago*; Inishdara, *Arenaria trinervia*; Namanfin, *Vicia sylvatica*; Strongbow I., *Agrimonia Eupatoria*; Cleenishmeen, *Eupatorium*. The vegetation is often remarkably luxuriant. *Campanula rotundifolia* was measured 2′ 8″ high; *Solidago*, 3-4 ft.; a plant of *Polygala vulgaris* bore over 100 upright stems a foot long, crowned with very large deep blue flowers; and magnificent snow-white tussocks of *Galium boreale* are abundant. *Caltha radicans* and *Circœa alpina* are common, also *Crepis paludosa* and *Lysimachia Nummularia*. *Thalictrum minus, Aster* sp., *Gentiana Amarella, Mimulus Langsdorffii, Centunculus, Chenopodium rubrum, Polygonum minus, Sagittaria, Sparganium minimum, Ophrys apifera, Carex Hudsonii, C. gracilis, Osmunda*, occur more sparingly. *Ranunculus scoticus* is on the shore at Castlecaldwell. *Potamogeton filiformis* grows at Kesh. *Callitriche autumnalis* is also present, and at Drumcose *Viola stagnina* (**91**), accompanied by hybrids with *V. canina*. On the smaller islands the woods are largely or entirely native, and the flora undisturbed. On Bilberry I., for instance, there is dense mixed wood of *Quercus, Fraxinus, Alnus, Betula pubescens, Populus tremula, Corylus, Cratægus, Ilex, Sorbus Aucuparia, Pyrus Malus (acerba). Euonymus, Rhamnus catharticus, Viburnum Opulus, Prunus spinosa;* with *Rosa canina, R. tomentosa, R. spinosissima, Myrica, Salix cinerea* and *S. aurita* round the margins.

The neighbourhood of Ely Lodge, on the W shore of the Lower Lake, has yielded good plants. *Monotropa, Polygonum Bistorta, Ophrys apifera* grow about the demesne. To the W, at **Carrickreagh**, ridges of bare limestone rise from the lake-shore and extend for some distance inland, covered with wood or scrub, with cutting going on continually. *Quercus sessiliflora* is sometimes dominant, sometimes *Betula pubescens*, sometimes *Corylus*, with much *Fraxinus, Viburnum Opulus, Prunus spinosa*. Ground vegetation with much *Primula vulgaris, Scilla non-scripta, Lysimachia nemorum, Rubus saxatilis, Epipactis latifolia* (in quantity, and up to 4 ft. in height), *Sesleria* (in open places), *Aquilegia, Arum, Pteris, Athyrium,* and sometimes *Blechnum*. On dry flat ground under oaks, on a thin skin of moss and humus over limestone rubble, *Phegopteris polypodioides* forms patches up to 20 ft. across, with fronds to 2 ft. high—the only habitat of the kind which I know in Ireland, its characteristic stations being wet chinks or ledges on the mountains. North-westward, the ground rises into the interesting hills described in **438**.

At the termination of Lower L. Erne the river falls abruptly at Belleek (*Juncus subnodulosus*, etc.) into a beautiful gorge, down which it foams to Ballyshannon (**444**) where it becomes tidal and soon enters the sea amid extensive sand-dunes, at the southern end of which lies the watering-place of Bundoran (**444**), which forms a good centre.

R. M. BARRINGTON: "Report on the Flora of the Shores of Lough Erne." Proc. Roy. Irish Acad., ser. 2, 4 (Science), 1–24. 1884. R. LL. PRAEGER: "In Camp on Lough Erne." Irish Nat. 1892, 110–114.

ENNISKILLEN.

437. The busy little town of Enniskillen is built on an island in the River Erne, between the lake-expansions of Upper and Lower L. Erne, and it forms a convenient centre for the exploration of these waters. In summer a steamer runs on the Lower Lake, and open boats are to be had in abundance. Railways in three directions assist the explorer. The surrounding country is undulating and fertile, with many lakelets and deep marshes, still unexplored. The limestone cliffs of Belmore Mountain, lying 6 miles W, and the woods and hills at Florencecourt (**430**), situated about the same distance to the SW, are good ground for the botanist. In the immediate neighbourhood of Enniskillen *Sium latifolium* has its Lough Erne head-quarters. *Nasturtium sylvestre, Veronica peregrina,* and *Lemna polyrrhiza* occur. *Caltha radicans* flourishes on the river-banks here and elsewhere. At Lisgoole, a couple of miles S of Enniskillen, *Leucojum æstivum* grows in damp meadows and *Dipsacus sylvestris* has an outlying station. Devenish, an island two miles down from Enniskillen, is of high archæological interest. From its pastures rises one of the finest of the round towers of Ireland, surrounded by the remains of a number of ecclesiastical buildings, pointing to an establishment of much bygone importance. Away to the west, there is a record for *Galium Cruciata* which is in need of verification—by a small lake at Colebrooke, 1869 (More: Recent Add.). Elsewhere in Ireland the plant is known only from Downpatrick (**477**).

THE POULAPHUCA AND CARRICK DISTRICT.

438. Between L. Macnean and Lower L. Erne, and W of Derrygonnelly, interesting hill-country extends, formed mainly of Yoredale sandstone and Upper Carboniferous limestone. The sandstones dip S, so that while they fringe the shores of the former lake they overlook L. Erne from a height of 1000 ft. or more (1135 ft.) perched on the top of imposing cliffs of limestone (fig. 26). The limestone fringes the whole area, and forms green foot-hills with

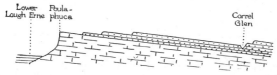

Fig. 26.—Diagrammatic Section across the Poulaphuca district: Upper Carboniferous Limestone capped by Yoredale Sandstone. Length of section, 3 miles; height, 1100 feet. Direction, N to S (left to right of sketch).

lakes between, and grey cliffs and grassy slopes, around
the dense heather which clothes the sandstone moors. The
sandstone on the higher grounds protrudes through the
peaty covering in a series of E and W scarps, often with
lakes at their bases; these parallel scarps run in straight
lines for considerable distances, and are backed by long
dip slopes. A number of rare northern plants haunt both
the limestone and the sandstone, the scarps of the latter in
particular. The most interesting spots on the limestone
area are the cliffs of Poulaphuca overlooking L. Erne, the
vicinity of Carrick Lake, and the cliff-walled hill of
Knockmore. *Saxifraga hypnoides, Circœa alpina, Hieracium lasiophyllum, Euphrasia salisburgensis* (**141**), *Taxus,
Sesleria* (**186**) are frequent or abundant on the limestone.
Poulaphuca yields in addition *Saxifraga aizoides, Sedum
roseum, Epilobium angustifolium, Lastrea montana*; and
Dryas is plentiful on Knockmore. About Carrick L. are
Juniperus communis, Eriophorum latifolium; at Bunna-
hone L. *Equisetum trachyodon* (**204**) grows on the shore.
Turning now to the moorland, the sandstone scarps, from
500 to 1100 ft. elevation, yield *Meconopsis, Pyrola secunda*
(in abundance), *P. media,* and *P. minor; Vaccinium Vitis-
Idœa, Listera cordata,* and *Asplenium viride,* all in
quantity; both species of *Hymenophyllum, Lastrea œmula*
and *Equisetum pratense.* The moorland lakes are fringed
with *Ranunculus scoticus* (**82**); *Lobelia* occurs sparingly.
It is interesting to note that conspicuously calcicole plants,
such as *Sesleria* (**186**), *Arabis hirsuta,* and *Asplenium
Ruta-muraria,* grow on the high sandstone scarps, far from
limestone; here they mix with such markedly calcifuge
species as *Vaccinium Vitis-Idœa, V. Myrtillus, Calluna,
Erica, Digitalis, Blechnum.*

One of the most charming spots in the district lies about
Carrick Lake, mentioned above. The lake is situated on
the junction of the limestone and the sandstone areas,
affording a picturesque contrast of scenery and vegetation.
On the limestone of Carrick Hill, which rises steeply from
its waters, grow *Euphrasia salisburgensis* (**141**) and other
calcicoles. Running up from the W end of the lake is
Correl Glen, a deep cut in the sandstone, in which a dense
Betula pubescens wood clothes masses of fallen rocks, deeply
covered with moss. Here *Vaccinium Vitis-Idœa* creeps
over stones and up tree-trunks at 400 ft. elevation, and
Trichomanes (**192**) and both species of *Hymenophyllum*

have been found amid the boulders, with *Pyrola secunda* on the low cliffs.

No hotels lie within the area, but it may be worked from Enniskillen, Belleek, or Garrison.

S. A. Stewart: "Report on the Botany of the Mountainous Portion of Co. Fermanagh, to the west of Lough Erne, and the adjoining district of Co. Cavan." Proc. Roy. Irish Acad., 2nd ser., **2** (Science), 531–544. 1882. R. Ll. Praeger: "Among the Fermanagh Hills." Irish Nat. 1904. 232–241.

THE COUNTY OF TYRONE.

439. The large hilly area which constitutes Tyrone is a complex of a great variety of rocks—mostly schist and gneiss in the N half and Old Red Sandstone, a little limestone, etc., in the S. Drainage is mostly towards the NW into the Foyle (**451, 454**), which is formed by the junction at Omagh, Newtownstewart, and Strabane of a number of fine streams descending from the uplands of Tyrone and Donegal and draining a large area of country. The Blackwater, flowing NE into Lough Neagh, fringes the county on the SE. L. Neagh itself forms part of the E boundary: the plants of that interesting lake are dealt with elsewhere (**463**). The highest hills are the Sperrin Mountains, which attain an elevation of 2240 ft. on the borders of Tyrone and Londonderry. In the NW, the long drowned valley of the Foyle permits the sea to flow up as far as Strabane, fringed with reed-beds and marshes which have not yet received due attention from the botanist: this region is referred to briefly below (**451, 454**). Tyrone is the headquarters of the American *Veronica peregrina*. It is a common plant there, and extends thence mainly SW as far as Galway.

The **Sperrin Mountains**, which are the culmination of a considerable area of hills in Tyrone and Londonderry, are among the least interesting uplands in Ireland. They form an E–W ridge with half a dozen summits exceeding 2000 ft. (Sawel 2240, Dart 2040), but the whole area is peat-covered, outcrops of the schistose rocks are small and very rare, and the higher grounds present miles of heavy, monotonous bog. One alpine plant, however, has here its only Irish habitat—*Rubus Chamæmorus*, of which a single patch, apparently never flowering, exists W of Dart (see Journ. Bot., 1892, 279, Irish Nat., 1892, 124, 1902, 317).

Otherwise the alpine plants are of little interest—*Saxifraga stellaris, Vaccinium Vitis-Idæa, Salix herbacea, Listera cordata, Carex rigida, Lycopodium alpinum.* *Bartsia viscosa* has one of its rare northern stations near The Six Towns, in Londonderry S of the Sperrins.

For the rest, the flora reflects the hilly conditions which prevail generally, as well as the northern position of the county. The rarer species include *Meconopsis* (Bally-skeagh Hill, old record), *Cardamine amara* (frequent), *Callitriche autumnalis, Circæa alpina* (frequent), *Crepis paludosa, Lobelia, Pyrola media, P. minor, Pinguicula lusitanica, Stachys officinalis, Carex limosa, C. aquatilis* (Finn R. near Strabane), *C. diandra, C. strigosa, C. lasiocarpa* (near Stewartstown), *Festuca sylvatica, Hymenophyllum tunbridgense, H. peltatum, Trichomanes* (**192**) (Strabane Glen), *Cystopteris fragilis, Lastrea Thelypteris, L. æmula* (frequent), *L. montana, Isoetes lacustris* (old record). *Equisetum trachyodon* (**204**) grows at Milltown Tullylagan, and *E. variegatum* on floating islands in Ardpatrick Lough near Stewartstown. Some other plants of this eastern region, which adjoins L. Neagh, are mentioned in **462**. The several tributaries of the Foyle are fine rocky streams in deep hill-rimmed valleys. So far very few plants have been found on their banks. *Hieracium stictophyllum* by the Mournebeg above Castlederg is one of the rarest. *Ophrys apifera* occurs between Pomeroy and Carrickmore. Among the plants which owe their presence to human influence are *Geranium phæum, Myrrhis, Peucedanum Ostruthium* (frequent), *Crepis biennis, Campanula rapunculoides, Lamium molucellifolium,* and probably *Leucojum æstivum,* of which a large patch occurs on a bog near Dungannon.

The visitor can stay at Strabane, Omagh, Dungannon, Cookstown and elsewhere.

H. C. HART: ''Notes on the Plants of some of the Mountain Ranges of Ireland.'' Proc. Roy. Irish Acad., 2nd ser., 4 (Science), 244–257. 1884. M. C. KNOWLES: ''Flowering Plants of Tyrone.'' Irish Nat. 1899, 83–84.

THE COUNTY OF DONEGAL.

440. Donegal, occupying the NW corner of Ireland, is mountainous or at least hilly over practically the whole of its large area (1870 sq. miles). Only towards the SE, over the low limestone country lying E of Donegal Bay and the fertile lands which lie along the rivers Finn and Foyle, is the county other than a rough rocky area, full of hills (to 2466 ft.) and lakelets, tiny fields and heather, the whole bounded by a coast-line of great extent and intricate nature, with noble sea-inlets here and there breaking the succession of creeks and islets. For purposes of convenience, the county is here broken up into half-a-dozen areas, but these do not mostly correspond to any change in the prevailing metamorphic or granitic rocks, or in the flora. The most "Atlantic" flora, along with the representatives of the Lusitanian and American groups, is found towards the NW, while the region the flora of which most approximates to that of the more typical Irish counties is the lower fertile part towards the SE.

Donegal as a whole forms a fairly natural botanical province. It is a highly picturesque area, owing its present form to a very ancient period of folding of the Earth's crust (5)—the same which has given us Scandinavia and the Highlands of Scotland, and in Ireland Connemara (384) and the Wicklow Mountains (261). This threw the surface into a series of high ridges. Subsequently denudation has reduced these, and left their roots exposed—ancient schists, gneisses, and quartzites which over the greater part of Ireland lie deeply buried under later deposits. In the NW there is a considerable exposure of intrusive granite, a handsome rock marked by its red felspar crystals. This is of later age than the ancient beds which it has invaded, but is still very old.

Fragments of deposits of subsequent epochs remain to show the great extent of the denudation—Old Red Sandstone over small areas in Fanet, Carboniferous sandstone on the summit of Slieve League (1900 ft.), and, as will be mentioned immediately, a little Carboniferous limestone. The main masses of rock still display in a striking degree the NE-SW trend of the original folding, which controls the direction of the present main ridges and valleys. A change is found in the SW corner of the county, at Donegal Bay, where the Carboniferous limestone which extends far to the S invades Donegal, providing a variety from the acid rocks and peaty soils which prevail elsewhere, and giving a habitat for a number of calcicole plants. But calcicole species are by no means confined to this limestone area; all round the coast they find a home on sea-sands, which are strongly alkaline from the plentiful admixture of broken shells of Mollusca and tests of Foraminifera with the grains of silica.

Donegal is botanically a rather isolated area, for though the ancient schists extend E over much of Londonderry and Tyrone, their surface is different, forming low grounds occupied by tillage or broad high hills deeply peat-covered. The county is moreover a peninsula; while its maximum SW–NE breadth, from Malinmore Head to Inishowen Head, is 84 miles, a parallel line only a few miles to the SE, from Donegal Bay to Lough Foyle, gives only 44 miles, and from tidal water to tidal water only 30 miles. The area is, then, three-quarters surrounded by sea; on account of the innumerable indentations of the margin and the number of islets, the coast-line (mostly low and rocky, but in places magnificently precipitous) is greatly extended; and the influence of the Atlantic is everywhere evident.

Donegal is a region of heather and gorse, of rocky ridges great and small, of little fields and stone walls, peaty lakelets, far-extending sea-inlets, innumerable rocky points and islets, and sandy bays. Crops occupy less than one-fifth of the area of the county, grass-land about one-third, barren mountain-land and peat-bog nearly 40 per cent. The tilled land is mostly in the form of small holdings dotted among larger areas of heather and rock, and cultivation is rarely seen at a greater height than 500 ft. above the sea. The search for fuel has denuded the county of most of its woods, and the peat has been cut down to the underlying rock, thus enhancing the poverty of the land. Windswept and sodden in winter, it is most beautiful throughout the rest of the year, blazing with gorse in spring, glowing with heather in summer and autumn, sparkling with blue-brown lakelets, and surrounded by the open ocean, the tumbled rugged surface giving to the traveller fresh vistas with every turn of the road.

441. Though much of the area is mountainous, lofty summits are absent, the highest points averaging about 2000 ft., with a maximum (in Errigal) of 2466 ft. The higher hills are widely scattered over the county. The prevalence of hill and bog is well shown by the fact that the following plants occur—most of them abundantly—in all the 8 districts into which, in his "Flora," H. C. Hart has divided the county :—

Saxifraga stellaris
Drosera rotundifolia
　　anglica
　　longifolia
Lobelia Dortmanna
Juniperus sibirica

Pinguicula vulgaris
　　lusitanica
Empetrum nigrum
Osmunda regalis
Selaginella selaginoides
Isoetes lacustris.

But it should be added that in Donegal all of these descend to sea-level except the first, which is not as yet recorded from less than 500 ft. Almost equally widely spread are

Sedum roseum
Crepis paludosa
Hieracium anglicum
　　iricum
　　umbellatum
Arctostaphylos Uva-ursi
Salix herbacea
Carex rigida
　　limosa

Carex lasiocarpa
Hymenophyllum peltatum
Cystopteris fragilis
Asplenium viride
Lastrea æmula
　　montana
Lycopodium alpinum
　　clavatum.

Utricularia intermedia is widespread over the W half of the area.

As a consequence of the general structure, rivers are usually short, and lakes, though very numerous, small. The Foyle, here broad and estuarine, forms with Lough Foyle the SE boundary of the county, and the Erne flows in a fine limestone gorge for some miles on the SW edge. Close by, Donegal touches the shore of Lower L. Erne, a much larger lake than any which are included within the county.

While its position in relation to the rest of Ireland would naturally lead to lines of communication running towards the NW, the topography of Donegal, due to the ancient folding, compels a prevalence of NE-SW direction in the railways and roads, especially the former; the hilly nature of the county leads to the use of narrow-gauge lines, so that railway travel in Donegal tends to be circuitous and slow; but the areas traversed are so interesting that this is of little account. These lines penetrate to the most remote parts of the county; Bundoran, Donegal town, Killybegs, Glenties, Burtonport, Gweedore, Creeslough, Letterkenny, Buncrana, Cardonagh are all approachable by rail, and from these and other places along the lines of route, every part of the county is easily reached. Hotels large and small are numerous, and places which are not accessible in the course of a day's walking are few.

Although the county includes the most northerly part of Ireland, the climate is mild, resembling that of Connemara and Kerry rather than that of the adjoining NE area. There is the same absence of extremes in summer and winter, the mean January temperature being 41°–43° F., and the July mean 58°–59° F. The minimum winter temperature, which probably is the factor which most affects vegetation, is practically the same in Donegal as it is in Kerry, Devon, and Cornwall. Rainfall is high—40 to 60 inches, and probably more at spots among the hills. Spring is the driest part of the year. With mild winters and absence of drought in summer, conditions are suitable for the presence of southern and mesophile plants: but wind remains a serious enemy. Where shelter is found, the number and variety of half-hardy shrubs and herbs which will flourish in Donegal is surprising.

442. The flora is rich as Irish areas go, and it combines in an interesting manner two special elements, derived respectively from the S and the N. Oceanic influence and edaphic factors together produce conditions repeating those of Kerry and Connemara, and so providing a suitable home for the Lusitanian element. The same may be said as regards the interesting American group, which, like the last, is so characteristic of Kerry and Galway-Mayo (see **37**). At the same time, we find in Donegal, as a result of geographical and climatic conditions, a high percentage of "Highland" and "Scottish" plants. As regards the former, the absence of alpine ground is compensated by the northern position of the county and the great exposure, so that many mountain species are found at lower elevations here than elsewhere in Ireland, mixing with the southern group. Of the southern plants, most of the species occurring both in Kerry and Galway or Clare extend into Donegal—*Saxifraga spathularis* (**114**), *Bartsia viscosa* (**142**), *Euphorbia hiberna* (**154**), *Trichomanes* (**192**); *Carum verticillatum* (**119**) also, which is absent from the Clare-to-Sligo area, and *Adiantum* (**193**), present there but absent from Kerry. Many of the most characteristic of the Lusitanian group, such as the three heaths of Connemara (*E. mediterranea, E. Mackaii, Dabeocia*), the *Arbutus* of Kerry and Sligo and the *Pinguicula grandiflora* of Kerry, do not extend so far N as Donegal (but see **145, 450**). Of the American group, Donegal possesses *Sisyrinchium angustifolium* (**159**), *Naias flexilis* (**172**), *Eriocaulon* (**173**), all found also in the Kerry and Galway regions. The Lusitanian and American plants in Donegal are, as a group, widely spread, but most of them are local in their range, and confined to small areas. There is none of the profuseness of Kerry or Connemara, and all are limited to the more "Atlantic" (NW) portions of the county. The "Highland" and "Scottish" plants, in which Donegal is rich, reach their Irish maximum in the N of Ireland, diminishing rapidly down the E coast, more slowly down the W coast, and very rapidly towards the centre. Several *Hieracia* are known in Ireland only from Donegal. Apart from the species belonging to the southern and American groups, the rarer Irish plants occurring in Donegal are mostly of northern range in Great Britain or Ireland or both; such are :—

Thalictrum alpinum	Saxifraga aizoides
Cochlearia grœnlandica	Ligusticum scoticum
Trollius europæus	Mertensia maritima
Crambe maritima	Oxyria digyna
Cardamine amara	Salix phylicifolia
Silene acaulis	Cryptogramme crispa.
Vicia lathyroides	

443. The alpine flora is rich despite the absence of lofty mountains. This is due in part to the northern and exposed position of the county, and in part to the abundance of suitable habitats at low elevations. The vegetation of the highest point in the county (Errigal, 2466 ft.), situated in N lat. 55° is (as is usual with Irish mountains), only very slightly alpine (so far as that term implies plants usually confined to high elevations), the summit list (2400–2460 ft.) being as follows :—

Potentilla erecta	Salix herbacea
Galium saxatile	Luzula sylvatica
Calluna vulgaris	Carex rigida
Vaccinium Myrtillus	Deschampsia flexuosa
Armeria maritima	Agrostis tenuis
Euphrasia officinalis	Festuca ovina
Rumex Acetosa	Lycopodium Selago.
Acetosella	

On the other hand, a number of "alpine" species descend in Donegal to sea-level or near it. The local range (in feet) of some of them is appended :—

Draba incana, 0–?250
Saxifraga oppositifolia,
  0–1850
Sedum roseum, 0–1900
Lobelia Dortmanna,
  0–1180

Arctostaphylos Uva-ursi,
  0–1900
Juniperus sibirica, 0–1750
Selaginella selaginoides,
  0–1800
Isoetes lacustris, 0–1500.

At the other end of the scale (alpines with a high lower limit) we have

Thalictrum alpinum, 900–
  2100
Dryas octopetala, 1300
Saussurea alpina, 900–
  2100
Polygonum viviparum,
  1310–1500

Oxyria digyna, 900–1800
Carex rigida, 1000–2400
Polystichum Lonchitis,
  1470–1850
Lycopodium alpinum,
  1400–2219.

The average elevation of plants of Watson's "Highland" type (of which some 30 or more occur, depending on how segregates are treated) is about 1000 ft. The depression of the heights at which alpine plants grow in W Ireland is well known; here, at the N extremity of the island, this is more pronounced than elsewhere. In Kerry, which however has much higher hills, as well as a much greater area of elevated ground, the average is nearly 1000 ft. higher.

The numerous lakes are generally of small size, with peaty margins and a calcifuge flora. In them occur some of the most interesting of the plants mentioned above, and other rarities besides—*Naias flexilis* (**172**) and *Eriocaulon septangulare* (**173**) of the American group, *Subularia, Lobelia Dortmanna* and *Isoetes lacustris* of the northern and mountain group, and so on. The Charophyta (**207**), which have been well worked by Canon Bullock-Webster, include *Chara muscosa, Nitella spanioclema,* and *N. batrachosperma.*

The maritime flora is fairly rich, its outstanding feature being the occurrence of distinctly northern plants, such as *Ligusticum* and *Mertensia.* Among the species widely distributed along the very extensive shore-line are *Cochlearia grœnlandica, Viola Curtisii, Spergularia rupicola, Crithmum, Euphorbia portlandica, Scirpus rufus, S. filiformis.* The more widespread plants of the adventive flora include *Myrrhis, Veronica peregrina, Lamium molucellifolium.*

Donegal, as will be observed, has, on account of the similarity of its climate and soil to those of the other two great western buttresses of Ireland—Kerry and Galway-Mayo—a considerable affinity to these areas as regards its flora : a more obvious affinity, indeed, than it has to the areas immediately adjoining. On the S rise the limestone highlands of Sligo (**422**), Leitrim (**429**), and Fermanagh (**438**), harbouring an interesting flora containing many species unknown in or uncharacteristic of Donegal, though they also are largely of northern type; while to the E, occupying the NE corner of Ireland, the basaltic area of Antrim and Londonderry extends (**455**). There also northern and Scottish plants are characteristic; but the climate is less Atlantic and colder, and the rocks yield a rich alkaline soil, with the result that again we have there a flora which has less affinity with that of Donegal than the proximity of the two areas might lead one to expect.

Literature.—Clear knowledge of the Donegal flora is due to Henry Chichester Hart, himself a Donegal man, who for over thirty years was a student of local botany, and after publishing a series of papers on the subject, combined the results of his work, in 1898, in his "Flora of the County Donegal" (Dublin : Sealy, Bryers, & Walker —out of print). This is a very complete account, and subsequent additions have been few; but some of them are of high interest, such as *Sisyrinchium angustifolium* (**172**), *Naias flexilis* (**173**), *Nitella spanioclema, N. batrachosperma, Chara muscosa* (**206**); such as they are, they will be found listed in the three Supplements to "Irish Topographical Botany." Some additional localities for rarer plants are scattered through the pages of the "Irish Naturalist" and other journals. Canon Bullock-Webster's contributions on the Characeæ (Irish Nat., 1918 1, 1919 7, 1920 1, 1921 55) are especially noteworthy.

THE DONEGAL BAY AREA.

444. As is usual where the hard crystalline rocks of W Ireland are replaced by Carboniferous limestone, Donegal Bay bites deeply into the land, and a band of usually low-lying and occasionally bare and fissured lime-stone, 5 to 7 miles in breadth, fringes the bay northward from Bundoran and then westward to St. John's Point. Much of these rocks belongs to the Calp or Middle Lime-stone, mostly shaly, but about Ballintra and St. John's Point the pure Lower Limestone yields a characteristic calcicole flora. Inland of this zone stretches everywhere the heathery rocky metamorphic country characteristic of Donegal. The limestone fringe is botanically a portion not of this county (to which it belongs politically) but is the extreme extension northward of the great limestone area of central Ireland, which, S of Donegal, in Sligo, abuts for a considerable distance on the Atlantic, as it does in Clare. Usually low in elevation and much covered with drift and peat, in Clare (**346**) and Sligo (**422**) the limestone rises into considerable hills with much exposure of the rock : it is to these regions, rather than that of the Central Plain, that the flora of the Donegal limestones is akin. Hart gives (Fl. Don. 27) a list of the plants which in Donegal flourish best in this area : of these the following are not found elsewhere in the county, having reached here their extreme NW limit in Ireland :—

Aquilegia vulgaris	Calamintha ascendens
Geranium sanguineum	Verbena officinalis
lucidum	Ophrys apifera
Poterium Sanguisorba	Juncus inflexus
Cornus sanguinea	Sesleria cœrulea (plentiful)
Carlina vulgaris	Arum maculatum
Origanum vulgare	

To these, *Euphrasia salisburgensis* (from Brown Hall), here at the N extremity of its Irish and European range (**141**), may now be added.

A few of the above, it should be said, are not always calcicole in Ireland, but all excepting the last in the list display a strong preference for an alkaline soil. The absence of *Arum* from almost the whole of Donegal is interesting. The rarest plant of the area is *Helianthemum Chamæcistus*, which, widespread in Britain, finds here its only Irish station on bare limestone near Ballintra, amid *Juniperus communis, Rubus saxatilis,* and abundance of *Sesleria* (Irish Nat. Journ., **5**, 76).

For the rest this Donegal Bay district is pretty and

attractive. From Ballyshannon and Bundoran one may work the interesting Ben Bulben region to the S (**422**) as from Belleek and Pettigoe one may explore Lower L. Erne (**435**) and the hills to the S of it (**438**). Nearer at hand, the limestone gorge of the Erne between L. Erne and the sea yields some good ground. *Juniperus communis* and *Galium boreale* are here, *Erodium moschatum* near Ballyshannon, *Eleocharis uniglumis* further down. At the N side of the Erne mouth is a curious dune association— *Hypericum pulchrum* dominant, dwarf *Pteris* subdominant, much *Linum catharticum* and *Thymus Serpyllum*, a little *Viola Riviniana* and *V. Curtisii*, *Gentiana campestris* and *G. Amarella*, *Erythræa Cantaurium*, and very little else. *Parnassia palustris* var. *condensata* is abundant in damp hollows. *Rosa rugosa* is an established escape on the stony beach close by, and *Cochlearia grœnlandica* occurs. Near Bundoran is *Viola Curtisii* and at Bundrowes River *Equisetum trachyodon* (**204**), *Carex gracilis*, *Aster* sp. (naturalized), etc.; *Rhynchospora fusca* (**177**) grows near Belleek; on the limestone N of Ballyshannon *Helianthemum Chamæcistus* (already mentioned), *Draba incana*, *Apium Moorei* (**118**), *Hieracium hibernicum* (Moynalt near Laghy, a plant elsewhere in Ireland known only from the Mourne Mountains), *Neottia*, *Juniperus communis*, *J. sibirica*, *Epipactis latifolia*, *Juncus subnodulosus*, *Carex diandra*, *Lastrea Thelypteris*, etc. *Cochlearia anglica* and *Limonium humile* are at Donegal town; the former also at Bally-shannon, where the aliens *Sedum album* and *Mimulus Langsdorffii* grow profusely on rocks at the picturesque falls of the Erne. *Hieracium opsianthum* is by the River Eglish, and *Spergularia rubra* by the shore at Mountcharles. Mr. Hart considered *Ulmus montana* native in limestone glens S of Donegal and on the older acid rocks around L. Eske.

Lough Eske above Donegal town is an attractive place from all points of view; its plants include *Caltha radicans*, *Elatine hexandra*, *Sagina subulata*, *Carum verticillatum* (**119**), *Lobelia*, *Cephalanthera ensifolia*, *Epipactis palustris*, *Festuca sylvatica*, *Hymenophyllum tunbridgense*, *H. peltatum*, *Trichomanes*, *Cystopteris fragilis*, *Lastrea œmula*, *L. montana*, *Asplenium viride*, *Equisetum trachyodon*. *Juniperus communis* is at Barnesmore, and there is an old record for *Subularia* at L. Carban. North of L. Eske, the **Croaghgorm** or **Bluestack Mountains** make a fine very rocky ridge, with several points of over 2000 ft. (Bluestack, 2219) and a few tarns and lakes. *Thalictrum alpinum* grows freely here, accompanied by most of the more widespread Donegal alpines; *Epilobium angustifolium* (in Ireland, as a native, a rather rare mountain plant) occurs also. Southward, near Pettigoe, is Lough Derg, a considerable lake and a famous place of pilgrimage, but to botanical pilgrims it offers but little. *Lobelia* and *Sparganium angustifolium* are almost the only plants in the lake, and on every side are miles of desolate bog-covered moorland. Westward, at Mountcharles, the hybrid sedge *Carex Bœnninghauseniana* has been found, also *Spergularia rubra,* and *Equisetum pratense.* Further W, the limestone makes its final appearance in and about a long narrow rib which projects far into the ocean (St. John's Point). This last refuge of the calcicole flora offers some limestone scarps which yield *Geranium sanguineum, Rubus saxatilis, Galium boreale, Sesleria, Asplenium Ruta-muraria,* growing under conditions of intense exposure. In the marshes is *Cladium,* two feet high and scarcely flowering, and on maritime rocks *Cochlearia grœnlandica.*

There are in this area plenty of places where the visitor may stay. Bundoran is a well-known watering-place; Ballyshannon is a good business town; and at Belleek, Pettigoe, Donegal, Killybegs, etc., hotels will be found. At Donegal town, the ruined abbey is interesting as the place where the famous ''Annals of the Four Masters,'' a remarkable chronology of Irish history from the earliest times, was compiled in 1632–6. Near by, the Jacobean castle of the O'Donels stands by the river.

H. C. HART: ''Report on the Flora of South-west Donegal.'' Proc. Roy. Irish Acad. 2nd ser., 4 (Science), 443–469. 1885. ''Further Report on the Flora of Southern Donegal.'' *Ibid.*, 568–579. 1886. ''Flora of the Croaghgorm Range, Co. Donegal.'' Journ. Bot. 1882, 198–200, also ''Flora of Donegal,'' 1898.

THE MALINMORE PROMONTORY.

445. West of the area last dealt with, a great mountain promontory formed of gneiss and quartzite runs out into the Atlantic. At its southern base is the little port of Killybegs, at its northern, Ardara and Glenties. Near its extremity, on the S side, Slieve League impends over the ocean.

Slieve League.—This imposing hill rises from the inland side to a knife-edge, which drops 1900 ft. into the Atlantic in a magnificent precipice, one of the finest things of its kind in Ireland. The northern face, which falls with equal steepness down towards the little Lough Agh, is the home of a remarkable assemblage of alpine plants. Here on the cliff or close by are

Thalictrum alpinum
Dryas octopetala
Sedum roseum
Saxifraga stellaris
    aizoides
    oppositifolia
Saussurea alpina
Hieracium anglicum
    iricum
Arctostaphylos Uva-ursi
Vaccinium Vitis-idæa
Polygonum viviparum
Oxyria digyna
Salix herbacea
Carex rigida
Juniperus sibirica
Polystichum Lonchitis
Asplenium viride
Lycopodium alpinum
Selaginella selaginoides
Isoetes lacustris.

Most of these occur well up on the cliff, their average height above the sea being 1325 ft. Nowhere else in Ireland is so large an assemblage of plants of Watson's Highland type to be found in so small an area. Slieve League, in addition to its alpine flora, yields some plants of interest :— *Hieracium anglicum* var. *cerinthiforme*, f. *Hartii*, which has here its only known station; *H. scoticum* and *H. sticto-phyllum* are by the Carrick River near by; the dwarf form of *Ophioglossum* previously mentioned (var. *polyphyllum*, 200) grows on Carrigan Head, and *Adiantum* in several places by the shore (very rare now and needing preservation).

About **Killybegs**, E of Slieve League, *Limonium humile* (**132**) and *Cochlearia anglica* grow, and *Sisyrinchium angustifolium* (**159**) has been recently found in several places : *Elymus* and *Althæa* (at Dunkineely) and *Hymeno-phyllum tunbridgense* occur. **Ardara** and its neighbour-hood yield *Draba incana, Sagina subulata, Callitriche autumnalis, Hieracium opsianthum, Eriophorum latifolium*; and *Cryptogramme* has one of its few and scattered stations on Alt Mountain close by. *Hieracium scoticum* has the second of its two Irish stations near Glenties, where also *H. orimeles* and *Festuca sylvatica* occur.

Further N, by the Barra River near Doochary Bridge,

Utricularia ochroleuca and *Carex aquatilis* (**181**) are found. Near **Narin** and Portnoo are the two Junipers and *Eriophorum latifolium*, and *Atropa* is naturalized on wild ground; small lakes are full of *Eriocaulon, Lobelia, Isoetes lacustris*; *Equisetum litorale* is in wet places. *Cochlearia grœnlandica* is common along the coast. The W coast from Slieve League to Slieveatooey (1515 ft.), which, like the former, drops into the Atlantic in a grand precipice but offers none of Slieve League's botanical attractions, is high and exceedingly wild. At the extremity of the promontory, at Glencolumkille, are the remains of an ancient ecclesiastical settlement, interesting as being associated with Saint Columbkille, the founder of Iona in 563, himself a Donegal man. To the S, opposite the islet of Rathlin O'Beirne, *Limonium binervosum* grows.

The visitor will find accommodation at Killybegs, Ardara, Narin, Portnoo, Glenties, etc.; the first and the last of these places are rail-heads, connecting with the broad-gauge system at Strabane.

H. C. HART: ''Report of the Botany of South-west Donegal.'' Proc. Roy. Irish Acad., 2nd ser., 4 (Science), 443–469. 1885. Also ''Flora of Donegal,'' 1898.

THE DUNGLOE-GWEEDORE-DUNFANAGHY AREA.

446. This region, lying beyond (N of) the great NE–SW trough which forms Glenveagh and Gweebarra and extends almost from sea to sea, is the most remote part of Donegal, reached by a series of long zig-zags of road, or a lengthy and tortuous railway journey. It contains much of the finest scenery in the county, as well as the highest point (**Errigal**, 2466 ft.), and also the two largest islands. The coastal region is broad, mostly low, very wild and barren, with an extremely dissected shore-line; the inland part presents high hills and deep valleys. Errigal itself, a beautiful quartzite cone, has very little botanical interest; its summit flora has been given already (**443**); an old record for *Epilobium angustifolium* needs confirmation. The other highest points, Slieve Snacht West (2240 ft.), the great ridge of Muckish (2197 ft.), and Dooish (2147 ft.) yield the alpine plants which might be expected, but not nearly in so great variety as Slieve League. **Muckish** however supplies a habitat for *Dryas* (**106**), and a very rare hybrid willow, *Salix Moorei* (*S. herbacea* × *phylicifolia*), known

PLATE 37.

HORN HEAD, CO. DONEGAL, 600 feet high.
The dark band is a sill of dolerite in the quartzite rocks.

R. Welch, Photo.

446.

PLATE 38.

BENEVENAGH, CO. LONDONDERRY,

1235 feet high. Basalt cliffs facing north, subtended by land-slides, with many alpine plants.

elsewhere only in Sweden (Journ. Bot., 1913, Suppl., 83–4). *Saxifraga spathularis* (**114**), absent from the Slieve League area, is widely spread here on the hills, at all elevations. Among the rarer *Hieracia* are *H. rubicundiforme, H. argenteum, H. opsianthum.*

An interesting and highly picturesque region is the chain of lakes which runs into the hills from **Gweedore,** by Dunlewy, terminating in the Poisoned Glen. Here, by the lowest lake (L. Nacung), grow *Elatine hexandra* and a curious large form of *Erica Tetralix* approaching *E. Mackaii,* and needing study; *Sisyrinchium angustifolium* (**159**) has recently been found at Dunlewy; *Cochlearia alpina, Trichomanes* (**192**), and *Cystopteris fragilis* are recorded from the **Poisoned Glen,** where also there is an old unverified record for *Euphorbia hiberna* (**154**). **Glenveagh** is a very fine straight V-shaped valley passing between high hills, its floor occupied by a long narrow lake, and its sides by a good deal of native woodland; the two species of *Hymenophyllum* grow here in great luxuriance—of *H. tunbridgense* a sheet of 8 ft. × 3 ft. has been measured—and *Juniperus sibirica* occurs, with very old *J. communis* and *Lastrea æmula.* The lower grounds nearer the ocean, where *Carex lasiocarpa* is frequent, form the Donegal headquarters of *Eriocaulon* (**173**), particularly that peculiar area N of Dungloe called **The Rosses,** in which lakelets are nearly as abundant as in the region W of Roundstone in Connemara. *Draba incana* is on sand-hills here, and at Kincashla *Arabis Brownii,* as are *Elymus* and *Juncus subnodulosus* further N about **Bunbeg,** where *Sedum roseum* occurs as a shingle plant on the islands, and *Cochlearia grœnlandica* (**87**) is on sea-rocks. *Thalictrum dunense* grows at Falcarragh, and *Althœa* is naturalized on an islet there. SW of Bunbeg, *Naias flexilis* (**172**) has been lately found, growing sparingly in L. Mullaghderg and abundantly in L. Ibby close by (Irish Nat., 1920, 55). In the N, **Horn Head** (plate 37) rises in a broken range of grand cliffs over 600 ft. high, beloved of sea-birds, but offering little to the botanist except magnificent scenery and *Epipactis palustris.* On the wind-swept summit the small form of *Ophioglossum* already referred to (var. *polyphyllum,* **200**) occurs again, also *Ligusticum* along the adjoining shore. *Calystegia Soldanella* is at Tramore. At Dunfanaghy, the best stopping-place for Horn Head, *Circœa alpina* is recorded.

Ards, near the head of Sheep Haven, yields *Neottia Orobanche rubra* (**143**), and other plants.

447. The following impression of the vegetation of a piece of the exposed coast is of interest:—"The coastal moor of Kincasslagh Head (in the neighbourhood of the Rosses) can for all practical purposes be regarded as lying at sea-level. The central part may rise some 100 feet, but even on a fine day great flecks of spray are driven inland by the west wind, so that the whole air is subject to a moist salty atmosphere . . . The soil consists of broad pockets of rather shallow peat lying among granite boulders or on glacial drift. The massive rocks which form the coastline of the low headland are worn too smooth by the sea to support much vegetation. Where there is any shelter though within reach of the waves, *Spergularia salina* Presl, *Osmunda regalis* L., and *Sedum roseum* Scop. are frequently found in curious proximity, while, lying immediately behind, *Plantago maritima* L. (a small form) *Carex extensa* Good. and *Triglochin maritimum* L. grow close together, the first mentioned associating freely everywhere with inland species; *Salicornia* even mingles with *Eriophorum*. As the ground rises slowly and rocks protrude *Juniperus nana* Willd. becomes plentiful, always procumbent and with every stem wedged into the narrow crevices of the granite. This species, together with a form of *Salix repens* L. (identical with that found by brackish pools in the dunes west of Dunfanaghy), and *Arctostaphylos Uva-ursi* Spr., densely carpet a shallow humus of peat and decomposed granite. *Empetrum nigrum* L., with much *Schœnus nigricans* L., becomes abundant as the peat deepens, and in wetter places *Eriophorum angustifolium* Roth, and *Glyceria fluitans* R. Br. But to return to the drier rocky situations; here we have various *Euphrasiæ*, *Rosa spinosissma* L., *Solidago Virgaurea* L. var. *cambrica* Huds., *Lotus corniculatus* L. var. *crassifolius* Pers. (densely matted and woody), handsome varieties of *Anthyllis Vulneraria* L., occasionally *Viola canina* L. var. *crassifolia*, the universal *Juncus squarrosus* L. and commoner heath-grasses rubbing shoulders, as it were, in close proximity. Salt-marsh species compete with alpines, peat plants with rock plants, yet the whole is a definite local association . . . The chief features governing this association are maritime atmosphere, heavy rains, and a prevalent west wind."[36]

[36] C. G. TRAPNELL in Irish Nat. Journ., 1, 73–4, 1926.

Gweedore is the best centre for this district, and small hotels are scattered along the coast.

H. C. HART: "On the Flora of North-western Donegal." Journ. Bot., 1879, 77, etc.; 1880, 271, etc. G. R. BULLOCK-WEBSTER: "The Characeæ of the Rosses, West Donegal." Irish Nat. 1918, 7–10. See also HART Fl. Donegal.

ARRANMORE.

448. Arranmore or North Arran is worthy of mention chiefly because it is one of the larger Irish islands, having an area of nearly 7 sq. miles. It stands close to the low-lying mainland, but is plateau-like, much of it about 500 ft. high (highest point 750 ft.). All the higher parts are *Calluna* moor over peat. The W coast is precipitous and very exposed; the E and S coast sloping, with a good deal of cultivation. Flora 316 species: outstanding plants are few. *Saxifraga "hirta,"* found by H. C. Hart in 1879 at Polldoo on the W coast, and refound there in 1933, proves to be the same as the *S. Drucei* which has since (Journ. Bot., 1918, 332) been recorded by E. S. Marshall as collected in 1886 near Torneady Point (in the extreme N) by H. M. Wallis, and also recently refound there. *Arctostaphylos Uva-ursi, Empetrum, Juniperus sibirica* are characterisitic of the heath area, where *Pinguicula lusitanica, Hymeno-phyllum peltatum* ¼ inch high, *Lastrea œmula, Selaginella* also occur. On the W cliffs *Sedum roseum* is very abundant down to spray-level. *Sparganium angustifolium* is unusually profuse in the several lakelets, with a good deal of *Lobelia* and *Isoetes lacustris*. Along the coast *Spergularia rupicola* is frequent, also *Cochlearia grœnlandica* (**87**), which hybridizes with *C. officinalis; Elymus* and *Scirpus rufus* are rare. *Ammophila arenaria* flourishes on the edge of cliffs where sand is wholly absent (see also **408**). *Populus tremula* (3 ft. high, increasing by suckers and forming a grove) grows in one place, also a single *Quercus* and a single *Betula pubescens*. Of alien plants, *Matricaria occidentalis* and *Mimulus Langsdorffii,* both doubtless recent arrivals, are the most noteworthy. Over the whole island, grazing is even a worse enemy to vegetation than the extreme exposure and poverty of soil.

H. C. HART: "On the Plants of (North) Aran Island, Co. Donegal." Journ. Bot. 1881, 19–23. R. LL. PRAEGER: "On the Flora of Arranmore, Co. Donegal." Irish Nat. Journ., 4, 50–54, 1932.

TORY ISLAND.

449. A more interesting island than the last, though still poorer botanically, is Tory, a mere shelf of rock three miles long and half-a-mile wide, lying in the full sweep of the Atlantic about 7 miles off shore. Great earthworks at its cliff-bound E end, the remains of an imposing promontory fort, show that it was inhabited in early times, and the stump of a round tower and other medieval relics point to a considerable ecclesiastical settlement a thousand years ago; protected coves on the S side allow a sparse present population to pursue their occupation of fishing. The peaty surface has been mostly stripped off for fuel, and starved agriculture further reduces the flora by killing out native plants and increases it by introducing weeds. The enumerated flora numbers 147 species, many of them occurring very sparingly. The few interesting plants include *Cochlearia grœnlandica*, *Spergularia rupicola*, *Ligusticum*, *Lamium molucellifolium*, *Sparganium angustifolium* and *S. minimum*, *Scirpus filiformis*, *Carex extensa*, *Isoetes lacustris*. In old days *Crambe* also grew there.

R. M. BARRINGTON: "The Plants of Tory Island, Co. Donegal." Journ. Bot. 1879, 263–270. R. LL. PRAEGER: "Notes on the Flora of Tory." Irish Nat. 1910, 189–192.

ROSGUILL, ROSAPENNA, FANET.

450. This region lies between the two long sea-inlets of Sheep Haven and Lough Swilly, and is itself dissected down the middle by the mazy island-studded Mulroy Bay, producing a delightful medley of heathy land and clear ocean water; the combination of hill and lakelet, wood and sea, sand-dune and rock, is very lovely. High hills are absent, Lough Salt Mountain alone exceeding 1500 ft., but Knockalla, 1200 ft., is the richer in plants. There is much rough ground, and a great variety of habitat, as well as a variety of rocks. The only Old Red Sandstone in Donegal occurs about Knockalla, but does not appear to be connected with the presence of several rare plants there. A remarkable band of primitive limestone, lying in the metamorphic strata, crosses the Fanet peninsula (the area E of Mulroy Bay) from E and W, but according to Hart its botanical effect is seen only in the absence of heather and its substitution by gorse.

In this district we have passed beyond the main area of the Lusitanian and American plants, but *Naias flexilis* (**172**) grows in Kindrum Lough (Irish Nat., 1917, 15), and the flora is generally interesting. The pretty Knockalla ridge, crossing Fanet from sea to sea, yields a group of alpine plants—*Vaccinium Vitis-Idœa, Salix herbacea* (both also on L. Salt Mountain), *Arctostaphylos Uva-ursi* (and elsewhere), *Carex rigida.* Other Fanet plants are *Erodium moschatum, Callitriche autumnalis, Hieracium Sommerfeltii, Centunculus, Limonium humile, Orobanche rubra, Euphorbia Paralias, Juniperus communis, Listera cordata* (frequent here as elsewhere in Donegal), *Potamogeton angustifolius, P. nitens* and *P. filiformis* (Kinny Lough), *Festuca sylvatica, Hymenophyllum tunbridgense, Nitella batrachosperma,* and *N. spanioclema* (both Lough Shannagh). A little N, at Ballyvicstocker, *Saxifraga hypnoides* has its only station, which is not above suspicion of introduction. *Pinguicula grandiflora,* introduced from Kerry into bogs at Carrablagh by H. C. Hart, was stated by him in 1898 to have established itself thoroughly. It cannot be found there now, but apparently its absence from Donegal is not due to unsuitability of climate or habitat. See also **145**. *Nuphar pumilum* was introduced by the same botanist to a mountain lake about a mile from Carrablagh (Fl. Don., 115). Its subsequent history is unknown. The well-known and curious "variety" of *Athyrium* called *Frizelliæ* has been found at the Seven Acres, Fanet.

On the Rosguill peninsula W of Mulroy Bay are *Limonium binervosum, Centunculus, Orobanche rubra, Mertensia, Potamogeton angustifolius*: also *Gentiana Amarella* and *Juncus subnodulosus,* elsewhere in Donegal known only on the limestone in the extreme SW. *Catabrosa aquatica* var. *littoralis* is at Tranarossan, *Chara aculeolata* in Rosapenna Lough, *Cuscuta Epithymum* near by, and *Ruppia maritima* near Rosguill House. To the S, at Glen Lough, *Salix phylicifolia* is found.

Further S, about **Rathmelton,** is a more fertile region with much pretty and interesting ground. *Orobanche Hederæ* is frequent; *Linaria repens,* possibly introduced, has here its only local station; *Trollius* (see also **451**), *Galium boreale, Caltha radicans,* and *Stachys officinalis* are at Lough Fern; *Cochlearia anglica* inhabits the muddy inlets at the head of Lough Swilly, as at Rathmelton and

Letterkenny; and near Kilmacrenan *Hieracium rubicundi-forme* grows. Another Kilmacrenan record, that for *Arctostaphylos alpina,* must be disallowed (see **213**); this plants remains an Irish desideratum.

West of Kilmacrenan there is very picturesque country, about L. Gartan and L. Akibbon. Here we are in the Irish headquarters of *Trollius,* which is confined to the basin of the Lennan River save for one station on L. Melvin in Fermanagh. With it about Loughs Gartan, Akibbon, and Fern are *Caltha radicans, Circæa alpina,* and *Potamogeton nitens.*

This district is much frequented by tourists and fishermen, and is well supplied with hotels, that at Rosapenna being widely known.

G. R. BULLOCK-WEBSTER: ''The Characeæ of Fanad, East [West] Donegal.'' Irish Nat. 1917, 1–5. *Ibid*: ''The Characeæ of the Rosses, West Donegal.'' Irish Nat. 1918, 7–10. H. C. HART: Fl. Donegal.

THE RAPHOE AREA.

451. From Letterkenny we pass S into a wide region lying behind the maritime areas which have been just dealt with, and draining inland (SE) into the River Foyle. This is much the most fertile part of Donegal, and correspondingly less attractive to the botanist. To the E is the Foyle, with much *Cochlearia anglica* and no doubt other plants. The remarkable sport *Fieldiæ* of *Athyrium Filix-fœmina* has been found in a lane at Newtowncunningham. To the W, the River Finn flows through upland country to the Foyle, and yields *Hieracium proximum* F. J. Hanb. and a good deal of *Carex aquatilis; Trollius* is by several of the streams, *Hieracium argenteum* near Stranorlar, *H. stictophyllum* by the Mournebeg R. near Ballybofey; and other *Hieracia, Lobelia, Pyrola media* and *P. minor* emphasize the northern character of the flora. The occurrence of *Asplenium viride* on old walls at Convoy only 200 ft. above sea-level is an instance of the tendency of ferns in Ireland to colonize unorthodox stations (see **75**).

The **River Foyle**, draining most of Tyrone and inland Donegal, and forming the county boundary, is broad and tidal from Strabane and Lifford past Londonderry to Culmore, where it enters Lough Foyle. It is formed of several rapid streams which descend from the uplands of

the two counties named to join near Strabane, the most
important being the Finn and the Mourne. The botany
of these tributary streams is but little known; that of the
tidal Foyle is referred to below (**454**).

INISHOWEN.

452. The peninsula of Inishowen, formed by the
convergence of the waters of Lough Swilly and Lough
Foyle, is a well-marked unit geographically but not
botanically. The N part, which includes Malin Head (the
most northern point of Ireland), possesses the characters
of the low exposed coastal region further W, the centre is
high and mountainous (Slieve Snacht, 2019 ft.), and in
the S the low neck which joins it to the mainland (and
which was below the sea in Neolithic times if not later,
making Inishowen an island, as its name implies) enjoys
considerable shelter and is largely farmland. Like the rest
of Donegal, this is a picturesque area, with a wild often
cliff-bound shore in the NW and NE, and the sheltered
waters of the two marine loughs on the other sides. A
railway from Londonderry to Carndonagh *via* Buncrana,
and buses to Moville and elsewhere, provide access to the
district.

The flora is of a mixed character, as might be expected.
Outposts of the Lusitanian element penetrate to this most
eastern portion of Donegal, for Lough Foyle forms the true
boundary between E and W in Ireland. *Euphorbia
hiberna* (**154**) grows along the Dunree River, and there
is an old record for *Saxifraga "umbrosa" "*at Knockglass,
Malin, close to the sea, and not more than 100 feet above
its level, rare and barren"* (DICKIE, Flor. Ulster, 1864),
which it would be very desirable to verify. *Bartsia viscosa*
(**142**), which grows in the Buncrana neighbourhood, is also
a southern plant, with a western range in Ireland similar
to that of the Lusitanians. On the other hand, *Ligusticum*
and *Mertensia,* which occur in a number of spots, are
characteristic northerners, with a range generally arctic.
Alpine plants are very local, and their main concentration
is not on the high Slieve Snacht group, but on Bulbein,
only 1630 ft. in elevation. Here grow *Saxifraga oppositifolia, Sedum roseum, Saussurea, Vaccinium Vitis-Idœa,
Polygonum viviparum, Salix herbacea, Listera cordata,
Carex rigida*; of these, the first, third, and last have here

their only Inishowen station. In addition, *Draba incana* grows on sea sands near Buncrana, *Silene acaulis* at 550 ft. on the cliffs of Dunaff Head, *Vicia lathyroides* near by, *Arctostaphylos Uva-ursi* at a low elevation near Dunree, *Malaxis* between Drumfries and Slieve Snacht, *Raphanus maritimus* in several stations in the N, *Lycopodium alpinum* near Slieve Snacht. A number of *Hieracia* are on Bulbein and Mintiaghs, including *H. rubicundiforme, H. argenteum, H. orimeles.* Other noteworthy plants are *Crambe,* formerly at Clonmany, now apparently gone—it is a shy and uncertain species; *Sagina subulata,* frequent in the N; *Circœa alpina,* Inishowen Head, *Orobanche Hederœ,* near Malin; *Glyceria distans, Festuca sylvatica* and *Elymus* are in the neighbourhood also, and *Equisetum pratense* at Culdaff. *Limonium paradoxum,* a recent split from *binervosum* (Pugsley in Journ. Bot., 1931, 44–7, fig.) grows near Malin Head, the most northerly point of the Irish mainland.

Hotels will be found at Buncrana, Carndonagh, Moville, Greencastle, and of course Londonderry.

H. C. HART: "On the Flora of Inishowen, Co. Donegal." Journ. Bot. 1883, 23, etc.; also HART: Fl. Donegal.

THE COUNTY OF LONDONDERRY.

453. The eastern two-thirds of Londonderry is included in the basaltic plateau of the North-east (**455**); the River Roe, flowing N through Dungiven and Limavady, roughly separates this area from the metamorphic region, the continuation of the Donegal rocks, which occupies western Derry. In the basaltic area the ground slopes E to the valley of the Bann (**461**); in the N and W it is high, a counterpart of the Antrim basaltic scarps, but facing W instead of E.

In the N, the great sandy triangle of Magilligan (**456**), almost blocking the entrance of Lough Foyle, is a notable feature. A basaltic cliff-range, beginning on the coast at Castlerock, continues W across the base of the triangle, and ends in the fine precipice of Benevenagh (1260 ft.). This is the extreme W outpost of the basaltic flora, and the occurrence of many alpine plants in addition (mostly at about 1000 ft.) adds to the interest of this fine hill. Eastern

Londonderry is dealt with below (**456**, **461**); the centre and W are less known.

S. A. STEWART and T. H. CORRY: ''A Flora of the North-east of Ireland, including the Phanerogamia, the Cryptogamia Vascularia and the Muscineæ.'' Belfast: Belfast Nat. Field Club. 1888. S. A. STEWART and R. LL. PRAEGER: ''Supplement to the 'Flora of the North-east of Ireland' of Stewart and Corry.'' Belfast, *ibid.* 1895. S. WEAR and R. LL. PRAEGER: ''A Second Supplement to and Summary of Stewart and Corry's 'Flora of the North-east of Ireland.' '' Belfast, *ibid.*, 1923. G. DICKIE: ''A Flora of Ulster and Botanist's Guide to the North of Ireland.'' 8vo. Belfast, 1864.

LONDONDERRY CITY.

454. The ancient city of Londonderry, standing picturesquely on a hill, still encircled by its ancient walls and half surrounded by the tidal Foyle, makes a convenient base for excursions. It forms the natural entrance to Donegal, since it takes in flank the series of NE and SW folds which determine the direction of the ridges and valleys of that county, and from it railways lead to every part of Donegal. In addition, the shores of Lough Foyle, the extensive sands of Magilligan, and the imposing cliff of Benevenagh with its alpine flora, are easily reached. Few uncommon plants are on record from the immediate vicinity of Londonderry—probably partly on account of incomplete working. Above the city *Cochlearia anglica* (**86**) grows abundantly by the Foyle, in three counties, accompanied often by *Mimulus Langsdorffii*. *Utricularia intermedia* (**144**), *Ceratophyllum demersum* and *Allium vineale* are recorded from the side of the river above the city, and their re-finding is desirable, as all are rare in the north-east. The American *Veronica peregrina* is naturalized and widespread on the Donegal side of the Foyle valley. Culmore on the Donegal bank is an old station for *Corydalis claviculata*; *Glyceria distans* is by the river also, with *Scirpus filiformis* and *S. rufus*, and *Ruppia maritima* at Carrickhugh. At Enagh Lough, below the city slightly above sea-level, *Elatine hexandra*, *Isoetes lacustris*, *Salix phylicifolia*, *Nitella translucens* have been found.

There is diversified hilly ground formed of metamorphic rocks in the area between Londonderry and the Sperrin Mountains (**439**), with some fine glens containing such plants as *Circæa alpina*, *Pyrola media*, *P. minor*, *Pinguicula* *lusitanica, Festuca sylvatica, Lastrea œmula, L. montana, Equisetum trachyodon.*

THE BASALTIC PLATEAU.

455. When we cross the valley of the Roe in the N or that of the Lagan in the S, we enter at once the area of Eocene basalts which occupies the E part of Londonderry and almost the whole of Antrim. These rocks were poured out as level sheets of lava covering a wide area. A collapse of the centre of the plateau along a N-S line, due no doubt to the extrusion of so much material, resulted in the formation of Lough Neagh (**463**) and of the wide valley of the Lower Bann (**461**) which extends from that lake to the northern ocean. On the E and N edges the plateau has been exposed to the onslaughts of the sea. At the present time, the basaltic area is scarped from Lough Foyle right round to Belfast, often with bold cliffs, the watershed being mostly close to the coast, with gentle slopes on the inland side, where the high moory hills drop down towards the fertile valley of the Bann. The basalts contain lime, and they weather into a deep, rich, clayey soil, so that both the scarps and the meadow-land provide habitats differing from those prevailing in the adjoining metamorphic and Silurian regions. As a result, there is a marked change of flora. The northern position of the area, and the fact that it lies nearest to Scotland and is the natural point of arrival for plant migrants from that country, tend to accentuate the difference when the Silurian area to the southward is compared. The difference with Donegal is not so striking, as it is in these combined areas that the northern and Scottish elements attain their maximum in Ireland.

It will be seen that the following short list of plants found in Antrim but not on the Silurian or granite rocks of Down is distinctly northern and Scottish in character (even though Down has mountains 1000 ft. higher than those of the basaltic plateau) :—

	Draba incana	Dryas octopetala
	Arenaria verna	Saxifraga Hirculus
	Sagina subulata	aizoides
A	Geranium pratense	hypnoides
A	sylvaticum	Parnassia palustris
	Vicia Orobus	Carum verticillatum

Galium boreale

Arctostaphylos Uva-
 ursi

B Pyrola secunda

A Melampyrum
 sylvaticum

A Salix phylicifolia

Taxus baccata

Malaxis paludosa

A Carex pauciflora
 aquatilis

A magellanica

B Equisetum pratense
 trachyodon.

Of these, those marked *A* are not found elsewhere in Ireland; those marked *B* are confined to N Ulster; and the remainder are mostly plants which range up the west of Ireland, generally on limestone, and which in Great Britain are mostly mountain species. The plants of the basaltic plateau and its picturesque scarps are treated under several headings below (**456–470**).

Literature : see **453**.

BENEVENAGH AND MAGILLIGAN.

456. A high cliff or scarp runs W and then S from Castlerock to Benbradagh near Dungiven, and some of the characteristic plants of the basaltic area (see **455**) are widespread here—*Arenaria verna, Epilobium angustifolium, Orobanche rubra.* Above Magilligan station *Meconopsis* grows on the cliffs with two very local trees, *Sorbus rupicola* and *Taxus.*

Benevenagh itself is a very beautiful hill with a lofty precipice subtended by extensive land-slides (plate 38). Here are two alpine plants unknown elsewhere in NE Ireland—*Silene acaulis* and *Saxifraga oppositifolia*; the former in profusion, with flowers of every shade from deep purple to white. Other plants are *Polygala vulgaris* var. *Ballii* (= *grandiflora*) (**93**) elsewhere known only from Ben Bulben in Sligo (**422**), *Cochlearia alpina, Draba incana, Dryas, Saxifraga sponhemica, Galium sylvestre, Hieracium pachyphylloides, H. cordigerum, Salix herbacea, Juniperus sibirica.* Several plants, normally maritime and lowland in Ireland, grow above the alpines at 1000–1100 ft.—*Cerastium semidecandrum, Erodium cicutarium, Plantago Coronopus.*

The extensive flat of **Magilligan** offers wide areas of both dry and damp sand, and of swampy ground. On the sands are *Viola Curtisii, Vicia lathyroides, Galium boreale, Hypochœris glabra* (extremely rare in Ireland), *Erythrœa* *littoralis* (only Irish station). From the cliffs, *Orobanche rubra* and *Draba incana* come down and join these. *Veronica peregrina* is a colonist in fields, with much *Lamium molucellifolium*. *Erica stricta* (**212**), escaped from a garden, has naturalized itself in two places on dunes near "The Umbra." In the wetter places are *Caltha radicans, Epipactis palustris, Lastrea Thelypteris* (both very rare in the NE), *Lemna gibba, Eleocharis uniglumis, Equisetum variegatum* (no doubt the var. *arenarium,* not seen recently). Towards their E end the mossy dunes are densely covered over many acres with *Rosa spinosissima* growing as a stoloniferous herbaceous plant, with annual unbranched stems a foot high, and never flowering, as the stems are killed each year by winter gales. Among it is much *Rubus saxatilis, Fragaria,* and *Primula vulgaris.* The damp flats among the dunes are full of *Parnassia palustris* var. *condensata,* growing 2 to 4 inches high. In autumn, when the *Rosa* turns all crimson and purple, and the *Parnassia* is still in flower, the colour effect is very striking. *Atriplex maritima* is at the E end of the long sandy beach at Downhill, where *Ligusticum* grows on rocks, also *Cochlearia grœnlandica, Scilla verna* and the introduced *Peucedanum Ostruthium.* Westward, *Zostera nana* is found on the L. Foyle shore near the mouth of the River Roe, and *Cochlearia anglica* is in the tidal portion of that river.

The western scarp of the basalts, culminating in Benbradagh (1536 ft.) and Carntogher (1521 ft.) is best worked from **Dungiven** in the Roe valley : there is a wide area of hilly ground here of which the botany is little known. On the hills are *Calamagrostis epigejos* (Formoyle), *Saxifraga hypnoides, Pyrola minor, Cryptogramme* (Clontygeragh, not seen recently). In the Roe valley near Dungiven grow *Hieracium pachyphylloides, Pyrola media, Salix Andersoniana, Hymenophyllum peltatum, Cystopteris fragilis, Equisetum trachyodon* (Ballyharrigan glen), *Lycopodium clavatum,* and lower down, about Limavady, *Circœa alpina* and *Lastrea œmula,* and the introduced *Anchusa sempervirens.* From Feeny, *Carum verticillatum* (**119**) is recently recorded.

H. C. HART : "Notes on the Plants of some of the Mountain Ranges of Ireland." Proc. Roy. Irish Acad., 2nd ser., **4,** 244–251. 1884. R. LL. PRAEGER : "Magilligan Plants." Irish Nat. 1917, 167–170. See also **453**.

THE COUNTY OF ANTRIM.

457. Three-quarters of the area occupied by the basaltic plateau of the NE, referred to above, is included in the county of Antrim. Only in the NW does the boundary of Antrim differ save to a minor degree from that of the basalt : there the volcanic rocks extend to cover the E portion of Londonderry, terminating in the fine hills of Benevenagh (**456**) and Benbradagh. The basaltic plateau is best considered as a single region (**455**), following which a more detailed treatment (**456–470**) is due to the different parts of this highly interesting and characteristic district.

PORTRUSH AND THE GIANT'S CAUSEWAY.

458. Portrush, one of the most popular watering-places in Ireland, stands on a basaltic promontory, with a rocky coast extending W, and to the E extensive dunes which give way to arched cliffs of white Chalk, which further on are replaced by basalt. Beyond these, when the sandy mouth of the River Bush is passed, the coast rises into the fine cliffs which overlook the Giant's Causeway and extend far beyond it. On the rocky coast both E and W of Portrush, *Sagina subulata, Spergularia rupicola, Scilla verna, Ligusticum* are characteristic. The sands at Portrush yield *Thalictrum dunense, Viola Curtisii, Vicia lathyroides, Euphorbia portlandica.* Several of these, with *E. Paralias* and *Atriplex maritima,* re-appear at Bushfoot, where the River Bush enters the ocean among sand-dunes below the falls at the little town of Bushmills. East of Portrush, where the great ruin of Dunluce Castle stands on its rock, we enter the region of *Geranium pratense,* in Ireland confined to the coast from this spot to the famous swinging bridge of Carrick-a-rede, 10 miles to the E. Inside this area it is often abundant, and is called locally "Flower of Dunluce." *G. sanguineum,* also found near Portrush, is not indigenous there. *Spergularia rubra,* very local in Ireland, is also a Dunluce plant.

The **Giant's Causeway** is a horizontal bed of basalt now projecting into the sea, which, owing to very slow cooling, has split up into a series of vertical jointed polygonal columns. It stands directly at the base of a grand cliff-wall cut out of the volcanic rocks. This neighbourhood yields *Cochlearia grœnlandica* (Runkerry), *Callitriche autumnalis* (Bush R.), *Ligusticum* (plate 39), *Lastrea œmula*; *Orobanche rubra* and *Arenaria verna* on the cliffs, *Mertensia* on the stony beach below them, *Carum verticillatum* (**119**) on the windy heath on their summit. Whitepark Bay, E of the Causeway headlands, is a lovely place, with half a mile of sand backed by Chalk cliffs. *Atriplex maritima* is here with *Orobanche rubra* and a profusion of *Geranium pratense*, and *Parnassia* is plentiful in its dwarf form *condensata*.

Eastward the cliffs rise again, and at Carrick-a-rede they are decked with *Epilobium angustifolium, Orobanche rubra, Juniperus* sp., *Lavatera* (**97**)—the last indigenous here, and on Sheep Island near by. At Kinbane, a promontory of Chalk which projects into the sea, *Hieracium flocculosum* has one of its few Irish stations.

There is plenty of hotel accommodation in the district— at Portrush, Portballantrae, the Causeway, Ballintoy, and Ballycastle.

Literature : see **453**.

RATHLIN ISLAND.

459. Rathlin is an attractive island—a tilted L-shaped high shelf of basalt, with a mostly precipitous coast-line, a heathy surface, several lakes, and the white Chalk showing under the dark volcanic rock on the side which faces Ballycastle. On the lofty cliffs towards the W innumerable sea-birds breed—gulls of several kinds, Guillemots, Razorbills, Puffins, Manx Shearwaters, Stormy Petrels. There is a fair amount of cultivation, the remainder, much of it once bogland, being wet rocky heath on a thin peaty soil, decked with amazing quantities of *Orchis maculata*.

The coast-line, which becomes low in the S, yields *Raphanus maritimus*, plenty of *Sagina subulata* and *Spergularia rupicola*, *Lavatera* (native on high sea-rocks), great abundance of *Asplenium marinum* and *Scilla verna, Ligusticum,* and luxuriant *Sedum roseum*. *Crambe* seems now extinct. Among the rocks are *Saxifraga hypnoides, Hieracium iricum, Orobanche rubra, Juniperus* sp., and *Populus tremula*—now the only native tree, the next largest native arborescent plants being *Salix aurita, S. cinerea, Prunus spinosa*, and Roses. On the heaths and boggy spots are *Hypericum elodes, Centunculus, Pinguicula lusitanica,*

LIGUSTICUM SCOTICUM,
Near Portstewart, Co. Londonderry.

W. J. Rankin, Photo.

PLATE 40.

FAIR HEAD, Co. ANTRIM, 636 feet.
Cliffs of columnar dolerite, with a scree of huge blocks of same.

Malaxis, Scirpus filiformis, Botrychium, Selaginella. Caltha palustris flourishes exceedingly, with flowers up to 3 inches across. In the lakes grow *Potamogeton prælongus, Ceratophyllum demersum, Nitella translucens.*

The flora, which numbers nearly 350 species of Phanerogams and Vascular Cryptogams, contains a smaller percentage of non-indigenous species than the mainland. Among the more interesting aliens are *Saponaria, Sempervivum tectorum, Conium, Myrrhis, Carum Carvi, Smyrnium, Inula Helenium, Tanacetum, Anchusa sempervirens*—almost all medicinal plants, or formerly used as such.

Motor-boats run from Ballycastle (6 miles). Previous to the advent of these the passage was very uncertain, on account of the great tide which sweeps past the island. Accommodation may be obtained.

S. A. STEWART: ''Report on the Botany of the Island of Rathlin, County of Antrim.'' Proc. Roy. Irish Acad., ser. 2 (Science), **4**, 82–104. 1884. R. LL. PRAEGER: ''The Flora of Rathlin Island.'' Irish Nat. 1893, 53. *Ibid.*: ''Three days on Rathlin Island, with notes on the Flora and Fauna.'' Proc. Belfast Nat. Field Club, 1889–90, 218–222. C. D. CHASE and others: ''Plants new to Rathlin Island.'' Irish Nat. Journ., **1**, 10. See also **453**.

BALLYCASTLE AND FAIR HEAD.

460. The extreme NE corner of Ireland, from Ballycastle to Cushendun, is good ground for the botanist. The uniform sheets of basalt give way here to a jumble of rocks mainly metamorphic (quartzite, etc.), with Devonian conglomerates in the S, Chalk and Keuper at Murlough Bay, a sill of intrusive basalt forming the magnificent columnar cliffs of Fair Head, and coal-bearing Carboniferous shales and sandstones near Ballycastle.

The coast is very bold and varied, and rises into heathery moorlands of 1000 ft. or more in elevation. **Fair Head** (plate 40), the finest feature of the Antrim coast, with its huge columns subtended by a talus of vast fallen blocks, and little lakes on its inland slope, is the home of *Meconopsis, Arenaria verna, Spergularia rupicola, Sagina subulata, Galium sylvestre, Pyrola media, Actostaphylos Uva-ursi* (not seen recently), *Orobanche rubra* (**143**), *Malaxis, Carex limosa, Cryptogramme, Hymenophyllum* *peltatum, Lastrea æmula, Isoetes lacustris. Cochlearia grænlandica* (**87**) occurs along the coast towards Ballycastle.

In the deep sheltered recess of **Murlough Bay**, where the sheets of Primroses (*P. vulgaris*) and Wild Hyacinths (*Scilla non-scripta*) mixed with *Lastrea æmula,* in rocky woods, are one of the loveliest sights in Ireland, *Draba incana* and *Saxifraga aizoides* grow on the steep slopes of Chalk and New Red Sandstone, with *Saxifraga hypnoides* and *Epilobium angustifolium*; all are found again at Torr Head to the S. On the coast from Torr past Runabay Head to Cushendun, and for some miles inland, ancient schists and other rocks of Dalradian age occupy the ground, but have no marked effect on the flora, save the loss of the characteristic plants of the basaltic scarps.

This lovely area can be worked from Cushendall, Cushendun, or **Ballycastle**, which lies in a bay to the E. At that watering-place *Thalictrum dunense, Erodium moschatum, E. maritimum, Vicia lathyroides, and Viola Curtisii* are yielding or have yielded to the exigencies of golf and other human pursuits. On the other hand, *Geranium pratense* and *Elymus*, recently introduced, have established themselves and are increasing. The glens behind Ballycastle will repay further investigation. *Carum verticillatum* (**119**) occurs at Carnsampson, W of the town. The great dome of Knocklayd (1695 ft.) is one of the several stations in Ireland for the elusive *Phegopteris Dryopteris*; no botanist now living has seen that fern growing in Ireland, though it has been found in five counties. *Cryptogramme*, almost equally rare, but not so elusive, is recorded from the same hill. *Salix Andersoniana* is there, and *Mercurialis perennis*.

Literature: see **453**.

THE VALLEY OF THE LOWER BANN.

461. The Bann, one of the larger Irish rivers, rises in the granite upland of the Mourne Mountains (**479**), and meanders across the Silurian rocks of County Down into Lough Neagh (**463**), emerging at the N end as a considerable stream to flow down the collapsed centre of the basaltic plateau to the Atlantic. The sheets of basalt which form the Antrim hills dip gently down into the valley of the Maine, which flows S into L. Neagh; a secondary ridge, flat-topped and boggy, succeeds on the W, separating for some distance the Maine from the Bann, which flows parallel to it, but N, towards the ocean. Beyond the Bann, the Derry portion of the broken-backed plateau (**12**) rises again towards the W. The broad trough thus formed is generally rich and fertile, but the upper part of the Maine valley is poor and boggy. In this latter region, about Killagan and Kellswater, *Andromeda, Carex limosa,* etc., are found; also *Carex aquatilis,* which fringes the river at intervals right down to Lough Neagh.

In the neighbourhood of Ballymena, a busy manufacturing town, are *Stachys officinalis* and *Equisetum litorale* (**201**), with *Crepis biennis* increasing in meadows and *Mercurialis perennis* at Galgorm. The remarkable hill of **Slemish** (1437 ft.), a volcanic neck, which rises conspicuously 8 miles E of Ballymena, harbours an outlying fragment of the flora of the basaltic scarps (**455, 467**)— *Sedum roseum, Hieracium iricum, Vaccinium Vitis-Idæa, Pyrola minor, Lastrea montana, Equisetum pratense,* etc. East of the Maine, in the Sixmilewater, *Ranunculus fluitans* (**80**) has its only Irish station, extending for some miles, and *Cephalanthera ensifolia* formerly grew at Muckamore (as also further S at Glenavy). *Potamogeton nitens* is in this stream; and along its banks *Allium oleraceum,* very rare in Ireland, grows freely, but is probably not native. From further up the Sixmilewater, near Ballyclare, *Lastrea Thelypteris* was recorded seventy years ago, but has not been seen since.

The ridge between the Bann and Maine runs N from Cranfield Point on Lough Neagh almost to Ballymoney, and in its higher part, between **Rasharkin** and **Dunloy**, flat wet peat-bogs extend, with occasional low scarps of basaltic rock. This is interesting ground. In the bogs are *Saxifraga Hirculus* (**116**), *Drosera anglica, Utricularia intermedia* (**144**), *Pinguicula lusitanica* (**146**), *Malaxis, Carex limosa. Eriophorum latifolium,* formerly found, has not been seen lately. The scarps yield *Arctostaphylos Uva-ursi, Pyrola media, P. minor.* There is an old record of *Poterium officinale* from Rasharkin.

462. As for the **Bann** itself and its valley, the rarest plant is *Spiranthes stricta* (**156**), which, widespread around L. Neagh, descends the river to Coleraine. The aquatic flora includes *Potamogeton panormitanus* (near Toome), also *Callitriche autumnalis, Apium Moorei, Cicuta, Potamogeton angustifolius*—all not infrequent. From the vicinity of **Coleraine**, where the river becomes tidal, several good plants are recorded which have not been seen for many years—*Subularia, Elatine hexandra, Ceratophyllum demersum, Pilularia.* Below Coleraine *Carum verticillatum* comes in in abundance, and *Scrophularia alata*, very rare in Ireland, occurs sparingly; the tidal banks yield *Cochlearia anglica, Aster* sp. (abundant), *Mimulus Langsdorffii, Carex gracilis, Scirpus rufus,* etc. On the extensive dunes which guard the river-mouth from Castlerock to Portstewart are *Viola Curtisii, Erodium maritimum* (old record), *Hypochœris glabra, Mertensia, Bartsia viscosa,* with *Sagina subulata, Cochlearia grœnlandica,* and *Ligusticum* on rocks.

On the Londonderry side of the river, as about **Kilrea,** there is rough ground with wooded glens which harbour *Circœa alpina, Carum verticillatum* (Kilrea), *Pyrola media, P. minor, P. secunda* (the last not seen here recently), *Melampyrum sylvaticum, Calamagrostis epigejos* (now missing, but grows a little further W), *Festuca sylvatica, Lastrea œmula.* Further S at Ardpatrick Lough near Stewartstown *Cladium* and *Equisetum variegatum* grow, both very rare locally. Other plants of the district are *Cardamine amara, Spergularia rubra* (by the Moyola River near Draperstown), *Poterium officinale* (Agivey, not seen recently), *Galium erectum, Hieracium strictum* (Garvagh), *Stachys officinalis, Salix laurina* (Castledawson), *Nitella translucens.*

West of Magherafelt, where Slieve Gallion rises to 1623 ft., there is a remarkable jumble of many geological formations, and much tumbled country, little known to the botanist, and needing exploration.

At Antrim (where there is a perfect Round Tower), Toome, Ballymena, Maghera, Kilrea, Garvagh, Cookstown, Ballymoney, Coleraine, etc., the explorer will find accommodation.

In several places along the river, notably under the bogs at Toome, thick beds of white diatomaceous earth or kieselguhr occur, the result of algal growth in former extensions of Lough Neagh. The material is cut and dried like peat, and is eventually ground to be used as an insulator.

Literature : see **453**.

LOUGH NEAGH.

463. Lough Neagh (153 sq. miles, 48 ft. above sea), the largest sheet of fresh water in Ireland and larger than any in Great Britain, is interesting in several respects. It was formed in Eocene times, through a local sinking of the Earth's surface caused by the great outpourings of basalt that now cover Antrim and much of Londonderry. That this sinking was considerable both in amount and in duration is shown by the great depth—1200 ft. or more—of the clays, laid down on the old lake-bottom, which underlie it and fringe the lough save on the N. The present lake is rectangular, almost devoid of islands, and preserving throughout its area a curiously uniform depth of 40 to 50 ft., except for a deep cut (102 ft.) in the NW corner, where the Bann leaves the lake. The water is rather alkaline, with a pH value of 7. Flooding to the extent of several feet, which is a common condition in winter, prevents the formation of acid peat on the marginal lands, which in consequence assume the condition of fen.[37] The shores are low, mostly sandy or stony, and often sloping gently up to an old scarp that tells of a former higher level. Along the S shore, a remarkable subaqueous cliff exists: the lake-bottom slopes gently from the shore till a depth of 3 to 6 ft. is reached, and then drops abruptly to 15 or 16 ft. This has been the subject of some discussion (Irish Nat. 1915, 8, 65), but its significance has still to be explained. The Tyrone shore is the wildest, with swamps, great reed-beds, rough woods and bogs, but it is not the richest for the botanist. The River Bann (**461–2**) enters the lough at its S side through miles of marshy meadows; emerging at the N end at Toome, it expands into the very shallow sandy Lough Beg, whence it flows, with occasional rapids, to meet tidal water at Coleraine, and to enter the sea between sand-dunes near Portstewart, 5 miles further down, after a course of 97 miles from its source. L. Neagh receives several considerable streams besides the Bann—from the SW the Blackwater, which comes down a limestone valley from Monaghan; in the N the Maine, and in the E the Sixmilewater, which drain much of central Antrim;

[37] J. SMALL: ''The Fenlands of Lough Neagh.'' Journ. Ecol. 1931, 383–388.

several streams also come from the W; all combine to make the Lower Bann (**461**) a considerable river.

Close to the lake on the E side is the shallow Portmore Lough, surrounded by swampy meadows where some of the rarer L. Neagh marsh plants still linger. Fed by rather limy springs, the conditions here are those of fen (see M. DUFF: "The Ecology of the Moss Lane region, Lough Neagh." Proc. Roy. Irish Acad., **39**, B, 477–496, pl. X–XIV. 1930).

Before the days of the motor car, much of the L. Neagh shore was rather inaccessible, especially the W side, where the towns—Dungannon, Cookstown, Magherafelt, etc.—lie far back from the lake. Elsewhere Lurgan, Antrim, and Toome form good bases.

464. The flora of L. Neagh offers several points of interest. In the first place, it includes three plants not known elsewhere in Ireland, and all until recent years unknown also in Great Britain: but each has now one station in the sister island. These are *Spiranthes stricta* (**156**) (long combined under the name *S. Romanzoffiana* with *S. gemmipara* of Cork and Kerry) which is spread all round the lake and along the Bann both above and below L. Neagh; in 1930 it was found on Colonsay, its first extra-Hibernian station in Europe (Journ. Bot. 1930, 346). Next comes *Carex fusca* (= *Buxbaumii*), very rare on L. Neagh and apparently now extinct—see **180**; a Scottish station (by a small loch in the Arisaig district of Inverness) was discovered in 1895 (Ann. Scott. Nat. Hist. 1895, 247–9). The third is *Calamagrostis neglecta* var. *Hookeri* (**185**), not uncommon around the lake, where it has long been known: this plant was lately added to the English flora, from a fen in Norfolk (Journ. Bot. 1915, 281, and Irish Nat. 1915, 170).

Then the Lough Neagh shores yield a group of plants which in Ireland at least are exclusively or almost exclusively maritime :—

Spergularia rupicola	Cerastium semidecandrum
Scirpus maritimus	arvense
Tabernæmontani	Viola Curtisii
Carex extensa	Erodium cicutarium
Plantago maritima	Trifolium arvense[38]

[38] Lately (Journ. Bot., 1934, 155) R. W. Butcher has recorded *Thalictrum arenarium* (*T. dunense* of the present book) from L. Neagh, this being the only inland station within his knowledge.

The sandy nature of the ground in some parts (inland sands are very rare in Ireland) may account for the presence of the last five, which in S England often grow on sandy tracts far from the sea. In Ireland, *V. Curtisii* is the only one occurring inland elsewhere save in rare stations open to suspicion; it grows by Lough Erne, and by Castlewellan Lake in Down. The five plants first on the list are more difficult to account for. Formerly it was suggested that they date from a period when submergence permitted the sea to invade the basin of L. Neagh (the lake is only 48 ft. above sea-level). But recent geological work appears to preclude this; Prof. Charlesworth has suggested (Geol. Mag. 1928, 212) that the re-advance of the Scottish ice in the late-Glacial times may have pushed before it the marine waters along the ice-front, and flooded L. Neagh for a while with sea-water : this might account not only for the maritime plants, but for certain relict animals found in the lake itself—the "Fresh-water Herring" or Pollan (*Coregonus pollan*) and the shrimp *Mysis relicta*, which have relationships with the sea.

465. The most interesting of the plants of Lough Neagh is undoubtedly *Spiranthes stricta* (**156,** plate 41), which has a wide range on cut-away bog, stony wet grassy lake-shores, and damp meadows along the Bann from above Portadown (Brackagh bog), all round the lake, and down the river to Coleraine.[39] It flowers at the end of July, but ripe seed has not been observed; its delicious vanilla scent is most attractive. The coloured plate in Irish Nat. Journal (**2,** pl. 1, reproduced in Proc. R. Irish Acad., **39,** B, pl. 1) gives an excellent idea of the plant, save that the pinkish colour of the flowers is erroneous : they are entirely greenish white.

The aquatic and marsh flora of the lake is rich, and was more extensive before drainage operations about 1855 lowered the level of the water by several feet.[40] Owing to this and other causes *Subularia, Elatine hexandra, E. Hydropiper* (**96**), *Sium latifolium, Carex fusca* (**180**), *C. elongata, C. lasiocarpa, Lastrea Thelypteris* appear to be now gone, and other local rarities, such as *Rhamnus catharticus, R. Frangula, Lathyrus palustris* (**103**), *Cicuta, Lobelia, Pilularia,* have shifted their ground, or at least are now

---

[39] See especially W. J. C. TOMLINSON: "The Occurrence of *Spiranthes Romanzoffiana* in Co. Antrim." Irish Nat. 1907, 311–314.
[40] See Irish Nat. 1920, 103–4.

known only from stations different from their former ones.
Among the other plants of the water and shores are :—

Ranunculus scoticus
Cardamine amara
Spergularia rubra
Callitriche autumnalis
Apium Moorei
Cicuta virosa
Galium boreale
Hieracium orarium
Mentha Pulegium
Polygonum laxiflorum
Salix laurina
Hydrocharis Morsus-ranæ
Typha angustifolia
Sagittaria sagittifolia

Butomus umbellatus
Lemna gibba
　　polyrrhiza
Potamogeton fluitans (=
　　　　lucens × natans)
　　angustifolius
　　obtusifolius
　　panormitanus[41]
　　filiformis
Scirpus pauciflorus
Carex Pseudo-Cyperus
　　gracilis
　　acutiformis
Isoetes lacustris

The foreshore is mostly grazed, which much reduces its
botanical decorativeness. Where this is not the case, it
shows in summer brilliant sheets of *Senecio Jacobœa, S.
aquaticus,* and their hybrid, *Mentha aquatica,* with much
M. sativa, Lysimachia vulgaris and *L. nemorum, Juncus
effusus, Lythrum Salicaria, Polygonum amphibium.* In
wetter places *Iris Pseudacorus* and beds of *Phragmites,
Eleocharis palustris, E. acicularis*; and in the water
Littorella and pondweeds, including *P. filiformis.* The
above example is taken from the Tyrone shore.

Especially on the W side there are extensive deposits
of silicious sand, forming in places low dunes. Their flora
is calcifuge and very poor, *Teesdalia* (Washing Bay only),
Filago minima, and *Senecio sylvaticus* being the only
plants other than very common ones.

At the S end of the lake, in N Armagh, there is an
extensive area of low peat-bog, known as The Moyntaghs,
only about 100 ft. above sea-level, still retaining much of
the flora lost to the area to the S and E by the cutting of
the peat for fuel. Here *Drosera anglica, D. longifolia,
Andromeda, Vaccinium Vitis-Idœa, Oxycoccus, Listera
cordata, Malaxis, Rhynchospora alba, Osmunda* find a
refuge (some, it will be observed, at unusually low levels).
Here also a tongue of Carboniferous limestone from the
Central Plain occupies the valley of the Blackwater, which

---

[41] Journ. Bot. 1919, 285.

PLATE 41.

SPIRANTHES STRICTA.

A group from the shores of Lough Neagh.

R. Welch, Photo.

465.

PLATE 42.

GLENARIFF, CO. ANTRIM,
Looking down towards the sea.

R. Welch, Photo.

discharges into the lake : this may account for the presence in Lough Neagh of a few of the plants characteristic of that area (see **240**), such as *Ranunculus circinatus, Sium latifolium, Lemna polyrrhiza, Potamogeton densus, Carex acutiformis,* which in the NE are found only here. *Galium boreale* is quite common on the lake-shores, as is *Lysimachia Nummularia,* and *Circæa alpina* occurs, also *Centunculus, Stachys officinalis, Carex strigosa, Mercurialis perennis.* A series of extraordinary sports of *Ophioglossum vulgatum* was observed at Shane's Castle (Irish Nat. 1898, 112, 3 figs.), where also *Cephalanthera ensifolia* grew formerly. *Subularia,* then unknown in Britain, was discovered by W. Sherard in L. Neagh in 1691, and figured by Plukenet in his "Phytographia," tab. clxxxviii. 1692, from specimens supplied by the finder.

466. A plant to be looked for is *Euphrasia salisburgensis,* whose present known distribution is over the limestones of the west (**141**). It was recorded from the border of L. Neagh by Newbould (see Colgan in Irish Nat. 1897, 106). Another desideratum is *Potamogeton sparganiifolius* (**168**), of which an unsatisfactory record exists (see Irish Nat. 1909, 84).

Some uncommon plants of doubtful standing are at least established here. Such are *Leucojum æstivum* (**161**) and *Sisyrinchium angustifolium* (**159**), near Antrim, both native in some parts of Ireland, but here open to question. *Teesdalia nudicaulis* also, at Washing Bay on the Tyrone shore, is slightly doubtful. An American *Aster* is abundant on shores and islets on the Tyrone side, where *Epilobium angustifolium* in quantity has colonized cut-away bogs. *Althæa, Barbarea intermedia, Lepidium campestre* are among the other aliens.

Literature : see **453**.

THE GLENS OF ANTRIM.

467. Under this head is included a long strip of coastal land, extending from Larne to beyond Cushendun, and also several types of ground—(1) the flattish wet moorland which forms the more elevated parts, from 1000 ft. to the highest point in Antrim (Trostan, 1811 ft.); (2) the basaltic eastern scarps and cliffs, where many of the more interesting plants have their home, about 1000 ft. to sea-level:

(3) the glens themselves, with sloping or cliffy sides, some native scrub, and a good deal of cultivation; (4) the sea-coast, stony, rocky or sandy.   Of the glens, which cut deeply into the high scarp which fronts the North Channel, the finest are **Glenariff** (plate 42), near the pretty village of Cushendall, and **Glendun**, a little further N, at the foot of which Cushendun lies. Glenariff is rather wide and flat-bottomed below, cliff-walled, rising steeply at the head and branching, with waterfalls and wooded ravines. Glendun is longer, with steeply sloping sides.

The district is remarkable for its beauty.   The great variety of rocks, which are frequently exposed in lofty cliffs, leads to alternations of black (basalt), white (Chalk), and red (Trias), which contrast with the green of the grasslands and the brown and purple of the moorlands. The great abundance of certain wild-flowers is also a characteristic.   In spring Primroses (*P. vulgaris*) in masses are everywhere (lingering on till July), with sheets of Wild Hyacinths (*Scilla non-scripta*), much *Orchis mascula,* and by the sea gay carpets of *Armeria, Silene maritima, Lotus corniculatus, Scilla verna* and so on.

The characteristic flora is that of the basaltic scarps and glens.   The following are frequent, many of them extending from Benbradagh or Benevenagh in Londonderry (**456**) right round to Belfast (**470**) :—

Arenaria verna.	Pyrola minor
Saxifraga hypnoides	Melampyrum sylvaticum
sponhemica	Orobanche rubra
Sedum roseum	Ulmus montana
Epilobium angustifolium	Festuca sylvatica
Circæa alpina	Cystopteris fragilis
Galium boreale	Lastrea æmula
Crepis paludosa	montana
Pyrola media	Equisetum pratense

More local are *Meconopsis* (near Garron Point), *Vicia Orobus* (plentiful on rough ground in the Ballygowan area SW of Larne), *Dryas* (Knockdhu), *Sorbus rupicola, Geranium sylvaticum* (distributed over about 20 sq. miles of country about Glenarm, its only Irish station), *Galium sylvestre, Hieracium flocculosum, H. farrense, H. pachyphylloides, H. strictum, Arctostaphylos Uva-ursi, Pyrola secunda, Calamintha ascendens* (once found in Glendun), *Ulmus montana, Salix Andersoniana, S. phylicifolia, Carex strigosa, Juniperus communis, J. sibirica, Hymenophyllum tunbridgense* (Glendun), *H. peltatum, Equisetum trachyodon* (Glenarm and Glenariff).

To descend to the shore-line (which would be often difficult of access owing to cliffs were it not that a raised beach of Neolithic age intervenes, allowing the famous Coast Road to pass between the cliff and the sea from Larne to Cushendall), *Raphanus maritimus, Viola Curtisii, Sagina subulata, Spergularia rupicola* are widespread on sands or rocks according to their wont. More local are *Crithmum, Ligusticum, Mertensia* (all about Garron Point), *Vicia lathyroides* (Cushendun), *Scirpus rufus, Erodium maritimum* (Glenarm, not seen recently).

Poterium officinale, strangely local as an Irish plant, grows in rough meadows by the sea N of Carnlough, where (as in neighbouring places) *Crepis biennis* is abundant and spreading. At Cushendun a unique rose, *R. canina* × *rugosa* (*R. Praegeri* W. Dod) (**110**) is found (close to the church), and *Mimulus moschatus* appears naturalized, as in Armagh and Wicklow. *Myrrhis, Anchusa sempervivens, Lamium molucellifolium* are among the more interesting local aliens. The native *Circœa alpina* is a frequent—and troublesome—garden weed in northern Ireland.

468. Garron Plateau.—On the high moorland, the best ground is the great flat area which lies behind Garron Point at about 1000–1200 ft. elevation—a treeless, roadless, and houseless stretch, tenanted mainly by breeding birds— Curlew, Dunlin, Redshank, Golden Plover, and colonies of Lesser Black-backed and Blackheaded Gulls. Here two sedges, *Carex pauciflora* (**178**) and *C. magellanica* have their only Irish station, both growing over a considerable area with *C. limosa; C. lasiocarpa* (very rare locally), more rarely *Saxifraga Hirculus* (**116**),[42] and abundance of *Drosera anglica,* also occur.

The vegetation of this northern mountain-moor is interesting, in comparison with that of bogland in other parts of the country (**52**). In a wet place where *Saxifraga Hirculus* grows, and which suggests a grown-over bog-pool,

[42] See Irish Nat., 1920, 96–99.

Hypnum revolvens occupies 75–80 per cent. of the ground with much

Saxifraga Hirculus	Juncus sylvaticus
Bellis perennis	Carex limosa
Menyanthes trifoliata	diversicolor
Prunella vulgaris	flava ;

a good deal of

Ranunculus Flammula	Cerastium vulgatum
Cardamine pratensis	Scirpus setaceus

and a little

Taraxacum vulgare	Carex panicea
Pinguicula vulgaris	Goodenowii

Near by, beyond the *Hypnum-Saxifraga* area, are

Linum catharticum	Pedicularis sylvatica
Sagina procumbens	Narthecium ossifragum
Drosera rotundifolia	Potamogeton polygoni-
anglica	folius
Epilobium palustre	Triglochin palustre
Galium saxatile	Schœnus nigricans
Erica Tetralix	Carex pulicaris
Oxycoccus quadripetala	inflata
Pinguicula lusitanica	Selaginella selaginoides

Where the ground is less wet, the vegetation consists of—

Calluna vulgaris	Potentilla erecta
(dominant)	Agrostis tenuis
Molinia cœrulea	Poa(?)pratensis
(sub-dominant)	Carex pulicaris
Scirpus cæspitosus	echinata
(sub-dominant)	panicea
Viola palustris	Goodenowii
Polygala vulgaris	inflata[43]

---

[43] See A. W. STELFOX and S. WEAR in Irish Nat. 1914, 229–250. Also R. LL. PRAEGER : ''Notes on Antrim Plants.'' Irish Nat. 1920, 95–105.

As to other less common species, *Pinguicula lusitanica* is widespread on the high grounds (as also low down), and *Malaxis* has been found on Slievenanee, where *Salix herbacea* grows on rocks.   *Lycopodium alpinum* is frequent on the drier parts of the plateau, *L. clavatum* rare; *Cryptogramme* occurs as half a dozen small clumps, widely separated.

Hotels will be found at Larne, Glenarm, Carnlough, Garron Point, Cushendall, Cushendun.

Literature : see **453**.

LARNE LOUGH AND ISLANDMAGEE.

469. Just as Strangford Lough lies parallel to the coast, opening through a narrow strait and cut off from the Irish Sea by a peninsula known as the Ards (**476**), so Larne Lough lies similarly, opening by a narrow entrance at Larne, and cut off from the North Channel by the peninsula of Islandmagee.   But Larne L. is the smaller and shallower of the two.   The salt-marshes and mud-banks yield *Cochlearia anglica* (**86**), *Limonium humile* (**133**), *Ruppia maritima, Scirpus rufus.   Raphanus maritimus* is on gravelly shores lower down, and at Larne and elsewhere.   *Ophrys apifera* grows on Chalk spoil-banks at Magheramorne, and in the glen at Redhall *Phyllitis Scolopendrium* and *Polypodium vulgare* var. *semilacerum* attain extraordinary luxuriance.   *Melilotus arvensis* is naturalized, and *Mercurialis perennis* looks native above Glynn.   The outer shore of the highly-tilled Islandmagee is precipitous, with high cliffs of basalt (The Gobbins) descending into the water.   *Ligusticum* grows here, and there are old records of *Mertensia* and *Atriplex maritima.   Picris echioides* and *Geranium phæum* are among the rarer Islandmagee aliens. The neighbourhood of Larne Lough is interesting and very pleasing, and forms an introduction to the splendid scenery of the Antrim coast.   Good accommodation at Whitehead and at Larne.

THE ENVIRONS OF BELFAST.

470. Belfast makes a good centre, for not only can any part of Down or Antrim or Lough Neagh be visited on a day's excursion by motor or rail, but the immediate vicinity yields a number of plants of interest—now reduced to a slight extent by the encroachments of the city. The hills which impend over Belfast from W to N (Divis rises to 1567 ft. , and the 1000-ft. contour is reached 3 miles from the City Hall), with a fine precipitous scarp on the Cave Hill, support many of the plants characteristic of the basaltic plateau : for instance, *Saxifraga hypnoides, S. sponhemica, Epilobium angustifolium, Circœa alpina, Pyrola media, P. minor, Melampyrum sylvaticum, Orobanche rubra* (**143**), *Juniperus* sp., *Carex strigosa, Festuca sylvatica, Hymenophyllum peltatum, Equisetum pratense, E. trachyodon* (**204**), *Lycopodium alpinum.* Also several *Hieracia* (**128**), of which *H. pachyphylloides* (on the Knockagh) is a rare plant, and G. C. Druce has recorded *H. killinense* from Whitewell (Irish Nat. Journ., **3**, 218).

The River **Lagan,** rising on Slieve Croob in the centre of Down, curves through that county and down a fertile valley at the base of the basaltic hills to enter the head of Belfast Lough. *Potamogeton prælongus* and four *Lemnæ* are in the river above Belfast; *Elatine Hydropiper* (**96**) and *Cicuta* formerly grew there, but have retreated further up, to Lisburn or towards Lough Neagh, which is connected with the Lagan by canal. *Centaurea Jacea,* collected by John Templeton near Drum Bridge nearly 150 years ago (see Britton in Bot. Exch. Club Rep., 1920, 164), has not been recognized in Ireland since. *Potamogeton Cooperi* (*crispus* × *perfoliatus*) is in the river at Magheralin, in a form which A. Bennett distinguishes as f. *hibernicus* (Journ. Bot., 1919, 17). In the Lagan Canal, *Acorus* is abundant (an old escape), with *Butomus* and *Carex gracilis.* *Carum verticillatum* (**119**) was found formerly on the river banks, as *Cardamine amara* is still. The visitor should not fail to see the fine earthwork surrounding a dolmen, known as the Giant's Ring, close to the river at Drumbo. In the deer-park on Cave Hill, *Adoxa* has been seen not many years ago in its only Irish station : it has been known there for over a century, and was formerly somewhat more widely spread. *Hypericum hirsutum,* curiously rare in Ireland, has—or had—one of its few stations near the same place.

Adventive plants (in most cases naturalized) about Belfast include *Barbarea intermedia* (frequent), *Teesdalia* (at Lambeg, probably came with sand from L. Neagh, one of its few possibly native stations in Ireland), *Papaver*

Argemone, Geranium phœum, G. sanguineum, G. pratense, Epilobium roseum, Galium erectum, Crepis biennis (common and spreading), *Lactuca muralis* (Dundonald), *Tragopogon porrifolium* (long established on railway banks), *T. pratense, Nymphoides peltatum, Mentha Nouletiana* (*nemorosa* × *viridis*),[44] *Acorus, Lemna polyrrhiza* (the last three about the Lagan), *Bromus britannicus* (Sydenham).

Literature: see **453**, also R. TATE: "Flora Belfastiensis," 12mo, Belfast, 1863.

BELFAST LOUGH.

471. The rapid growth of Belfast and consequent reclamation have driven out most of the old salt-marsh flora at the head of the lough, but *Cochlearia anglica* (**86**) still clings to the Lagan and the Connswater. The alien *Zannichellia polycarpa* has gone from its original British station in Victoria Park. *Ruppia maritima* still survives at Holywood, whence *Trigonella* and *Scilla verna* have departed. A little lower down the lough, on the County Down side, *Spergularia rupicola, Limonium humile, Atriplex littoralis, Scilla verna, Eleocharis uniglumis, Scirpus rufus, S. filiformis,* still persist. *Crithmum* begins at Carnalea, *Atriplex maritima* and *Vicia lathyroides* about Groomsport. *Ligusticum* is on the Copelands—the only rare plant which these islands yield. The Japanese *Rosa rugosa* is naturalized on the shore at Craigavad, where it is drenched with sea-water during storms. A colony of *Spartina Townsendii* was planted in 1929 by the Belfast Harbour Commissioners on the foreshore near Tillysburn (Proc. R. Irish Acad., **41**, B, 121–2, 1932); the ground has now been reclaimed, and the grass re-planted. The Antrim shore of the lough is not favourable for plants, and little is seen till Whitehead is reached, where *Trifolium striatum* has its most northerly Irish station.

The lover of history will stop at the old town of Carrickfergus, much more ancient than Belfast, dominated by its great 13th century castle built on a volcanic dyke, and still displaying much of its town walls, gay in spring with *Cheiranthus.*

[44] Journ. Bot. 1926, 282.

THE NORTH-EASTERN SILURIAN AREA.

472. Two lines, joining Slieve Bane in Roscommon with Drogheda on the E and with Belfast on the NE, cut off an area of over 2500 sq. miles, extending as a wedge from the E coast into the centre of Ireland, and formed almost entirely of slates and grits of the Silurian Period—Ordovician along the NW margin, Gotlandian elsewhere. This tract includes almost the whole of Down (**473**) and Louth (**489**), the greater part of Armagh (**482**) and Monaghan (**483**), and much of Cavan (**484**). The surface has a character of its own, being low and hummocky, with many small lakes; the soil is generally light and fertile: and the characteristic flora strongly calcifuge. The chief interruptions to the continuity of the slates are two considerable granite intrusions in southern Down (the more recent of which forms the Mourne Mountains (**479**) and Carlingford Mountains (**490**)), and an outlier of Carboniferous limestone in the area where Meath, Louth, Cavan, and Monaghan meet.

Compared with the Central Plain of limestone, which adjoins on the S, there is a marked abundance of many plants which throughout the latter area usually haunt only patches of non-calcareous rocks. Such are:—

Ranunculus hederaceus
Lepidium heterophyllum
Raphanus Raphanistrum
Spergula vulgaris
Hypericum humifusum
Cytisus scoparius
Lotus uliginosus
Lathyrus montanus
Sedum anglicum
Peplis Portula
Galium saxatile
Gnaphalium uliginosum
Chrysanthemum segetum
Jasione montana

Digitalis purpurea
Teucrium Scorodonia
Polygonum Hydropiper
Rumex Acetosella
Potamogeton poly-
            gonifolius
    obtusifolius
Carex binervis
Deschampsia flexuosa
Nardus stricta
Blechnum Spicant
Athyrium Filix-fœmina
Equisetum sylvaticum

On the other hand, if one comes N into this area from the Central Plain, the following plants (with others), which are usually abundant on the limestone there, become rare or are absent, and if present mostly occur where some lime is found :—

Ranunculus circinatus
Geranium lucidum
Poterium Sanguisorba
Parnassia palustris
Myriophyllum verticillatum
Carlina vulgaris
Leontodon hispidus
Primula veris
Gentiana Amarella

Epipactis palustris
Anacamptis pyramidalis
Orchis morio
Ophrys apifera
Juncus inflexus
    subnodulosus
Potamogeton coloratus
Carex acutiformis
Chara aculeolata

The richest part botanically of the Silurian area is the County of Down; this is due largely to its extensive and varied coast-line. Also one finds there a curious outlying group of calcicole species to the south of Strangford Lough (**475**) arising from the scattering by ice from the north of calcareous debris from a now small outlier of Carboniferous limestone at Castle Espie, near the head of this same sheet of water. Louth, similarly with a varied coast-line (**491**), has a more varied flora than the inland parts of the region.

Entering the Silurian area from the basaltic region of Antrim, we find a change somewhat similar to that just described, for the basalts harbour a good many lime-loving species. Many of the characteristic Antrim plants are in the Silurian area found, if present at all, where calcareous sea-sands or other substrata supply a suitable habitat. Such are :—

Arenaria verna
Cerastium arvense
Geranium lucidum
Saxifraga hypnoides
Parnassia palustris

Galium boreale
Orobanche rubra
Ulmus montana
Juncus inflexus

THE COUNTY OF DOWN.

473. As stated above (**472**) Down is botanically the richest portion of the extensive Silurian region of NE Ireland. While the vegetation is generally that of the area mentioned, several causes combine to increase the flora :—1. The extensive and varied coast-line; (2) the calcareous strip already mentioned; (3) the high granite mass of the Mourne Mountains. Each of these is dealt with separately below. Among the characteristic plants which are widespread in addition to the list in **472**, *Cicuta* may be mentioned, and also the colonists *Barbarea intermedia*, *Myrrhis*, *Peucedanum Ostruthium*, *Galium erectum*, *Anchusa sempervirens*, *Lamium molucellifolium*. All of these display in Ireland an increase towards the NE— shown most strongly in *Peucedanum*, which is confined to that region, least strongly in *Galium erectum*.

Central and eastern Down (as around Ballynahinch, Dromore, Banbridge), not dealt with below, is a hummocky well-tilled area, with little waste land (save about Slieve Croob, 1755 ft.) and in the hollows small lakes and marshes, which are the home of many of the scarcer plants. It is drained by the upper waters of the Bann (**461–2–3**) and Lagan (**470**); few plants are recorded from it. The lakelets of the centre yield *Elatine hexandra*, *E. Hydropiper* (L. Briclan; this last also haunts the Lagan Canal, and at various times has been found in places extending from Belfast by Lough Neagh to Newry). *Rubus Lettii* is frequent in central Down, and *R. regillus* occurs at Gillhall. *Hieracium cinderella* is a rare Hawkweed growing on walls at Rowallane, near Saintfield, where *H. scanicum* and *H. grandidens* also occur, and *Rubus morganwgensis*. Other uncommon plants are *Callitriche autumnalis*, *Galium sylvestre* (in a suspicious station at Lenaderg), *Hieracium lucidulum* (Gilford), *Rubus hesperius* (Banbridge), *Ceratophyllum demersum*, *Typha angustifolia*, *Butomus*, *Potamogeton nitens*, *P. prælongus* (L. Aghery), *P. obtusifolius*, *Nitella translucens*; and *Acorus* is an old introduction at Ballynahinch. Before the bogs were destroyed for fuel, *Rhamnus Frangula*, *Cephalanthera ensifolia*, and *Carex limosa* grew, chiefly about **Ballynahinch**, where *Centunculus*, *Lastrea æmula*, and *Osmunda* still persist. In the S, *Spergularia rubra* has one of its few local stations at Lough Island Reavy, and *Viola Curtisii*, seldom found save on the coast (see **464**), is on the shore of Castlewellan Lake. *Crepis biennis* is an increasing colonist, and *Mercurialis perennis*, once introduced, seldom departs. *Geranium Endressi* is abundant in Hillsborough demesne. The most striking absentee is *Parnassia*, which has a very wide distribution in Ireland save in the SW.

R. LL. PRAEGER: ''Official Guide to County Down and the Mourne Mountains.'' 8vo. Belfast. 1898 (Botany, pp. 25–27, etc.). 2nd ed., 1900. See also **453**.

NORTH DOWN.

474. The district lying between Belfast and Donaghadee, including Bangor and Newtownards, though highly tilled, offers some heathy ground at Conlig, Carngaver, and above Holywood, where *Crepis paludosa, Pyrola media, P. minor, Pinguicula lusitanica, Lastrea montana, Phegopteris polypodioides* find a home, and glens yield *Lastrea æmula*: *Asplenium Adiantum-nigrum* var. *acutum* (**195**) has here its only Ulster station. Peat-bogs west of Donaghadee, such as the Cotton Moss, now nearly cut away, yield—or till lately yielded—*Drosera anglica, Andromeda, Pinguicula lusitanica, Carex limosa,* and no doubt other plants in old days. Human activities also account probably for the disappearance of *Geranium sanguineum* from the bluffs at Crawfordsburn and of *Carex strigosa* from the glen at the same place, but a similar cause has added to the local flora *Papaver hybridum, Geranium phæum, G. pusillum, Valerianella carinata, Crepis biennis, Matricaria occidentalis, Allium oleraceum, Hordeum murinum,* etc., all of which appear now established. *Butomus* has recently appeared by the railway SW of Holywood, perhaps bird-sown. *Rosa rugosa,* thrown out from a garden, is spreading on the gravelly shore at Craigavad (Irish Nat. 1924, 9). *Poterium officinale,* very rare in N Ireland, is at Donaghadee, and *Nitella translucens* at Conlig and Clandeboye. *Rosa hibernica* of Templeton (**109**) (usually ascribed to Smith[45]) has its original station at Tillysburn near Belfast, but the area is being rapidly built over now. It is a hybrid, *R. dumetorum* × *spinosissima*. *Cardamine amara,* in Ireland an entirely Ulster species, occurs in several places in N Down.

Literature: see **453**.

STRANGFORD LOUGH.

475. In the east of County Down the hummocky surface dips gently down under the sea, making a sheltered almost enclosed island-studded inlet twenty miles in length, opening to the Irish Sea through a deep narrow strait filled

---

[45] See J. BRITTEN in Irish Nat. 1907, 309–310.

with swirling water save at ebb and flood. This land-locked lough is attractive especially to the ornithologist, for on the low gravelly islets great numbers of sea-birds breed—terns of several kinds, gulls, Oyster-catchers, Ringed Plover, Mergansers and other sorts of duck. In many places the botanist has to pick his steps carefully among the innumerable eggs that are scattered on the shingle or on the fringe of dried *Zostera,* or to avoid treading on a sitting Merganser among the coarse grass.

Much the most interesting plant of the islands is *Glyceria Foucaudii* (**188**), which occurs in great profusion in many places. It haunts low reefs where the Glacial drift is covered by an inch or two of stones or gravel, where it often forms a dense belt 30 ft. in width, looking like a waving field of corn two feet in height. It is the lowest plant on the beach. Above it is a zone of *Atriplex* spp. and *Aster Tripolium,* and then *Agropyron repens.* *Ruppia maritima* grows at the head of the lough, *Zostera nana* at Greyabbey, mixed with *Zannichellia.* *Cochlearia anglica* (**86**) and *Limonium humile* (**133**) are common, *Scirpus rufus* and *Glyceria distans* frequent. *Falcaria vulgaris* (first recorded as *Ammi majus*) has formed a colony at an unspecified point (Irish Nat. 1913, 18; 1914, 20). *Juncus diffusus* is on the Quoile estuary. An out-crop of Carboniferous limestone at Castle Espie, by the shore near Comber, is of importance botanically—see **472, 477**. A local outpouring of basalt forms the conspicuous Scrabo Hill, near Newtownards, but is not colonized by any of the plants of the basaltic plateau on the N.

Literature: see **453**.

THE ARDS.

476. The Ards is a long narrow strip of land lying between Strangford L. and the Irish Sea, and forms the most easterly part of Ireland. It is typical County Down country—low and hummocky, the Silurian slates covered by drift yielding a light fertile soil; and highly tilled, so that the coast and the little lakes and marshes are the refuge of most of the rarer plants. The outer shore is sandy, with a good deal of low glaciated rock, and the more interesting part is in the S, from Ballyhalbert to Ballyquintin. Among the seaside species are

Thalictrum dunense	Ligusticum scoticum
Raphanus maritimus	Mertensia maritima
Crambe maritima	Atriplex portulacoides
Viola Curtisii	maritima
Spergularia rupicola	Chenopodium rubrum
Erodium maritimum	Scilla verna
Trifolium striatum	Scirpus rufus
filiforme	filiformis
Vicia lathyroides	Glyceria Foucaudii
Crithmum maritimum	

Ligusticum reaches here its southern limit in Ireland (at Burial Island); *Potamogeton coloratus* (characteristic of the Central Plain), *Glaucium,* and *Atriplex portulacoides,* here attain their most northerly Irish stations. The lakelets yield *Ceratophyllum demersum, Hydrocharis, Typha angustifolia* in several places. *Geranium columbinum* has here its East Coast northern limit. *Hieracium lucidulum* is on walls at Greyabbey. *Valerianella rimosa* and *Lamium molucellifolium* in fields are the most noteworthy of a large adventive flora. *Cuscuta Epithymum* is on sandy warrens at Kirkiston. *Lavatera* is quite likely an original native, but all possible present habitats are in the zone of human activities. There is an old record of *Erodium moschatum* (**98**) from Portaferry.

In this district also (near Mountstewart) was found the curious sport of *Ulex europœus* known as var. *strictus, U. hibernicus,* or Irish whin—a single plant, from which all those in cultivation have been derived by cuttings—see **99**.

Fertility of soil is often reflected in the architecture of the past as well as of the present, so it is not surprising to find in this area, at Greyabbey, the beautiful ruins of one of the larger Irish ecclesiastical establishments. The abbey here, built of the local Bunter sandstone, was founded in 1193 by Affreca, wife of John de Courcy, and colonized by Cistercian monks from Holm Cultram in Cumberland.

Hotels will be found at Donaghadee, Newtownards, and Portaferry.

R. LL. PRAEGER: "Botanizing in the Ards." Irish Nat. 1903, 254–265. See also **453**.

THE DOWNPATRICK-ARDGLASS DISTRICT.

477. This almost peninsular area is interesting because here occurs that scattering of calcareous detritus already mentioned (**472**), forming a soil suitable for certain calcicole species characteristic of the Central Plain, in this place quite isolated; and other noteworthy plants are not infrequent. Of the Central Plain plants, *Stellaria glauca* occurs in one marsh, *Potamogeton coloratus* in three places, *Chara aculeolata* in two; *Juncus inflexus, J. subnodulosus, Carex diandra,* and *C. lasiocarpa* are widespread, the last three forming a close floating felt which fringes or sometimes obliterates the numerous lakelets lying in hollows of the hummocky country. *Anacamptis pyramidalis* along the coast owes its presence at least in part to limy sea-sands, and possibly the same rather than calcareous Glacial drift accounts for *Ophrys apifera* at Killard Point, where *Orobanche rubra* also grows, and *Cuscuta Epithymum.* A small lake S of Strangford furnishes an isolated low-level station for *Lobelia Dortmanna.* The four species of red poppy—*P. Rhœas, P. dubium, P. Argemone, P. hybridum*—grow here in a profusion rare in Ireland. *Barbarea verna* and *Erodium moschatum* (**98**) are at Strangford.

The coast, from Killard to St. John's Point, is mostly rocky, with a shallow inlet at Killough, and its flora includes *Cochlearia anglica* (**86**), *Raphanus maritimus, Spergularia rupicola* in abundance, *Erodium moschatum* (**98**), *Trigonella, Trifolium striatum, Crithmum, Artemisia maritima, Limonium binervosum, L. humile* (**133**), *Pinguicula lusitanica* (**145**), *Atriplex maritima, A. portulacoides, Scilla verna, Juncus subnodulosus, Eleocharis uniglumis, Scirpus rufus*; an old Ardglass record for *Trifolium scabrum* is quite possibly correct, since its usual concomitants on the Irish E coast, *T. striatum, T. filiforme,* and *Trigonella* all occur there. These locally rare Leguminosæ usually occupy low glaciated bosses of slate, growing in a close sward of *Thymus, Anthyllis, Trifolium dubium, Medicago lupulina, Scilla verna.*

In the east of the area, the ancient town of **Downpatrick** stands on a hill among the marshes of the river Quoile, which becomes tidal a mile lower down and flows through a widening island-studded estuary into Strangford Lough (**475**). Several plants very rare in Ireland are found here, notably *Hottonia* in the stream (also further up at Crossgar) and *Galium Cruciata* on the great Norman motte: the standing of both as natives is uncertain. *Nasturtium sylvestre* has here its second NE station; *Apium Moorei* (**118**), *Ceratophyllum demersum, Lemna gibba,* and *Carex limosa* occur, also the alien *Barbarea verna. Erinus alpinus* is naturalized on the walls of Downpatrick gaol. This is a pretty and attractive district. The extensive marshes of the Quoile have never been thoroughly explored, as might be done during a period of drought. They were subject to tidal influence until the construction of lock-gates about a century ago. To the archæologist the area is of great interest; megalithic monuments are frequent, together with old castles, abbeys, and churches, some of the last associated with Saint Patrick.

The little fishing town of Ardglass, with its old castles, forms an excellent base, and Downpatrick is within an hour's rail of Belfast.

R. LL. PRAEGER: "Some Plants of the North-east Coast." Irish Nat. 1902, 200–210; "Some County Down Plants." *Ibid.*, 1918, 116–8. See also **453**.

NEWCASTLE AND DUNDRUM BAY.

478. From Newcastle, which is a popular watering-place, a semi-ellipse of sandy coast extends N and E to St. John's Point, broken only by the curious hammer-headed shallow Inner Bay of Dundrum. In the Newcastle section of this area are found *Thalictrum dunense, Viola Curtisii, Sagina ciliata, Spergularia rupicola, Erodium maritimum, E. Ballii* (one of Jordan's splits, described from Irish material, see Journ. Bot., 1920, 26; 1928, 363), *Ornithopus perpusillus* (at its N limit in Ireland), *Carlina, Campanula rapunculoides, Centunculus, Mertensia* (both N and S of Newcastle), *Atriplex portulacoides, A. littoralis, A. maritima, Euphorbia Paralias.* Many of these re-appear in the sandy stretches E of the Inner Bay, about Ballykinler, along with *Teesdalia* (also at Murlough), *Geranium phæum, Vicia lathyroides, Carlina, Euphorbia portlandica* and other plants. The profusion of the beautiful *Mertensia* in some spots near St. John's Point, with groves of *Thalictrum dunense* and *Echium vulgare,* and slopes starred with *Anacamptis pyramidalis,* alone make this coast worth a visit.

About the **Inner Bay** are *Cochlearia anglica* (**86**), *Apium Moorei* (**118**), *Centunculus, Limonium humile, Juncus subnodulosus, Zostera nana, Carex Hudsonii*; with *Erodium moschatum* (**98**) near the village of Dundrum, which is dominated by the great Norman castle built by John de Courcy in the 12th century.

The most interesting part of the dunes is that which lies south of Murlough House. Here *Hypericum perforatum, Rosa spinosissima, Erica cinerea, Teucrium Scorodonia,* and *Hippophae* (the last introduced) are in turn dominant. *Carlina,* very rare in NE Ireland, grows a dozen to the square yard in places, and most of the plants above-mentioned occur. In hollows, parallel ridges of gravel (raised beach) are sometimes carpeted over an area of an acre or two with the moss *Racomitrium canescens* to the exclusion of almost all other plants. An old record for *Equisetum hyemale* from here was thought by A. G. More (Recent Add.) to be possibly referable to the Wicklow-Wexford *E. Moorei,* but the plant has not been refound.

Newcastle, Dundrum, and Ardglass make suitable bases for the exploration of this area.

Literature : see **453**.

THE MOURNE MOUNTAINS.

479. An outburst of volcanic activity in Eocene times has left its memorial in the Mourne Mountains, a fine group of granite hills occupying an oblong area in S Down, with the Silurian rocks lapping their flanks (fig. 27). At the NE end, at **Newcastle** (**478**), the hills descend steeply into the Irish Sea, and at the SW end, at Rostrevor, into Carlingford Lough (**481**). A dozen summits exceed 2000 ft. (Slieve Donard is 2796 ft.) and especially in the E part, where most of the higher hills are, the valleys are deep, with steep slopes. The deepest and longest glen is that of the Kilkeel River (the Silent Valley), in which a great reservoir has recently been constructed to supply Belfast with water. A much more important stream, the Bann (**461–3**), rises on the N side of the range in the plateau-like moorland called the Deer's Meadow. There are some fine cliffs, hung with ferns and mosses, and in places the weathering of the granite has produced fantastic pinnacles, jointed so as to

PLATE 43.

MOURNE MOUNTAINS, CO. DOWN.

Weathered pinnacles of granite on Slieve Commedagh, amid slopes of *Calluna*.

R. Welch, Photo.

479.

PLATE 44.

THE BOYNE AT BEAUPARC, CO. MEATH.

River gorge in Carboniferous limestone, with luxuriant planted woods.

resemble a series of boulders precariously poised one on another (plate 43). The Silurian rocks have in places been forced up to considerable elevations, and there is frequently a sharp contrast between the deep peat and heather which

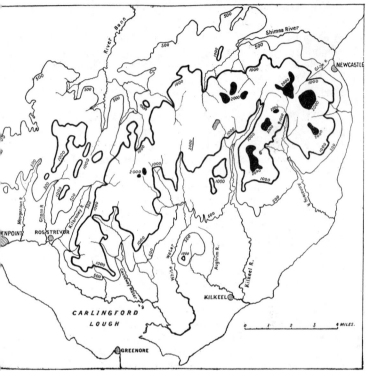

FIG. 27.—THE MOURNE MOUNTAINS.

cover the granite, and the less acid lighter soil and grassy vegetation of the slates, whose outcrops also yield a richer flora than the granites. There are a few small tarns.

Among the plants of the high grounds are

A	Saxifraga stellaris	Salix herbacea
A	Sedum roseum	*A* Juniperus sibirica
	Saussurea alpina	*A* Cryptogramme crispa
A	Lobelia Dortmanna	Lycopodium alpinum
	Vaccinium Vitis-Idæa	*A* Isoetes lacustris

but these are not particularly alpine in habitat, six of them (marked *A*) descending to 1200 ft. or less. Other plants of mountain habitat are *Meconopsis* (valley above Rostrevor), *Epilobium angustifolium* (only on high cliffs of Slieve Bingian and Eagle Mountain), *Cystopteris fragilis*, *Lycopodium clavatum*, and a number of *Hieracia* (**128**), including *H. Stewartii* (R. Bann above Hilltown, and in Tollymore Park, endemic), *H. hibernicum* (Broughnamaddy, elsewhere only in S. Donegal) *H. flocculosum* (Spinkwee R.), *H. sagittatum* (Pigeon Rock Mtn. and Eagle Mtn., only Irish stations), *H. senescens* (Tollymore Park, only Irish station), *H. argenteum*, etc.

480. *Pinguicula lusitanica* has a wide range vertically and horizontally, as has *Crepis paludosa*. An interesting relict is *Carex lasiocarpa*, rare locally, which occupies wet bog (a former lake) at the head of the valley of the Rowantree River, 1000 ft. elevation. The vegetation of some portions of the range has recently been the subject of special study.[46] On the lower grounds the most attractive spot is **Tollymore Park** near Newcastle, richly wooded, with a beautiful stream, the Shimna River, and a flora including *Circæa alpina*, *Pyrola minor*, *Mimulus moschatus* (naturalized) and quantities of *Festuca sylvatica*, *Lastrea æmula*, *L. montana*, *Hymenophyllum peltatum*, and some rare *Hieracia* (*ante*). *Drosera longifolia* has very isolated colonies by the Kilkeel River, now in part destroyed by the reservoir in the Silent Valley. By the infant Bann above Hilltown *Equisetum litorale* (**201**) occurs in several places, and a curious form of *E. hyemale* with alternate lateral sessile cones. A form of *Hymenophyllum peltatum* from Cove Mountain produced in cultivation fronds which continued to grow at the tip in successive seasons, making in three years fronds 9 inches in length. The *Filix minor longifolia* of Ray's "Synopsis," Plunkenet's "Phytographia," Petiver's "Almagestum," and the Sherardian and Sloanean herbaria, formerly attributed to *Asplenium Adiantum-nigrum* var. *acutum*, from "a dark cave among

---

[46] J. I. ARMSTRONG, J. CALVERT, and C. T. INGOLD: "The Ecology of the Mourne Mountains, with special reference to Slieve Donard." Proc. Roy. Irish Acad., 39, B, 440–452. 1930. ARMSTRONG, INGOLD, and K. C. VEAR: Vegetation Map of the Mourne Mountains, Co. Down, Ireland. Journ. Ecol., 1934, 439–444.

the mountains of Mourne" is a barren plumose form of *Athyrium Filix-fœmina* close to that known to fern-growers as *kalothrix*. It has not been refound. One of the more noticeable aliens is *Rhododendron ponticum,* originally introduced as undergrowth in a wood at Kinnahalla near Hilltown. The wood was cut some years ago, and the *Rhododendron* has spread in thousands over some acres of stony hillside, making a wonderful display in May.

The Mournes form a beautiful and compact group of hills, with some fine cliffs and crag-topped hills, and provide excellent walking and climbing. The visitor can stay at Newcastle, Bryansford, Hilltown, Rostrevor, or Kilkeel: the first three being best for the mountains. At Newcastle there is the additional attraction of the great stretches of dunes to the N (**478**), while at Rostrevor one is on the margin of lovely Carlingford Lough (**481**). The shore, which swings round in a semicircle between these points, is generally stony, backed by a cliff of Boulder-clay, and offers little to the botanist; but *Mertensia* and *Glyceria Foucaudii* occur.

S. A. STEWART and R. LL. PRAEGER: "Report on the Botany of the Mourne Mountains, Co. Down." Proc. Roy. Irish Acad., 3rd ser., 2, 335–380. 1892. R. LL. PRAEGER: "Official Guide to County Down and the Mourne Mountains." 1898. H. C. HART: "Notes on the Plants of some of the Mountain Ranges of Ireland." Proc. Roy. Irish Acad., 2nd ser. (Science), 4, 238–244. 1885. See also 453.

CARLINGFORD LOUGH.

481. The Newry River, an inconsiderable stream, enters below Newry a long straight narrow arm of the sea, which at Warrenpoint widens into the hill-embosomed waters of Carlingford Lough. This is a lovely place, with much interest for the botanist. Where the river reaches tidal waters, at **Newry,** *Cochlearia anglica* (**86**), *Limonium humile* (**133**), *Lemna gibba, Scirpus filiformis,* etc., become immediately abundant. In the canal just above the town *Callitriche autumnalis* occurs, and *Subularia* grew formerly, and may turn up again; *Elatine Hydropiper* (**96**) has here its largest colony in the British Isles, extending along the canal from Newry for fifteen miles northward. *Nasturtium sylvestre,* possibly alien, is at Sugar Island, Newry.

Linaria repens, in Ireland rare and of uncertain standing, grows near the canal locks a few miles down, and further, at the castle-guarded bend at Narrow-water, the foreshore is covered with *Limonium humile, Atriplex portulacoides, Glyceria distans.* On the hill just above the ferry, on the Louth side, *Lastrea œmula, L. montana* and *Hymenophyllum tunbridgense* inhabit a small glen. *Spergularia rubra,* a very local plant, is scattered along the railway from Newry to Omeath; *Rubus Lettii* grows about Newry, and *Raphanus maritimus* at Omeath. At the pleasant watering-place of Warrenpoint, where the river widens into the lough, *Spergularia rupicola* begins, and *Tragopogon porrifolium* has established itself.

About **Rostrevor,** famous for its beauty and mild climate, few rare plants occur save those of the hills which rise steeply from the lough shore (**479**); *Lactuca muralis, Campanula rapunculoides,* and the New Zealand *Acœna Sanguisorbœ,* are among the alien population. The gravelly point of Killowen, further down, supports colonies of *Lepidium campestre* and *Linaria repens* with its hybrid with *L. vulgaris (L. sepium).* The entrance of the lough is impeded by reefs of Carboniferous limestone, a small outlier from the Central Plain, with extensive sands and gravels on both sides. On the Down shore, about Greencastle and Cranfield Point, are *Raphanus maritimus, Dipsacus sylvestris, Mertensia, Linaria sepium, Atriplex maritima, Euphorbia Paralias, E. portlandica,* etc. Most of these are repeated on the Louth side, about Greenore, along with *Silene noctiflora, Erodium moschatum,* and near Ballagan *Typha angustifolia.* The visitor to this area should not miss the picturesque old town of **Carlingford,** on the lough shore at the foot of the fine mountain of the same name (**490**), with its ancient Norman castle and Dominican priory hung with *Cheiranthus* and *Kentranthus.* One can stay at Newry, Warrenpoint, Rostrevor or Carlingford.

THE COUNTY OF ARMAGH.

482. In Armagh, as in Down and Monaghan, Silurian slates, with a rather low undulating fertile surface, occupy much of the ground : but a band of Carboniferous limestone runs through the ancient city of Armagh, with Triassic beds to the north of it, and Eocene clays bordering L. Neagh (**463**) and basalts to the NE ; in the S of the county the granite mass which culminates in Slieve Gullion (**490**) rises to 1893 ft.

The richest area for the botanist is the N. Excluding the remarkable flora of the shores of L. Neagh, which is dealt with separately (**463–466**), the plants occurring here, in the area including Armagh, Loughgall, Portadown and Lurgan, include among native species *Spergularia rubra* (on railway), *Apium Moorei* (**118**), *Stachys officinalis, Spiranthes stricta, Potamogeton angustifolius, Lycopodium clavatum, Chara aculeolata* (which with *Anacamptis pyramidalis* and *Juncus inflexus* shows the influence of the limestone) ; also *Hydrocharis, Sagittaria, Sparganium minimum, Lemna polyrrhiza, Carex Pseudo-Cyperus, Equisetum litorale* (**201**) ; and of aliens *Papaver Argemone, Barbarea intermedia, B. verna, Lepidium campestre, Silene noctiflora, Epilobium roseum, Myrrhis, Crepis biennis, Hieracium lucidulum* (walls at Armagh), *Lamium molucellifolium.* An unusual alien is *Mimulus moschatus*, which is naturalized by the Bann at Whitecoat above Portadown. A rare hybrid pondweed, *Potamogeton Bennettii* (*crispus* × *pusillus*, or *crispus* × *obtusifolius*), which may rather be *P. Lintoni* (*crispus* × *Friesii*), grows in the canal below Caledon in the extreme NW (see Journ. Bot., 1919, 16). The Silurian and granite area of the centre and S yield (apart from mountain plants, see **490**), *Lepidium heterophyllum, Callitriche autumnalis, Crepis paludosa, Lobelia Dortmanna, Potamogeton obtusifolius, Isoetes lacustris, Nitella translucens.* In old days *Subularia* grew in the canal at Newry, probably derived from Lough Neagh : but it has not been seen for some time. *Elatine Hydropiper*, also long since recorded from Newry, has recently been found to occupy the canal thence northward for fifteen miles. In marshes by the railway along the E margin *Cicuta* grows abundantly, with *Butomus,* etc. ; *Festuca sylvatica* is in the demesne at Tanderagee, where also *Mercurialis perennis* is naturalized.

Armagh, situated in a very fertile area, has long been a place of importance. For six hundred years (300 B.C.– 332 A.D.) it was the residence of the Kings of Ulster, the site being the great earthwork of Emania, now called Navan Fort. Later, St. Patrick founded a church on the site of the present cathedral : and Armagh became and still is the metropolitan see of all Ireland.

Armagh and Lurgan form the best centres for the exploration of the north of the county, and Newry of the south.

R. Ll. Praeger: ''The Flora of County Armagh.'' Irish Nat. 1893, 11, etc.

THE COUNTY OF MONAGHAN.

483. The Silurian rocks continue SW from Down and Armagh across Monaghan, and occupy most of that county except the N, giving a fertile surface much like the rest of the Silurian area, and a similar but rather poorer flora. The N portion is the more interesting part.

Monaghan, the county town, is on the S edge of a band of Carboniferous limestone which runs ENE from the N edge of the Central Plain. One of the few rarer calcicole plants which follow the limestone here is *Chara aculeolata*; *Juncus subnodulosus* occurs; and *Sium latifolium*, in the Finn River near Clones, is probably also on the limestone. Northward the ground rises into barren hills (Slieve Beagh, 1190 ft.) of Millstone Grit and Yoredale beds, with a heathery flora of little interest. Here there are miles of barren peat with so limited and monotonous a flora, that *Oxycoccus, Empetrum,* and *Lycopodium Selago* form welcome finds. The flora of the Silurian area to the S includes *Callitriche autumnalis, Apium Moorei, Cicuta, Crepis paludosa, Lobelia Dortmanna, Potamogeton nitens, P. Friesii, P. obtusifolius, Lastrea æmula, Isoetes lacustris, Nitella flexilis,* and *N. mucronata*—the last in its only Irish stations in several lakes near Carrickmacross, where *C. aculeolata* also grows. The rare var. *nidifica* of *Nitella flexilis* is in Annaghmakerig Lough near Newbliss. There are many other lakes, most of them unexplored, in the S half of the county, Muckno Lake near Castleblayney being the largest. The alien flora of the county includes a good deal of *Myrrhis, Crepis biennis, Mercurialis perennis. Veronica peregrina* and *Acorus Calamus* have colonies at Monaghan, and *Aster* sp. at Glasslough. This western part of the Silurian region has not hitherto yielded much to its few botanical explorers.

G. R. BULLOCK-WEBSTER: ''Characeæ from Co. Monaghan.'' Irish Nat. 1902, 141–6. 183.

THE COUNTY OF CAVAN.

484. From the point of view of the botanist, Cavan is merely a geographical expression, for the county is the reverse of a natural area—shaped somewhat like a saucepan, with the handle formed half of Millstone Grit and Yoredale rocks and half of Carboniferous limestone, and the rest mainly of Silurian slates. It is a watery region, and botanical interest centres around the lakes. In the middle of the county, the mazy Lough Oughter (**434**), on the Erne, lies on the limestone, as does also Lough Sheelin (**487**) in the S. L. Gowna (**434**, also on the Erne), and L. Ramor (**488**), are on the slates. The NW arm of the county, stretching up to L. Macnean, is hilly and wild. Here the Shannon has its source, W of the high hills of Cuilcagh (2188 ft., **427**) and Tiltinbane (1949 ft.). This area, as also the mass of slate country in the SE, is not well known. The rarer plants, according to present knowledge, are grouped along the course of the Erne, which is dealt with above (**433**). E Cavan, around Cootehill, Bailieborough, Ballyjamesduff, Kingscourt, is an agricultural district with many small hills and lakes. It is very little known botanically.

THE COUNTY OF LONGFORD.

485. The small county of Longford lies on the E bank of the Shannon on the N edge of the Limestone Plain. It is only here and there of botanical interest. In the NE the mazy Lough Gowna, the uppermost of the tortuous lakes of the Erne basin (**432, 433**), introduces us to the curious water-logged country that characterizes the course of that river. In the SW, on the Shannon, Lough Ree and the Inny mouth below Ballymahon afford some good plants (**379**). There is an old record for *Hymenophyllum tunbridgense*—rare in central Ireland—''in the neighbourhood of Longford,'' which seems an unlikely station. About Longford town *Apium Moorei* is in the canal and a few local aliens are found—*Valerianella rimosa, Crepis biennis, Campanula rapunculoides, Lamium molucellifolium.*

The Royal Canal crosses the SW of the county, yielding as usual *Equisetum variegatum.*

In the N, the Silurian slates which prevail over Down, Armagh, and Monaghan extend across Longford also, and rise to 912 ft. in Carn Clonhugh. A calcifuge flora prevails there; the only uncommon plant recorded from the hill is *Lycopodium clavatum.*

THE COUNTY OF WESTMEATH.

486. The County of Westmeath, a limestone area, lying in the centre of Ireland just N of Offaly (King's County), offers two areas of interest to the botanist—a series of lakes among low but conspicuous hills (Knock Eyon, 707 ft., etc.) in the centre, and, on the W border, Lough Ree (**379**) one of the large expansions of the Shannon. For the rest we are here in typical Central Plain country (**240**) with the usual wide areas of pasture, peat-bog, and marshy land lying on calcareous drift which rests on the Carboniferous limestone. The broad band of Silurian slates which stretches SE from Down half across Ireland has been nearly left behind in Longford and the calcicole flora is in full possession.

THE WESTMEATH LAKES.

487. **Mullingar,** the county town, lies in the centre of the lake district, and makes a good base for their exploration—L. Ennell, L. Owel, L. Iron, L. Derevaragh, while further NW, about Castlepollard, are other lakes, of which L. Lene is the largest. The area N and NW of Mullingar is rich in rare plants. Of these, *Pyrola rotundifolia,* found on several of the bogs, is unknown elsewhere in Ireland. *Chara tomentosa,* common in the lakes, is elsewhere in Ireland known only from the Shannon; it is absent from Great Britain. *Galium uliginosum, Carlina, Crepis paludosa, Hydrocharis, Juncus subnodulosus,*

Sparganium minimum, Carex diandra, C. Hudsonii, Lemna polyrrhiza, Chara aculeolata, occur in a number of places; *Lathyrus palustris* (**103**), *Drosera anglica, Cicuta, Chenopodium rubrum, Ophrys apifera, Sagittaria, Potamogeton*

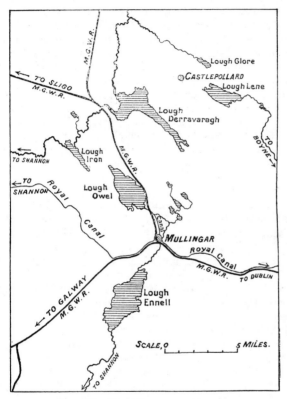

Fig. 28.—The Westmeath Lakes.

coloratus, P. angustifolius, P. Friesii, Carex paradoxa (**179**), *C. lasiocarpa, Lastrea Thelypteris*, are more local. *Ceratophyllum demersum, Potamogeton decipiens* and *P. nitens* are in L. Derevaragh; *P. prælongus* in L. Owel. *Chara denudata* (**207**) has its only Irish station in Brittas Lake, whence also *Caltha radicans* is recorded; *Nitella tenuissima* (**207**), which occurs in L. Owel and in pools in the Scraw Bog, is known elsewhere in Ireland only from E Galway. This "Scraw Bog," ½ mile E of L. Owel, is very interesting. It is a long narrow lake on the limestone, entirely covered by a thick floating felt of vegetation, so that it can be traversed on foot from end to end by the venturesome. Over the greater part a caricetum prevails formed of *C. lasiocarpa* and *C. diandra*, with *C. dioica*, *C. pulicaris* and *C. limosa*, *Juncus subnodulosus*, *Cladium*, *Menyanthes*, *Equisetum limosum*; on low mossy tussocks, or among clumps of *Salix cinerea* and *S. aurita*, *Pyrola rotundifolia* has its headquarters, growing in great profusion with *Oxycoccus quadripetala*; but it is not confined to the tussocks. In drains are *Hydrocharis*, *Nitella tenuissima*, *Potamogeton coloratus*, *Chara aculeolata*. The presence of the last two, which are distinctly calcicole, is probably due to water from the surrounding area which passes across the "bog" in a large drain, or else to limy springs. Lisclogher Bog yields *Saxifraga Hirculus* (**116**) and *Pinguicula lusitanica* (**146**). Other local plants are *Sium latifolium*, *Utricularia intermedia* (**144**), *Typha angustifolia*, *Eriophorum latifolium* (2 miles W of Mullingar), *Carex limosa*, *C. aquatilis* (**181**), *C. strigosa* (Killucan), *Festuca sylvatica*, *Lastrea æmula*, *Equisetum variegatum*. An old record by F. J. Foot for *Lamium Galeobdolon* from L. Ennell needs verification; it is very local in Ireland. *Pyrola minor* has been found both within this area (at Ballynagall) and further S near Tyrrellspass. *Sorbus porrigens* is in woods at Knockdrin.

It will be seen that this group of lakes yields a galaxy of rare plants equalled by very few localities in Ireland : almost all are hydrophytes or pertain to marshes or bogs. Our knowledge of them is largely due to an active botanist—the late H. C. Levinge of Knock Drin—who had his residence in the middle of the area. The variety of plants here is to be accounted for in some measure by variety of conditions. Thus L. Owel, fed by limestone springs, has very clear highly calcareous water of a pale green hue, showing the lime-incrusted rocks on the bottom; while L. Derevaragh, on the River Inny, a few miles away, has brown more acid water due to the miles of bog through which this stream flows, and which fringe its N and W shores. The lakes do not lie within a single catchment area, the waters of the northern ones reaching the Shannon by the Inny, while the southern ones discharge into the Shannon much further S by means of the Brosna.

The **Inny**, rising in Cavan near **Lough Sheelin** (which yields *Lathyrus palustris, Potamogeton filiformis, Lastrea æmula,* etc.), flows through that lake and the smaller L. Kinale, and thence down a wide bog-filled valley to the Westmeath lakes. There is plenty of little-known country here, best explored from Granard, a town built on a hill, its tall spire and great Norman motte dominating the country far and wide.

At Coolatore near Moate, *Sarracenia purpurea* is naturalized on a peat-bog (see also **380**).

Westmeath as a whole is typical eastern Central Plain country, being more fertile, better wooded, and more undulating than the tracts W of the Shannon; and the lake district dealt with above is an area of much charm and high botanical interest.

H. C. LEVINGE: ''The Plants of Westmeath.'' Irish Nat. 1894, 1895, 1896. E. F. LINTON and W. R. LINTON: ''Westmeath Plants.'' Journ. Bot. 1896, 119–122.

MEATH AND THE BOYNE.

''See, how swift Boyn to Trim cuts out his way!
  See, how at Drogheda he joyns the Sea!''
SPENSER.

488. The botany of the large county of Meath is for all purposes the botany of the Boyne. The two main branches of that river, the **Boyne** itself and the Blackwater, rising respectively in Kildare and in Cavan, soon enter Meath and flowing, the first NE and the second SE, unite at Navan and continue E to the Irish Sea, draining between them almost the whole county. This is for the most part a fertile and well-populated area. The grazing lands of Meath are the richest in Ireland; the ground is undulating and well wooded save in the W, where portions resemble the Central Plain. In its lower course the river offers much charming scenery (plate 44). Its long estuary and sand-rimmed mouth yield many good plants. Limestone prevails save locally in the N and E, where slates of the Silurian period come to the surface.

From the upper reaches of the Boyne few plants are on record. The **Blackwater** appears richer. **Lough Ramor** or Virginia Water, a considerable lake in the Silurian region across the Cavan border, through which this stream passes, yields *Rubus Lettii, Polygonum minus, P. laxiflorum*; and *Carex aquatilis* (**181**) is abundant in the river below the lake. Smaller lakes and streams within this northern Silurian area yield *Hottonia* (stream flowing into Newcastle Lake near Kingscourt; its only possibly native station in Ireland is in County Down (**473**)), *Callitriche autumnalis, Hydrocharis, Cicuta, Carex diandra, C. limosa*; the last occurs also with *Chara aculeolata* on the limestone at Lough Crew in the W, close to the Bronze Age cemetery of Slievenacalliagh, remarkable for its wealth of inscribed stones.

At **Trim** and **Bective** the Boyne is already a considerable stream, with reedy banks, which enhance and are enhanced by the fine architectural remains—castles, churches, abbeys—of those historic places. The plants of this neighbourhood include *Dipsacus sylvestris, Blackstonia, Acorus, Epipactis palustris, Potamogeton angustifolius, Galeopsis intermedia* (apparently very rare in Ireland) and the introduced *Myrrhis*. About and below **Navan**, the county town, are *Apium Moorei, Calamintha ascendens, Galium uliginosum, Atropa Belladonna* (river-cliff near Slane), *Ceratophyllum demersum* (not seen recently), *Acorus, Poa palustris* (only Irish station; it extends some miles down stream from Navan, and its habitat does not suggest introduction). *Carex Hudsonii, C. gracilis, C. lasiocarpa, Potamogeton prælongus, P. decipiens* are here too, and about the town *Hordeum murinum*.

Above **Drogheda**, where the river becomes tidal, *Cochlearia anglica* is characteristic, and *Dipsacus sylvestris, Calamintha ascendens, Verbena, Butomus,* and *Lemna gibba* occur. Five miles of estuary give way to sand-dunes at the river-mouth, where the rare southern snail, *Helix pisana*, occurs in great profusion. This is good ground, as about Mornington and Baltray, with *Sisymbrium Sophia, Silene noctiflora, Cerastium arvense, Melilotus arvensis, Epilobium roseum, Fœniculum, Artemisia maritima, Centunculus, Cuscuta Trifolii, Salvia horminoides, Chenopodium rubrum, Atriplex maritima, Euphorbia portlandica, Allium vineale, Lemna gibba, Hordeum nodosum*; also

Linum bienne, Erodium maritimum, Trifolium fragiferum, Dipsacus sylvestris, Limonium humile, L. binervosum, Atriplex portulacoides, Euphorbia Paralias, Festuca uniglumis. These last nine occur again at the mouth of the Nanny River, at **Laytown**, a few miles S; at the last-mentioned place *Zostera nana* also grows, with *Senecio erucifolius* (here at the N limit of its very limited Irish range, save for an old unverified station between Drogheda and Dundalk), *Picris echioides, Campanula rapunculoides, Scirpus filiformis.* Miscellaneous plants of the Boyne area include *Rubia peregrina* (Monasterboice, very rare inland), *Valerianella rimosa* (frequent), *Andromeda* and *Carex limosa* on bogs, *Lamium molucellifolium, Hydrocharis* (N and centre), and *Lastrea Thelypteris* (Drumconrath).

The rich valley of the Boyne has been from very early times a centre of population and of culture, as witness the royal cemetery of Brugh-na-Boinne, of Bronze Age date, the finest monument of its kind in Europe, of which the leading structures are the imposing tumulus of Newgrange, and the lesser ones of Dowth and Knowth, containing chambers with great stones richly inscribed with spirals and other patterns of unknown meaning. A whole series of such tumuli, of smaller size, stands on Slievenacalliagh near **Oldcastle** (from which Lough Sheelin (**487**) and Lough Ramor (**488**) can be worked). In the SE of the county is Tara, long the residence of the Irish kings. Monuments of more recent date—castles, abbeys, high crosses and so on—at Trim, Kells, Monasterboice, Mellifont, Drogheda, and elsewhere, remain to testify to the importance of the district in early Christian times.

The explorer will find accommodation at Drogheda, Slane, Navan, Trim, Oldcastle, Virginia, etc.

THE COUNTY OF LOUTH.

489. The smallest of Irish counties, with an area of 316 sq. miles; but it offers much variety to the botanist—the fine hills of the Carlingford range (**490**) in the N, a diversified coast-line of gravel, sand, mud and rock (**491**), and a mixture of Silurian slates (mainly) with limestone and volcanic rocks.

CARLINGFORD MOUNTAIN AND SLIEVE GULLION.

490. N and E of Dundalk, an upheaval of Eocene times has produced a mountainous area, of which the main mass, the Mourne Mountains in Co. Down, are described above (**479**). They are cut off by the deep and picturesque inlet of Carlingford Lough from their westward extension, across which the Great Northern Railway climbs between Dundalk and Goraghwood on its way from Dublin to Belfast.   The highest summits here are Carlingford Mountain (1935 ft.), a fine rugged hill overlooking the lough of the same name, and Slieve Gullion (1893 ft.), out to the NW.   In the hilly region between, Clermont Carn rises to 1674 ft.   Botanical interest centres in Carlingford Mountain.   Here grow *Sedum roseum, Crepis paludosa, Hieracium hypochœroides, Pinguicula lusitanica, Salix herbacea, Lastrea œmula, Hymenophyllum peltatum, Phegopteris polypodioides*; and *Cryptogramme* has one of its widely scattered stations on rocks near the summit. *Phegopteris Robertiana* (**199**), planted on the hill in 1878, was still flourishing in 1889.   *Hymenophyllum tunbridgense* occurs with *Lastrea montana* and *L. œmula* near sea-level at Narrow-water (**481**), and *Circœa alpina* further to the SW.

Slieve Gullion, though only 42 ft. lower than Carlingford Mountain, has a flora almost devoid of alpine plants, but one unexpected alien occurs—the S European *Mentha Requienii,* which fringes numerous rills on the N slope at about 1000 ft. : elsewhere in Ireland it is naturalized in Cork (**298**).   *Crepis paludosa, Vaccinium Vitis-Idœa, Phegopteris polypodioides* and *Selaginella* are the only species growing on the hill which suggest mountain conditions, and all are in Ireland frequently lowland.   At its N base, near Camlough, *Nitella flexilis* occurs, *Geranium columbinum* is at Adavoyle, and to the S, at Flurry Bridge, Ravensdale, the rare hybrid *Asplenium Clermontæ* (*A. Ruta-muraria* × *Trichomanes*) was once found (**196**). *Nitella translucens* is at Windy Gap.

THE COAST OF LOUTH.

491. The coast between Drogheda and Dundalk presents
considerable variety of ground and of flora. N of the mouth
of the Boyne a Silurian ridge terminates seaward in the
bold promontory of **Clogher Head**. Here grow *Spergularia rupicola, Trigonella* (not seen recently), *Trifolium
striatum* (the last two are characteristic species of the S
and E coasts), *T. fragiferum, Crithmum, Artemisia
maritima, Limonium binervosum, Atriplex portulacoides,
Scilla verna.* Entering Dundalk Bay, we find extensive
sandy beaches backed by low ground, with muddy inlets
about Dundalk town. This region yields some of the foregoing species, with *Silene noctiflora, Cuscuta Trifolii,
Atriplex maritima, Limonium humile. Mertensia* is also
recorded. The muddy shores near Dundalk support some
of the foregoing, with the addition of *Cochlearia anglica,
Artemisia maritima, Lemna polyrrhiza, Eleocharis uniglumis.* At Dundalk railway station the introduced *Sedum
rupestre* is unusually abundant and wild-looking on rock-
cuttings, and a little to the north the tall inflorescences of
an alien *Petasites* form a conspicuous object in spring.

The hinterland, low-lying Silurian country, presents a
good deal of marsh, bog and stream. At **Castlebellingham**
Orobanche Hederæ finds its NE limit, and *Geranium
columbinum, Blackstonia, Epipactis palustris, Juncus
subnodulosus, Sparganium minimum,* etc., occur; *Osmunda*
has its largest E coast colony in Braganstown bog. *Drosera
anglica, Cicuta, Andromeda, Hydrocharis* in plenty, *Lemna
gibba, Potamogeton coloratus, P. angustifolius, Carex
diandra, Lastrea Thelypteris* are in the **Ardee** district. At
Termonfeckin, *Melilotus arvensis* and *Campanula rapunculoides* are reported as naturalized; there is an old record
of *Lactuca muralis* from Collon, and *Valerianella rimosa*
is a cornfield colonist.

Drogheda, Dundalk, Ardee, and Carlingford form the
best centres for the exploration of Louth.

INDEX.

NOTE.—Principal references are in italic type. The last number under each name gives the PAGE on which the name originally appeared in the Census List, if it appeared at all. This Census List has been omitted; see note on page viii.